Connections, Curvature, and Cohomology

Volume I

De Rham Cohomology of Manifolds and Vector Bundles

This is Volume 47 in
PURE AND APPLIED MATHEMATICS
A Series of Monographs and Textbooks
Editors: PAUL A. SMITH AND SAMUEL EILENBERG
A complete list of titles in this series appears at the end of this volume

Connections, Curvature, and Cohomology

Werner Greub, Stephen Halperin, and Ray Vanstone

DEPARTMENT OF MATHEMATICS
UNIVERSITY OF TORONTO
TORONTO, CANADA

VOLUME I

De Rham Cohomology of Manifolds and Vector Bundles

ACADEMIC PRESS New York and London 1972

ACADEMIC PRESS, INC.
111 Fifth Avenue, New York, New York 10003

United Kingdom Edition published by
ACADEMIC PRESS, INC. (LONDON) LTD.
24/28 Oval Road, London NW1

LIBRARY OF CONGRESS CATALOG CARD NUMBER: 79-159608

AMS (MOS) 1970 Subject Classifications: 53-00, 55C20, 55C25,
55F10, 55F25, 58A05, 58A10, 58C05, 58C15

PRINTED IN THE UNITED STATES OF AMERICA

Respectfully dedicated to the memory of

HEINZ HOPF

Contents

Chapter IV Calculus of Differential Forms

Chapter V De Rham Cohomology

Chapter VI Mapping Degree

Chapter VII Integration over the Fibre

Chapter VIII Cohomology of Sphere Bundles

Preface

This monograph developed out of the Abendseminar of 1958–1959 at the University of Zürich. It was originally a joint enterprise of the first author and H. H. Keller, who planned a brief treatise on connections in smooth fibre bundles. Then, in 1960, the first author took a position in the United States and geographic considerations forced the cancellation of this arrangement.

The collaboration between the first and third authors began with the former's move to Toronto in 1962; they were joined by the second author in 1965. During this time the purpose and scope of the book grew to its present form: a three-volume study, ab initio, of the de Rham cohomology of smooth bundles. In particular, the material in volume I has been used at the University of Toronto as the syllabus for an introductory graduate course on differentiable manifolds.

During the long history of this book we have had numerous valuable suggestions from many mathematicians. We are especially grateful to the faculty and graduate students of the institutions below. Our exposition of Poincaré duality is based on the master's thesis of C. Auderset, while particular thanks are due to D. Toledo for his frequent and helpful contributions. Our thanks also go to E. Stamm and the Academic Press reviewer for their criticisms of the manuscript, to which we paid serious attention. A. E. Fekete, who prepared the subject index, has our special gratitude.

We are indebted to the institutions whose facilities were used by one or more of us during the writing. These include the Departments of Mathematics of Cornell University, Flinders University, the University of Fribourg, and the University of Toronto, as well as the Institut für theoretische Kernphysik at Bonn and the Forschungsinstitut für Mathematik der Eidgenössischen Technischen Hochschule, Zürich.

The entire manuscript was typed with unstinting devotion by Frances Mitchell, to whom we express our deep gratitude.

A first class job of typesetting was done by the compositors.

A. E. Fekete and H. Rummler assisted us with the proof reading; however, any mistakes in the text are entirely our own responsibility.

Finally, we would like to thank the production and editorial staff at Academic Press for their unfailing helpfulness and cooperation. Their universal patience, while we rewrote the manuscript (ad infinitum), oscillated amongst titles, and ruined production schedules, was in large measure, responsible for the completion of this work.

Werner Greub
Stephen Halperin
Ray Vanstone

Toronto, Canada

Introduction

The purpose of this monograph is to develop the de Rham cohomology theory, and to apply it to obtain topological invariants of smooth manifolds and fibre bundles.

In the de Rham theory, the real cohomology algebra of a smooth manifold is constructed by means of the calculus of differential forms, which, in turn, is the natural global version of the usual differential calculus in $\mathbb{R}^n$. Because of this approach, our principal technique is the following one:

First, establish local analytic properties in $\mathbb{R}^n$.

Second, piece these together to prove global results about differential forms on a manifold.

Finally, pass to cohomology to obtain topological properties.

This interplay between local and global phenomena is of fundamental importance in the book, and leads to the major results of the later chapters. The Euler–Poincaré–Hopf theorem is a prime instance; it states that on a compact manifold the index sum of a vector field equals the alternating sum of the Betti numbers.

Although the final results are largely taken from algebraic topology, with only one exception, no formal algebraic topology (simplices, homology and homotopy groups, etc.) is included in the text, nor is any prior knowledge of the subject assumed. (In the proof of the de Rham theorem in article 7, Chapter V, simplicial complexes are introduced; the subsequent development, however, is independent of this article.)

The contents are organized as follows: In the first four chapters we introduce manifolds and vector bundles and develop both the differential and integral calculus of differential forms. This is applied in Chapters V and VI to yield the basic properties of de Rham cohomology. In particular, Poincaré duality and the theory of mapping degree are presented as applications of integration.

In Chapter VII the partial integral, $\int f(x, y)\, dy$, appears as an intrinsic analytic operator associated with smooth fibre bundles. This fibre integral of Chern and Spanier is the main tool in Chapters VIII–X, which culminate in the Euler–Poincaré–Hopf formula and the Lefschetz fixed point theorem. (A more detailed description of the contents follows below.)

A plentiful supply of problems of varying difficulty accompanies the text. They introduce a considerable amount of additional material; however, they are never used in the proofs in the text.

All the material in this exposition is either in the literature or is well-established folklore. Nevertheless we have not attempted to associate with the theorems the names of their discoverers, except where this is already common usage.

This monograph is intended for graduate students in mathematics, especially those interested in global analysis or differential geometry. In particular, it could be used as a text or reference for an introductory course on manifolds. It presupposes a solid background in linear and multilinear algebra, and in the calculus of several real variables. The reader should also be familiar with elementary facts about rings and modules, as well as the rudiments of point set topology.

Aside from these prerequisites, and two individual quotations, the book is completely self-contained. One such quotation (Sard's theorem) is developed in the problems, while the other (existence of simple covers) occurs in article 7, Chapter V.

Every chapter consists of a number of articles which are further divided into sections. The sections, theorems, propositions, and lemmas are individually and consecutively numbered within each chapter.

In general, the reader should follow the order of presentation. However, sec. 1.2 contains only point set topology and could be omitted; the reader would then take Theorem I of that section for granted. Moreover, the detailed exposition of vector bundles in Chapter II has been placed at the beginning so that it could be used as needed. The reader might omit articles 2, 4, and 5 of this chapter, and return to them only when necessary for reference.

This volume will be followed by volume II (Lie groups and the Chern–Weil theory of characteristic classes) and volume III (cohomology of principal bundles and homogeneous spaces).

Chapter 0. This is a summary of the algebra, analysis, and point set topology which is used throughout the book. Notation and definitions are fixed, and (with the exception of really basic material) all the results to be quoted later are explicitly stated, with references.

Chapter I. Basic Concepts. Manifolds and fibre bundles are defined in this chapter. In article 1 the basic properties of topological manifolds are established; the fundamental result is the "finite atlas" theorem (Theorem I, sec. 1.2).

Its proof is based on the following observation (Proposition II, sec. 1.1): If a basis for the topology of a manifold is closed under finite unions and countable disjoint unions, then it contains every open set. The same technique is used later to establish the Poincaré isomorphism (sec. 5.12), the Künneth isomorphism (sec. 5.19), and the Thom isomorphism (sec. 9.1).

Smooth manifolds and smooth maps are introduced in article 2. The examples (spheres, products, etc.) and concepts (homotopy, partitions of unity) reappear throughout the book.

Finally, smooth fibre bundles, (E, π, B, F), are defined in article 3; π, the *projection*, is a smooth map from the *total space* E to the *base space* B; and for each $x \in B$ the *fibre over* x, $\pi^{-1}(x)$, may be identified with the smooth manifold, F (typical fibre).

Chapter II. Vector Bundles. Vector bundles and bundle maps are defined in article 1. Of particular importance is the construction principle (sec. 2.5) and its application to pull-backs (sec. 2.6). Article 2 is devoted to extending the basic constructions of linear and multilinear algebra to vector bundles.

In article 4 we discuss orientations, Riemannian metrics, and complex structures in vector bundles.

In article 5 it is shown that the module of cross-sections in a vector bundle is finitely generated and projective (theorem of Swan). The corollaries of this result are quoted extensively in Chapter III. The existence of a Riemannian metric is used to show that every vector bundle is a direct summand of a trivial bundle.

Chapter III. Tangent Bundle and Differential Forms. In article 1 the tangent bundle of a smooth manifold is defined; the derivative of a smooth map appears as a bundle map between the corresponding tangent bundles. The inverse function theorem is translated into bundle language in article 2; its applications to submanifolds are cited frequently in the following chapters.

Vector fields on a manifold are introduced in article 3 as cross-sections in the tangent bundle. It is shown that the module of vector fields is canonically isomorphic to the module of derivations in the ring of smooth functions. This article also contains the Picard theorem for ordinary differential equations, restated in the terminology of vector fields.

Differential forms (defined in article 4) are among the fundamental analytic objects treated in the book. They are the cross-sections in the exterior algebra bundle of the dual of the tangent bundle, and they form a graded anticommutative algebra. In particular, differential forms of degree n are used in article 5 to study orientations on n-manifolds.

Chapter IV. Calculus of Differential Forms. Article 1 deals with the differential calculus on manifolds. Three basic operators are introduced: the substitution operator (sec. 4.1), the Lie derivative (sec. 4.2), and the exterior derivative (sec. 4.3). The first maps p-forms to $(p - 1)$-forms by evaluation on a given vector field. The second differentiates a p-form in the direction of a vector field, while the third generalizes the notion of gradient to differential forms of higher degree.

In article 2 we consider the derivative and integral of a 1-parameter family of differential forms—with respect to the (real) parameter. This is employed later to prove Stokes' theorem (article 4) and to study homotopy properties (sec. 5.2).

The integral is a canonical linear function in the space of compactly supported n-forms on an oriented n-manifold. It is constructed in article 3 by glueing together local Riemann integrals via a partition of unity. It is shown that the basic properties of the Riemann integral continue to hold.

In article 4 Stokes' theorem is established for the annulus and the n-ball. The general form of the theorem for manifolds-with-boundary (as well as the definition of these manifolds) is left to the exercises.

Chapter V. De Rham Cohomology. The exterior derivative converts the algebra of differential forms on a manifold into a graded differential algebra. The corresponding cohomology is called the de Rham cohomology algebra.

In article 1 it is shown that the de Rham cohomology satisfies the dimension, homotopy, disjoint union, and Mayer–Vietoris axioms. In article 2 various examples (retracts, Poincaré lemma, cohomology of S^n, and $\mathbb{R}P^n$) are discussed. In article 3 everything is done again (with the appropriate modifications) for differential forms with compact carrier.

In article 4 the integral is used to establish the Poincaré duality theorem for a smooth orientable manifold. This theorem is applied in article 5 (sec. 5.13 and 5.14) to determine the nth de Rham cohomology space for any n-manifold (orientable or nonorientable). In sec. 5.15 the duality theorem is used to show that a compact manifold has finite-dimensional de Rham cohomology.

The de Rham cohomology of the product of two manifolds is computed in article 6 (Künneth theorems). In article 7 one version of the de Rham theorem is established. The results of this article are not quoted elsewhere in the book.

Chapter VI. Mapping Degree. The degree of a smooth map between compact, connected, oriented n-manifolds is defined in article 1. It is shown that it is always an integer (Theorem I, sec. 6.3).

In article 3 we define the local degree of a smooth map between oriented n-manifolds at an isolated point. If both manifolds are compact and connected, then the degree of the map is the sum of the local degrees (Theorem II, sec. 6.13). The definitions and results of article 3 depend on a canonical linear map introduced in article 2.

In article 4 the smooth version of the Hopf theorem is proved: Two smooth maps $\psi, \varphi : S^n \to S^n$ which have the same degree are smoothly homotopic. This result is applied in Chapter VIII, Proposition VII, sec. 8.10.

Chapter VII. Integration over the Fibre. This chapter deals with a general smooth bundle $\mathscr{B} = (E, \pi, B, F)$. The notion of an orientation of $\mathscr{B}$ is defined in article 2. In article 3 it is shown that in the case of a vector bundle this definition coincides with that given in sec. 2.16.

The fibre integral in an oriented bundle, $\mathscr{B} = (E, \pi, B, F)$, is defined in article 5; it is a surjective linear map from the forms with fibre compact support on E to the forms on B. The derivation of its fundamental properties (commuting with the exterior derivative, naturality, Fubini theorem) is the object of this article.

Chapter VIII. Cohomology of Sphere Bundles. In article 1 the fibre integral is applied to an oriented r-sphere bundle $\mathscr{B} = (E, \pi, B, S^r)$ to obtain the Gysin sequence and the Euler class, $\chi_{\mathscr{B}} \in H^{r+1}(B)$. The Euler class is a fundamental global invariant associated with the sphere bundle; together with $H(B)$, it determines the cohomology of E (sec. 8.4).

In article 3 we consider r-sphere bundles, where $\dim B = r + 1$, and E is oriented. Then, to every isolated singularity a of a cross-section, σ, an integer, $j_a(\sigma)$, called the index of σ at a, is assigned.

If B is compact and oriented and $\dim B = r + 1$, the Euler class, $\chi_{\mathscr{B}}$, can be integrated over B to yield a real number. The main theorem of this chapter (article 4) is a fundamental global-local result. It states that

$$\int_B^{\#} \chi_{\mathscr{B}} = \sum_a j_a(\sigma),$$

where σ is any cross-section with finitely many singularities. In article 5, finally, it is shown that such cross-sections always exist.

Chapter IX. Cohomology of Vector Bundles. In this chapter we consider vector bundles $\xi = (E, \pi, B, F)$. In article 1 it is shown that in an oriented vector bundle the fibre integral induces an isomorphism from the fibre-compact cohomology of E to the cohomology of B. The inverse of this isomorphism is the Thom isomorphism, Th. The cohomology class $Th(1)$ is a fundamental global invariant of the vector bundle, which is intimately related to the Euler class of the sphere bundle associated with ξ via a Riemannian metric (article 2).

In article 3 the index of a cross-section at an isolated zero is defined (for vector bundles whose fibre dimension is equal to the base dimension). With the aid of the Thom class, $Th(1)$, this index is expressed as an integral over the base. Finally the theorems of Chapter VIII are applied to show that the index sum of a cross-section with finitely many zeros is the integral of the Euler class over the base.

Chapter X. The Lefschetz Class of a Manifold. In this chapter the results of Chapters VIII and IX are applied to the tangent bundle of a compact manifold. The goal of article 1 is to prove that the index sum of a vector field with finitely many zeros equals the Euler–Poincaré characteristic of the underlying manifold.

Articles 2 and 3 deal with coincidence theory. Two smooth maps $\varphi : M \to N, \psi : M \to N$ have $a \in M$ as a coincidence point if $\varphi(a) = \psi(a)$. If M and N are oriented n-manifolds, we associate an integer $L_a(\varphi, \psi)$ with each isolated coincidence point a (article 3). On the other hand, if M and N are compact n-manifolds, the Lefschetz number $L(\varphi, \psi)$ is defined by

$$L(\varphi, \psi) = \sum_{p=0}^{n} (-1)^p \operatorname{tr}(\varphi^p \circ \tilde{\psi}^p),$$

where $\varphi^p : H^p(N) \to H^p(M)$ is the map induced by φ and $\tilde{\psi}^p$ is the Poincaré dual of ψ^{n-p} (article 2).

The chapter closes with the coincidence theorem in article 3 which states that if two maps φ, ψ between compact oriented n-manifolds have only finitely many coincidence points, then

$$\sum_a L_a(\varphi, \psi) = L(\varphi, \psi).$$

If $N = M$ and $\psi = \iota$, this result reduces to the Lefschetz fixed point theorem.

Contents of Volumes II and III (Tentative)

Chapter 0

Algebraic and Analytic Preliminaries

§1. Linear algebra

0.0. Notation. Throughout this book ι_X denotes the identity map of a set X. When it is clear which set we mean, we write simply ι. If U_{α_i} $(i = 1,..., r)$ are subsets of X, then $U_{\alpha_1\alpha_2\cdots\alpha_r}$ denotes their intersection. The empty set is denoted by $\varnothing$.

The symbols $\mathbb{N}$, $\mathbb{Z}$, $\mathbb{Q}$, $\mathbb{R}$, and $\mathbb{C}$ denote the natural numbers, integers, rationals, reals, and complexes.

0.1. We shall assume the fundamentals of linear and multilinear algebra. We will consider only real vector spaces (except for the occasional complex space).

A pair of dual vector spaces is denoted by E^*, E and the scalar product between E^* and E is denoted by $\langle\ ,\ \rangle$. If $F \subset E$, then

$$F^\perp = \{y^* \in E^* \mid \langle y^*, F\rangle = 0\}.$$

The dual of a linear map $\varphi: E \to F$ is denoted by φ^*. A direct sum of spaces E^p is denoted

$$\sum_p E^p \qquad \text{or} \qquad \bigoplus\nolimits_p E^p.$$

The determinant and the trace of a linear transformation $\varphi: E \to E$ are denoted respectively by $\det \varphi$, $\operatorname{tr} \varphi$.

A *determinant function* in an n-dimensional vector space is a nonzero skew-symmetric n-linear function. Every nonzero determinant function Δ_E in a real vector space defines an *orientation*.

Given two vector spaces E and F, we shall denote by $L(E; F)$ the space of linear maps $E \to F$. $L(E; E)$ will also be denoted by L_E. Finally if $E_1, ..., E_p$, and F are vector spaces, $L(E_1, ..., E_p; F)$ denotes the space of p-linear maps $E_1 \times \cdots \times E_p \to F$.

The group of linear automorphisms of a vector space E will be denoted by $GL(E)$.

A *Euclidean space* is a finite-dimensional real space, together with a positive definite inner product (also denoted by $\langle\ ,\ \rangle$). A *Hermitian space* is a finite-dimensional complex space together with a positive definite Hermitian inner product (also denoted by $\langle\ ,\ \rangle$).

If F is a real vector space, make $F^{\mathbb{C}} = \mathbb{C} \otimes F$ into a complex space by setting

$$\beta(\alpha \otimes x) = \beta\alpha \otimes x, \qquad \beta, \alpha \in \mathbb{C}, \quad x \in F.$$

$F^{\mathbb{C}}$ is called the *complexification* of F.

If $\langle\ ,\ \rangle$ is a positive definite inner product in F, then

$$\langle \alpha \otimes x, \beta \otimes y \rangle_{\mathbb{C}} = \alpha\bar{\beta}\langle x, y\rangle, \qquad \alpha, \beta \in \mathbb{C}, \quad x, y \in F$$

defines a Hermitian metric in $F^{\mathbb{C}}$.

An *indefinite inner product* in a finite-dimensional real vector space E is a non degenerate symmetric bilinear function $\langle\ ,\ \rangle$. If E_+ is a maximal subspace in which $\langle\ ,\ \rangle$ is positive definite, then $E = E_+ \oplus E_+^{\perp}$. The integer

$$\dim E_+ - \dim E_+^{\perp}$$

is independent of the choice of E_+, and is called the *signature* of $\langle\ ,\ \rangle$.

The symbol $\otimes$ denotes tensor over $\mathbb{R}$ (unless otherwise stated); for other rings R we write $\otimes_R$.

0.2. Quaternions and quaternionic vector spaces. Let H be an oriented four-dimensional Euclidean space. Choose a unit vector $e \in H$, and let $K = e^{\perp}$; it is a three-dimensional Euclidean space. Orient K so that, if e_1, e_2, e_3 is a positive basis of K, then e, e_1, e_2, e_3 is a positive basis of H.

Now define a bilinear map $H \times H \to H$ by

$$pq = -\langle p, q\rangle e + p \times q, \qquad p, q \in K$$

$$pe = p = ep, \qquad p \in H,$$

where $\times$ denotes the cross product in the oriented Euclidean space K. In this way H becomes an associative division algebra with unit element e. It is called the *algebra of quaternions* and is denoted by $\mathbb{H}$. The vectors of $\mathbb{H}$ are called *quaternions* and the vectors of K are called *pure quaternions*.

Every quaternion can be uniquely written in the form

$$p = \lambda e + q = \lambda + q, \qquad \lambda \in \mathbb{R}, \qquad q \in K.$$

λ and q are called the *real part* and the *pure quaternionic part of* p. *The conjugate* $\bar{p}$ *of a quaternion* $p = \lambda e + q$ is defined by $\bar{p} = \lambda e - q$. The map $p \to \bar{p}$ defines an automorphism of the algebra $\mathbb{H}$ called *conjugation*. The product of p and $\bar{p}$ is given by $p\bar{p} = |p|^2 e = |p|^2$.

Multiplication and the inner product in $\mathbb{H}$ are connected by the relation

$$\langle pr, qr \rangle = \langle p, q \rangle \langle r, r \rangle, \qquad p, q, r \in \mathbb{H}.$$

In particular,

$$|pr| = |p| \, |r|, \qquad p, r \in \mathbb{H}.$$

A *unit quaternion* is a quaternion of norm one. A pure unit quaternion q satisfies the relation $q^2 = -e$. If (e_1, e_2, e_3) is a positive orthonormal basis in K, then

$$e_1 e_2 = e_3, \qquad e_2 e_3 = e_1, \qquad e_3 e_1 = e_2.$$

0.3. Algebras. An *algebra* A over $\mathbb{R}$ is a real vector space together with a real bilinear map $A \times A \to A$ (called *product*). A *system of generators* of an algebra A is a subset $S \subset A$ such that every element of A can be written as a finite sum of products of the elements of S.

A *homomorphism* between two algebras A and B is a linear map $\varphi: A \to B$ such that

$$\varphi(xy) = \varphi(x)\,\varphi(y), \qquad x, y \in A.$$

A *derivation* in an algebra A is a linear map $\theta: A \to A$ satisfying

$$\theta(xy) = \theta(x)y + x\theta(y).$$

A derivation which is zero on a system of generators is identically zero. If θ_1 and θ_2 are derivations in A, then so is $\theta_1 \circ \theta_2 - \theta_2 \circ \theta_1$.

More generally, let $\varphi: A \to B$ be a homomorphism of algebras. Then a *φ-derivation* is a linear map $\theta: A \to B$ which satisfies

$$\theta(xy) = \theta(x)\,\varphi(y) + \varphi(x)\,\theta(y).$$

A *graded algebra* A over $\mathbb{R}$ is a graded vector space $A = \sum_{p \geqslant 0} A^p$, together with an algebra structure, such that

$$A^p \cdot A^q \subset A^{p+q}.$$

If

$$xy = (-1)^{pq}\, yx, \qquad x \in A^p, \quad y \in A^q,$$

then A is called *anticommutative*. If A has an identity, and $\dim A^0 = 1$, then A is called *connected*.

If A and B are graded algebras, then $A \otimes B$ can be made into a graded algebra in two ways:

(1) $(x_1 \otimes y_1)(x_2 \otimes y_2) = x_1x_2 \otimes y_1y_2$
(2) $(x_1 \otimes y_1)(x_2 \otimes y_2) = (-1)^{q_1p_2}\, x_1x_2 \otimes y_1y_2$

where $x_1, x_2 \in A$, $y_1, y_2 \in B$, $\deg y_1 = q_1$, $\deg x_2 = p_2$. The first algebra is called the *canonical tensor product* of A and B, while the second one is called the *anticommutative* or *skew tensor product* of A and B. If A and B are anticommutative, then so is the skew tensor product.

An *antiderivation* in a graded algebra A is a linear map $\alpha: A \to A$, homogeneous of *odd* degree, such that

$$\alpha(xy) = \alpha(x)y + (-1)^p\, x\alpha(y), \qquad x \in A^p, \qquad y \in A.$$

If α_1 and α_2 are antiderivations, then $\alpha_2 \circ \alpha_1 + \alpha_1 \circ \alpha_2$ is a derivation. If α is an antiderivation and θ is a derivation, then $\alpha \circ \theta - \theta \circ \alpha$ is an antiderivation.

The *direct product* $\prod_\alpha A_\alpha$ of algebras A_α is the set of infinite sequences $\{(x_\alpha) \mid x_\alpha \in A_\alpha\}$; multiplication and addition is defined component by component. The *direct sum* $\sum_\alpha A_\alpha$ is the subalgebra of sequences with finitely many nonzero terms.

0.4. Lie algebras. A *Lie algebra* E is a vector space (not necessarily of finite dimension) together with a bilinear map $E \times E \to E$, denoted by $[\ ,\ ]$, subject to the conditions

$$[x, x] = 0$$

and

$$[[x, y], z] + [[z, x], y] + [[y, z], x] = 0, \quad x, y, z \in E \qquad \text{(Jacobi identity).}$$

A *homomorphism of Lie algebras* is a linear map $\varphi: E \to F$ such that

$$\varphi([x, y]) = [\varphi(x), \varphi(y)], \qquad x, y \in E.$$

0.5. Multilinear algebra. The tensor, exterior, and symmetric algebras over a vector space E are denoted by

$$\otimes E = \sum_{p\geqslant 0} \otimes^p E, \qquad \wedge E = \sum_{p\geqslant 0} \wedge^p E, \qquad \vee E = \sum_{p\geqslant 0} \vee^p E.$$

(If $\dim E = n$, $\wedge E = \sum_{p=0}^n \wedge^p E$.)

If F is a second space, a nondegenerate pairing between $E^* \otimes F^*$ and $E \otimes F$ is given by

$$\langle x^* \otimes y^*, x \otimes y\rangle = \langle x^*, x\rangle\langle y^*, y\rangle, \qquad x^* \in E^*,\ y^* \in F^*, \quad x \in E,\ y \in F.$$

If E or F has finite dimension, this yields an isomorphism $E^* \otimes F^* \cong (E \otimes F)^*$. In particular, in this case $(\otimes^p E)^* \cong \otimes^p E^*$.

Similarly, if $\dim E < \infty$, we may write $(\wedge^p E)^* = \wedge^p E^*$, $(\vee^q E)^* = \vee^q E^*$ by setting

$$\langle x^{*1} \wedge \cdots \wedge x^{*p}, x_1 \wedge \cdots \wedge x_p\rangle = \det(\langle x^{*i}, x_j\rangle)$$

and

$$\langle y^{*1} \vee \cdots \vee y^{*p}, y_1 \vee \cdots \vee y_p\rangle = \operatorname{perm}(\langle y^{*i}, y_j\rangle),$$

where "perm" denotes the permanent of a matrix.

The algebras of multilinear (resp. skew multilinear, symmetric multilinear) functions in a space E are denoted by

$$T(E) = \sum_{p \geqslant 0} T^p(E), \qquad A(E) = \sum_{p \geqslant 0} A^p(E)$$

and

$$S(E) = \sum_{p \geqslant 0} S^p(E).$$

The multiplications are given respectively by

$$(\Phi \otimes \Psi)(x_1, \dots, x_{p+q}) = \Phi(x_1, \dots, x_p)\, \Psi(x_{p+1}, \dots, x_{p+q})$$

$$(\Phi \wedge \Psi)(x_1, \dots, x_{p+q}) = \frac{1}{p!q!} \sum_{\sigma \in S^{p+q}} \epsilon_\sigma \Phi(x_{\sigma(1)}, \dots, x_{\sigma(p)})\, \Psi(x_{\sigma(p+1)}, \dots, x_{\sigma(p+q)})$$

and

$$(\Phi \vee \Psi)(x_1, \dots, x_{p+q}) = \frac{1}{p!q!} \sum_{\sigma \in S^{p+q}} \Phi(x_{\sigma(1)}, \dots, x_{\sigma(p)})\, \Psi(x_{\sigma(p+1)}, \dots, x_{\sigma(p+q)}).$$

Here S^p denotes the symmetric group on p objects, while $\epsilon_\sigma = \pm 1$ according as the permutation σ is even or odd.

If $\dim E < \infty$, we identify the graded algebras $T(E)$ and $\otimes E^*$ (resp. $A(E)$ and $\wedge E^*$, $S(E)$ and $\vee E^*$) by setting

$$\Phi(x_1, \dots, x_p) = \langle \Phi, x_1 \otimes \cdots \otimes x_p\rangle, \qquad \Phi \in \otimes^p E^*$$

$$\Psi(x_1, \dots, x_p) = \langle \Psi, x_1 \wedge \cdots \wedge x_p\rangle, \qquad \Psi \in \wedge^p E^*$$

and

$$X(x_1, ..., x_p) = \langle X, x_1 \vee \cdots \vee x_p \rangle, \qquad X \in \vee^p E^*.$$

A linear map $\varphi: E \to F$ extends uniquely to homomorphisms

$$\otimes\varphi: \otimes E \to \otimes F, \qquad \wedge\varphi: \wedge E \to \wedge F, \qquad \vee\varphi: \vee E \to \vee F.$$

These are sometimes denoted by $\varphi_\otimes$, $\varphi_\wedge$, and $\varphi_\vee$.

To each $x \in E$ we associate the *substitution* operator $i(x): A(E) \to A(E)$, given by

$$(i(x)\Phi)(x_1, ..., x_{p-1}) = \Phi(x, x_1, ..., x_{p-1}), \qquad \Phi \in A^p(E), \quad p \geq 1,$$
$$i(x)\Phi = 0, \qquad \Phi \in A^0(E),$$

and the multiplication operator $\mu(x)$: $\wedge E \to \wedge E$ given by

$$\mu(x)(a) = x \wedge a, \qquad a \in \wedge E,$$

$i(x)$ is an antiderivation in $A(E)$ and is dual to $\mu(x)$.

§2. Homological algebra

0.6. Rings and modules. Let R be a commutative ring. If M, N are R-modules, then the tensor product $M \otimes_R N$ is again an R-module (cf. [1, p. AII–56] or [2, §8, Chap. 3]). If Q is a third R-module and if $\varphi: M \times N \to Q$ is a map satisfying the conditions

(1) $\varphi(x + y, u) = \varphi(x, u) + \varphi(y, u)$
(2) $\varphi(x, u + v) = \varphi(x, u) + \varphi(x, v)$

and

(3) $\varphi(\lambda x, u) = \varphi(x, \lambda u)$

for $x, y \in M$, $u, v \in N$, $\lambda \in R$, then there is a unique additive map $\psi: M \otimes_R N \to Q$ such that

$$\varphi(x, u) = \psi(x \otimes u), \qquad x \in M, \quad u \in N$$

(cf. [1, Prop. I(b), p. AII–51] or [2, §8, Chap. 3]). If (iii) is replaced by the stronger

$$\varphi(\lambda x, u) = \lambda\varphi(x, u) = \varphi(x, \lambda u), \qquad x \in M, \quad u \in N, \quad \lambda \in R,$$

then ψ is R-linear.

The R-module of R-linear maps $M \to N$ is denoted by $\mathrm{Hom}_R(M; N)$. $\mathrm{Hom}_R(M; R)$ is denoted by M^*. A canonical R-linear map

$$\alpha: M^* \otimes_R N \to \mathrm{Hom}_R(M; N)$$

is given by

$$\alpha(f \otimes u)(x) = f(x)u, \qquad x \in M, \quad u \in N, \quad f \in M^*.$$

A module M is called *free* if it has a basis; M is called *projective* if there exists another R-module N such that $M \oplus N$ is free. If M is projective and finitely generated, then N can be chosen so that $M \oplus N$ has a finite basis.

If M is finitely generated and projective, then so is M^*, and for all R-modules N, the homomorphism α given just above is an isomorphism. In particular, the isomorphism

$$M^* \otimes_R M \xrightarrow{\cong} \mathrm{Hom}_R(M; M)$$

specifies a unique tensor $t_M \in M^* \otimes_R M$ such that

$$\alpha(t_M) = \iota_M .$$

It is called the *unit tensor for M*.

A *graded module* is a module M in which submodules M^p have been distinguished such that

$$M = \sum_{p \geqslant 0} M^p.$$

The elements of M^p are called *homogeneous of degree p*. If $x \in M^p$, then p is called the *degree of x* and we shall write $\deg x = p$.

If M and N are graded modules, then a gradation in the module $M \otimes_R N$ is given by

$$(M \otimes_R N)^r = \sum_{p+q=r} M^p \otimes_R N^q.$$

An R-linear map between graded modules, $\varphi: M \to N$, is called *homogeneous of degree k*, if

$$\varphi(M^p) \subset N^{p+k}, \qquad p \geqslant 0$$

An R-linear map which is homogeneous of degree zero is called a *homomorphism of graded modules*.

A *bigraded module* is a module which is the direct sum of submodules $M^{p,q} (p \geqslant 0, q \geqslant 0)$.

An *exact sequence of modules* is a sequence

$$\cdots \longrightarrow M_{i-1} \xrightarrow{\varphi_{i-1}} M_i \xrightarrow{\varphi_i} M_{i+1} \longrightarrow \cdots,$$

where the φ_i are R-linear maps satisfying

$$\ker \varphi_i = \operatorname{Im} \varphi_{i-1} .$$

Suppose

$$\begin{array}{ccccccccc} M_1 & \xrightarrow{\varphi_1} & M_2 & \xrightarrow{\varphi_2} & M_3 & \xrightarrow{\varphi_3} & M_4 & \xrightarrow{\varphi_4} & M_5 \\ \alpha_1 \downarrow \cong & & \alpha_2 \downarrow \cong & & \alpha_3 \downarrow & & \alpha_4 \downarrow \cong & & \alpha_5 \downarrow \cong \\ N_1 & \xrightarrow[\psi_1]{} & N_2 & \xrightarrow[\psi_2]{} & N_3 & \xrightarrow[\psi_3]{} & N_4 & \xrightarrow[\psi_4]{} & N_5 \end{array}$$

is a commutative row-exact diagram of R-linear maps. Assume that the maps α_1, α_2, α_4, α_5 are isomorphisms. Then the *five-lemma* states that α_3 is also an isomorphism.

On the other hand, if

$$\begin{array}{ccccccccc} & & 0 & & 0 & & 0 & & \\ & & \downarrow & & \downarrow & & \downarrow & & \\ 0 & \longrightarrow & M_{11} & \longrightarrow & M_{12} & \longrightarrow & M_{13} & \longrightarrow & 0 \\ & & \downarrow & & \downarrow & & \downarrow & & \\ 0 & \longrightarrow & M_{21} & \longrightarrow & M_{22} & \longrightarrow & M_{23} & \longrightarrow & 0 \\ & & \downarrow & & \downarrow & & \downarrow & & \\ 0 & \longrightarrow & M_{31} & \longrightarrow & M_{32} & \longrightarrow & M_{33} & \longrightarrow & 0 \\ & & \downarrow & & \downarrow & & \downarrow & & \\ & & 0 & & 0 & & 0 & & \end{array}$$

is a commutative diagram of R-linear maps with exact columns, and if the middle and bottom rows are exact, then the *nine-lemma* states that the top row is exact.

An algebra over R is an R-module A together with an R-linear map $A \otimes_R A \to A$. In particular if M is any R-module, the tensor, exterior, and symmetric algebras over M are written $\otimes_R M$, $\wedge_R M$ and, $\vee_R M$. If M is finitely generated and projective, there are isomorphisms, $(\otimes_R^p M)^* \cong \otimes_R^p M^*$, $(\wedge_R^p M)^* \cong \wedge_R^p M^*$, $(\vee_R^p M)^* \cong \vee_R^p M^*$, defined in exactly the same way as in sec. 0.5.

0.7. Differential spaces. A *differential space* is a vector space X together with a linear map $\delta: X \to X$ satisfying $\delta^2 = 0$. δ is called the *differential operator in* X. The elements of the subspaces

$$Z(X) = \ker \delta \qquad \text{and} \qquad B(X) = \operatorname{Im} \delta.$$

are called, respectively, *cocycles* and *coboundaries*. The space $H(X) = Z(X)/B(X)$ is called the *cohomology space* of X.

A *homomorphism* of differential spaces $\varphi: (X, \delta_X) \to (Y, \delta_Y)$ is a linear map for which $\varphi \circ \delta_X = \delta_Y \circ \varphi$. It restricts to maps between the cocycle and coboundary spaces, and so induces a linear map

$$\varphi_{\#} : H(X) \to H(Y).$$

A *homotopy operator* for two such homomorphisms, φ, ψ, is a linear map $h: X \to Y$ such that

$$\varphi - \psi = h \circ \delta + \delta \circ h.$$

If h exists then $\varphi_{\#} = \psi_{\#}$.

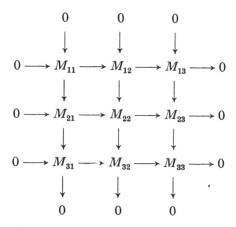

Suppose

$$0 \longrightarrow X \xrightarrow{f} Y \xrightarrow{g} Z \longrightarrow 0$$

is an exact sequence of homomorphisms of differential spaces. Every cocycle $z \in Z$ has a preimage $y \in Y$. In particular,

$$g(\delta y) = \delta z = 0$$

and so there is a cocycle $x \in X$ for which $f(x) = \delta y$. The class $\xi \in H(X)$ represented by x depends only on the class $\zeta \in H(Z)$ represented by z. The correspondence $\zeta \mapsto \xi$ defines a linear map

$$\partial\colon H(Z) \to H(X)$$

called the *connecting homomorphism* for the exact sequence. The triangle

$$\begin{array}{ccc} H(X) & \xrightarrow{\quad f_\# \quad} & H(Y) \\ & {}_{\partial}\nwarrow \qquad \swarrow_{g_\#} & \\ & H(Z) & \end{array}$$

is exact.

If

$$\begin{array}{ccccccccc} 0 & \longrightarrow & X & \longrightarrow & Y & \longrightarrow & Z & \longrightarrow & 0 \\ & & \downarrow{\scriptstyle \varphi} & & \downarrow{\scriptstyle \psi} & & \downarrow{\scriptstyle \chi} & & \\ 0 & \longrightarrow & X' & \longrightarrow & Y' & \longrightarrow & Z' & \longrightarrow & 0 \end{array}$$

is a row-exact diagram of differential spaces, then

$$\partial' \circ \chi_\# = \varphi_\# \circ \partial$$

(∂, ∂' the connecting homomorphisms).

0.8. Graded differential spaces and algebras. A graded space $X = \sum_{p \geqslant 0} X^p$ together with a differential operator δ homogeneous of degree $+1$ is called a *graded differential space*. In such a case the cocycle, coboundary, and cohomology spaces are graded:

$$Z^p(X) = Z(X) \cap X^p, \qquad B^p(X) = B(X) \cap X^p$$

and

$$H^p(X) = Z^p(X)/B^p(X).$$

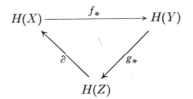

A *homomorphism of graded differential spaces* is a homomorphism of differential spaces, homogeneous of degree zero.

Now assume that X has finite dimension and let $\varphi: X \to X$ be a homomorphism of graded differential spaces. Let

$$\varphi^p: X^p \to X^p \qquad \text{and} \qquad (\varphi_\#)^p: H^p(X) \to H^p(X)$$

be the restrictions of φ and $\varphi_\#$ to X^p and $H^p(X)$. The *algebraic Lefschetz formula* states that

$$\sum_{p \geq 0} (-1)^p \operatorname{tr} \varphi^p = \sum_{p \geq 0} (-1)^p \operatorname{tr} (\varphi_\#)^p.$$

In particular, if $\varphi = \iota$, we obtain the *Euler-Poincaré formula*

$$\sum_{p \geq 0} (-1)^p \dim X^p = \sum_{p \geq 0} (-1)^p \dim H^p(X).$$

A *graded differential algebra* A is a graded algebra together with an antiderivation, δ, homogeneous of degree one such that $\delta^2 = 0$. In this case $Z(A)$ is a graded subalgebra and $B(A)$ is a graded ideal in $Z(A)$. Thus $H(A)$ becomes a graded algebra. It is called the *cohomology algebra* of A. If A is anticommutative, then so is $H(A)$.

A *homomorphism of graded differential algebras* $\varphi: A \to B$ is a map which is a homomorphism of graded differential spaces and a homomorphism of algebras. It induces a homomorphism between the cohomology algebras,

$$\varphi_\# : H(A) \to H(B).$$

Next let A and B be graded differential algebras and consider the skew tensor product $A \otimes B$. Then the antiderivation in $A \otimes B$, given by

$$\delta(x \otimes y) = \delta x \otimes y + (-1)^p x \otimes \delta y, \qquad x \in A^p, \quad y \in B,$$

satisfies $\delta^2 = 0$. Thus $A \otimes B$ becomes a graded differential algebra. The tensor multiplication between A and B induces an isomorphism

$$H(A) \otimes H(B) \xrightarrow{\cong} H(A \otimes B)$$

of graded algebras. It is called the *Künneth isomorphism*.

§3. Analysis and topology

0.9. Smooth maps. Let E, F be real, finite dimensional vector spaces with the standard topology. Let $U \subset E$ be an open subset. A map $\varphi: U \to F$ is called *differentiable* at a point $a \in U$ if for some $\psi_a \in L(E; F)$

$$\lim_{t \to 0} \frac{\varphi(a + th) - \varphi(a)}{t} = \psi_a(h), \qquad h \in E.$$

In this case ψ_a is called the *derivative of* φ *at* a and is denoted by $\varphi'(a)$. We shall write

$$\varphi'(a; h) = \varphi'(a)h = \psi_a(h), \qquad h \in E.$$

If φ is differentiable at every point $a \in U$, it is called a *differentiable map* and the map

$$\varphi': U \to L(E; F)$$

given by $a \mapsto \varphi'(a)$ is called the *derivative of* φ. Since $L(E; F)$ is again a finite dimensional vector space, it makes sense for φ' to be differentiable. In this case the derivative of φ' is denoted by φ''; it is a map

$$\varphi'': U \to L(E; L(E; F)) = L(E, E; F).$$

More generally, the kth derivative of φ (if it exists) is denoted by $\varphi^{(k)}$,

$$\varphi^{(k)}: U \to L(\underbrace{E, \dots, E}_{k \text{ terms}}; F).$$

For each $a \in U$, $\varphi^{(k)}(a)$ is a symmetric k-linear map of $E \times \cdots \times E$ into F. If all derivatives of φ exist, φ is called *infinitely differentiable*, or *smooth*.

A smooth map $\varphi: U \to V$ between open subsets $U \subset E$ and $V \subset F$ is called a *diffeomorphism* if it has a smooth inverse.

Assume now that $\varphi: U \to F$ is a map with a continuous derivative such that for some point $a \in U$

$$\varphi'(a): E \xrightarrow{\cong} F$$

is a linear isomorphism. Then the *inverse function theorem* states that there are neighbourhoods U of a and V of $\varphi(a)$ such that φ restricts to a diffeomorphism $U \xrightarrow{\cong} V$.

We shall also need the basic properties of the Riemannian integral of a compactly supported function in $\mathbb{R}^n$ (linearity, transformation of coordinates, differentiation with respect to a parameter). The theory extends to vector-valued functions (integrate component by component).

Finally, we shall use the Picard existence and uniqueness theorem for ordinary differential equations as given in [3, p. 22].

0.10. The exponential map. Let E be an n-dimensional real or complex vector space and let $\sigma: E \to E$ be a linear transformation. It follows from the standard existence theorems of differential equations that there is a unique smooth map $\tau: \mathbb{R} \to L_E$ satisfying the linear differential equation

$$\dot{\tau} = \sigma \circ \tau$$

and the initial condition $\tau(0) = \iota$. The linear transformation $\tau(1)$ is called the *exponential of* σ and is denoted by $\exp \sigma$.

In this way we obtain a (nonlinear) map $\exp: L_E \to L_E$. It has the following properties:

(0) $\exp 0 = \iota$.
(1) If $\sigma_1 \circ \sigma_2 = \sigma_2 \circ \sigma_1$, then $\exp(\sigma_1 + \sigma_2) = \exp \sigma_1 \circ \exp \sigma_2$.
(2) $\exp(k\sigma) = (\exp \sigma)^k$, $k \in \mathbb{Z}$.
(3) $\det \exp \sigma = \exp \operatorname{tr} \sigma$.
(4) If a Euclidean (Hermitian) inner product is defined in the real (complex) vector space E and if σ^* denotes the adjoint linear transformation, then

$$\exp \sigma^* = (\exp \sigma)^*.$$

(All these properties are easy consequences of the uniqueness theorem for solutions of differential equations.)

Relations (0) and (1) imply that $\exp \sigma$ is an automorphism with $(\exp \sigma)^{-1} = \exp(-\sigma)$. In particular, if σ is self-adjoint, then so is $\exp \sigma$ and if σ is skew (resp. Hermitian skew), then $\exp \sigma$ is a proper rotation (resp. unitary transformation) of E.

In terms of an infinite series we can write

$$\exp \sigma = \sum_{p=0}^{\infty} \frac{1}{p!} \sigma^p.$$

0.11. General topology. We shall assume the basics of point set topology: manipulation with open sets and closed sets, compactness, Hausdorff spaces, locally compact spaces, second countable spaces, connectedness, paracompact spaces, normality, open coverings, shrinking of an open covering, etc.

The closure of a subset A of a topological space X will be denoted by $\bar{A}$. If A and B are any two subsets of X, we shall write

$$A - B = \{x \in A \mid x \notin B\}.$$

A neighbourhood of A in X will always mean an *open* subset U of X such that $U \supset A$.

An *open cover* of X is a family $\mathcal{O}$ of open sets whose union is X. It is called *locally finite* if every point has a neighbourhood which meets only finitely many elements of $\mathcal{O}$. $\mathcal{O}$ is called a *refinement* of an open cover $\mathcal{U}$ if each $O \in \mathcal{O}$ is a subset of some $U \in \mathcal{U}$. X is called *paracompact* if every open cover of X has a locally finite refinement.

A *basis* for the topology of X is a family $\mathcal{O}$ of open sets such that each open subset of X is the union of elements of $\mathcal{O}$. If $\mathcal{O}$ is closed under finite intersections, it is called an *i*-basis. If X has a countable basis, it is called *second countable*.

Chapter I

Basic Concepts

§1. Topological manifolds

1.1. n-manifolds. An *n-dimensional topological manifold* (or simply a topological n-manifold) is a Hausdorff space M with a countable basis which satisfies the following condition:

Every point $a \in M$ has a neighbourhood U_a which is homeomorphic to an open subset of an n-dimensional real vector space E.

In this case we write $\dim M = n$.

A *chart* for a topological n-manifold M is a triple (U, u, V) where U is an open subset of M, V is an open subset of an n-dimensional real vector space E, and $u: U \to V$ is a homeomorphism. Because the chart (U, u, V) is determined by the pair (U, u), we will usually denote a chart by (U, u).

An *atlas* on an n-manifold M is a family of charts $\{(U_\alpha, u_\alpha) \mid a \in \mathscr{I}\}$, where $\mathscr{I}$ is an arbitrary indexing set, such that the sets U_α form a covering of M:

$$M = \bigcup_{\alpha \in \mathscr{I}} U_\alpha .$$

An atlas is called countable (or finite) if the index set is countable (or finite).

Proposition I: Every topological n-manifold M admits a countable atlas $\{(U_i, u_i, \mathbb{R}^n) \mid i \in \mathbb{N}\}$, where the closures $\bar{U}_i$ are compact.

Corollary I: A compact n-manifold admits a finite atlas with each member of the covering homeomorphic to $\mathbb{R}^n$.

Corollary II: Every topological n-manifold M admits a countable open cover $\{U_i \mid i \in \mathbb{N}\}$ such that

(1) U_i is a finite union of open sets each of which is homeomorphic to $\mathbb{R}^n$
(2) $\bar{U}_i$ is compact
(3) $\bar{U}_i \subset U_{i+1}$, $i \in \mathbb{N}$.

Next let $\mathcal{O}$ be any basis for the topology of M and let $\mathcal{O}_{\mathrm{f}}$ be the collection of open subsets $O \subset M$ of the form

$$O = \bigcup_{i=1}^{k} O_i, \qquad O_i \in \mathcal{O}, \qquad 1 \leqslant k < \infty.$$

Then $\mathcal{O}_{\mathrm{f}} \supset \mathcal{O}$ is a basis for the topology of M. Next let $\mathcal{O}_{\mathrm{s}}$ be the collection of open subsets $O \subset M$ of the form

$$O = \bigcup_{i=1}^{\infty} O_i, \qquad O_i \in \mathcal{O},$$

where the O_i are *disjoint*. Then $\mathcal{O}_{\mathrm{s}} \supset \mathcal{O}$ is a basis for the topology of M.

Proposition II: The basis $((\mathcal{O}_{\mathrm{f}})_{\mathrm{s}})_{\mathrm{f}}$ contains every open subset of M.

Proof: Let U be an arbitrary open subset of M and write

$$U = \bigcup_{i=1}^{\infty} K_i,$$

where K_i is compact and $K_i \subset K_{i+1}$. Next construct an open cover $\{U_i \mid i \in \mathbb{N}\}$ of U so that $U_i \subset U$ and

(1) $\bigcup_{i=1}^{k} U_i \supset K_k$

(2) $U_i \in \mathcal{O}_{\mathrm{f}}$ and $\bar{U}_i$ is compact

(3) $\bigcup_{i=1}^{k} U_i \supset \bigcup_{i=1}^{k-1} \bar{U}_i$

and

(4) $U_i \cap U_j = \varnothing$ unless $j = i - 1$ or $i + 1$.

Condition (1) shows that the U_i give a cover of U. Let

$$V_1 = \bigcup_{i \geqslant 0} U_{2i+1}, \qquad V_2 = \bigcup_{i \geqslant 1} U_{2i}.$$

By conditions (2), (4), V_1 and V_2 belong to $(\mathcal{O}_f)_s$. Hence

$$U = V_1 \cup V_2 \in ((\mathcal{O}_f)_s)_f .$$

Q.E.D.

Proposition III: A topological n-manifold, M, has the following properties:

(1) An open subset of M (with the induced topology) is again an n-manifold
(2) M is connected if and only if M is pathwise connected
(3) M is paracompact.

Corollary: For every open covering $\{U_\alpha \mid \alpha \in \mathcal{I}\}$ of M, there is a *shrinking* $\{V_\alpha \mid \alpha \in \mathcal{I}\}$; i.e., there is an open covering $\{V_\alpha \mid \alpha \in \mathcal{I}\}$ such that $\bar{V}_\alpha \subset U_\alpha$.

1.2. Dimension theory. In this section we develop some elementary results of dimension theory to prove

Theorem I: Let $\mathcal{O}$ be an open covering of a topological manifold M. Then there exists a refinement $\{V_{ij}\}$, where $j \in \mathbb{N}$ and i runs through a *finite* set, such that for each i

$$V_{ij} \cap V_{ik} = \varnothing, \qquad j \neq k.$$

To prove this theorem we need the following definitions and results. An open covering of a topological space X is said to have *order* $\leqslant p$ if the intersection of every $p + 1$ elements of the cover is empty. X is said to have *Lebesgue dimension* $\leqslant p$ if every open cover has a locally finite refinement of order $\leqslant p + 1$. We write this as *dim* $X \leqslant p$. If *dim* $X \leqslant p$, *dim* $X \nleqslant p - 1$, we say *dim* $X = p$.

Proposition IV: Every topological n-manifold M satisfies

$$\dim M \leqslant 7^n.$$

Remark: It can, in fact, be shown that *dim* $M = n$ (cf. [8]), but we shall not need this.

Lemma I: If *dim* $X \leqslant m$ $(m \geqslant 1)$, then $\dim(X \times \mathbb{R}) \leqslant 7m$.

Proof: Let $\mathscr{O}$ be any open cover of $X \times \mathbb{R}$. For each $n \in \mathbb{Z}$, choose an open cover $\mathscr{W}_n$ of X with the following property: If $W \in \mathscr{W}_n$ and $t \in [n, n+2]$, then for some $\epsilon > 0$ and $O \in \mathscr{O}$, $W \times (t - \epsilon, t + \epsilon) \subset O$. We may assume that each $\mathscr{W}_n$ is locally finite and of order $\leqslant m + 1$ (since $\dim X \leqslant m$).

By considering open sets of the form $W \times (t - \epsilon, t + \epsilon)$ $(W \in \mathscr{W}_n)$, obtain a locally finite open covering of $X \times (n, n+2)$ of order $\leqslant 2(m+1)$. These open coverings together provide an open covering of $X \times \mathbb{R}$ of order $\leqslant 4(m+1) \leqslant 7m + 1$.

Q.E.D.

Corollary: $\dim \mathbb{R}^n \leqslant 7^n$.

Lemma II: Let X be a normal space with a countable basis. Suppose U and V are open sets such that $\dim U \leqslant n$, $\dim V \leqslant n$ and $X = U \cup V$. Then $\dim X \leqslant n$.

Proof: Choose disjoint open sets $U', V' \subset X$ such that

$$(X - V) \subset U' \subset U \qquad \text{and} \qquad (X - U) \subset V' \subset V.$$

Let $\mathscr{O}$ be an open covering of X. By refining $\mathscr{O}$ if necessary we may assume that $\mathscr{O}$ is of the form

$$\mathscr{O} = \mathscr{O}_U \cup \mathscr{O}_{V'},$$

where

$$\mathscr{O}_U = \{O_k \mid k \in \mathbb{N}\}$$

is a locally finite open covering of U of order $\leqslant n + 1$, and $\mathscr{O}_{V'}$ is an open covering of V'.

Set $\mathscr{O}_V = \{O_k \cap V \mid k \in \mathbb{N}\}$. Then $\mathscr{O}_V \cup \mathscr{O}_{V'}$ is an open covering of V. Let $\mathscr{W}$ be a locally finite refinement of this covering of order $\leqslant n + 1$. Then $\mathscr{W}$ is the disjoint union of $\mathscr{W}^{(1)}$ and $\mathscr{W}^{(2)}$, where $\mathscr{W}^{(1)}$ consists of those open sets contained in V' and $\mathscr{W}^{(2)}$ consists of the others.

We denote the elements of $\mathscr{W}^{(1)}$ (resp. $\mathscr{W}^{(2)}$) by W_α (resp. W_β). Thus each W_β is contained in some O_k. Hence $\mathscr{W}^{(2)}$ is the disjoint union of the subcollections $\mathscr{W}_k^{(2)}$ given by

$$\mathscr{W}_k^{(2)} = \{W_\beta \mid W_\beta \subset O_k,\ W_\beta \not\subset O_i,\ i < k\}.$$

Now define open sets Y_k by

$$Y_k = (O_k \cap U') \cup \bigcup_{\beta_\nu} W_{\beta_\nu},$$

where the union is taken over those β_ν such that $W_{\beta_\nu} \in \mathscr{W}_k^{(2)}$. Set

$$\mathscr{Y} = \{Y_k \mid k \in \mathbb{N}\}.$$

We show that $\mathscr{Y}^{(1)} = \mathscr{Y} \cup \mathscr{W}^{(1)}$ is a locally finite refinement and has order $\leqslant n + 1$.

First note that since the O_k cover U, the Y_k cover U'. On the other hand, the Y_k contain all the W_β, and so the W_α together with the Y_k cover V (since $\mathscr{W}$ covers V). Since $X = U' \cup V$, it follows that $\mathscr{Y}^{(1)}$ is a cover of X.

Next observe that $Y_k \subset O_k$ and so $\mathscr{Y}$ refines $\mathscr{O}$. But $\mathscr{W}$ (and hence $\mathscr{W}^{(1)}$) also refines $\mathscr{O}$. Thus $\mathscr{Y}^{(1)}$ refines $\mathscr{O}$.

To show that $\mathscr{Y}^{(1)}$ has order $\leqslant n + 1$, let

$$x \in Y_{k_1} \cap \cdots \cap Y_{k_p} \cap W_{\alpha_1} \cap \cdots \cap W_{\alpha_q}.$$

We distinguish two cases.

Case I: $x \in U'$. Then $q = 0$ and $x \in Y_{k_1} \cap \cdots \cap Y_{k_p} \subset O_{k_1} \cap \cdots \cap O_{k_p}$. Hence $p \leqslant n + 1$ and so $p + q \leqslant n + 1$.

Case II: $x \notin U'$. Then for each k_i there is an element $W_{\beta_i} \subset \mathscr{W}_{k_i}^{(2)}$ such that $x \in W_{\beta_i}$. Moreover, the W_{β_i} are necessarily distinct. Thus

$$x \in W_{\beta_1} \cap \cdots \cap W_{\beta_p} \cap W_{\alpha_1} \cap \cdots \cap W_{\alpha_q};$$

i.e., x is in $p + q$ distinct elements of $\mathscr{W}$. It follows that $p + q \leqslant n + 1$.

Distinguishing between the same two cases and using the fact that $\mathscr{O}_U$ and $\mathscr{W}$ are locally finite, we see that $\mathscr{Y}^{(1)}$ is locally finite.

Q.E.D.

Lemma III: If a manifold M has a basis O_α such that for each α, *dim* $O_\alpha \leqslant p$, then for every open subset O of M

$$\dim O \leqslant p.$$

Proof: Clearly, if a space X is the *disjoint* union of open subsets with *dim* $\leqslant p$ then *dim* $X \leqslant p$. On the other hand, Lemma II implies that if a manifold Q is a *finite* union with $dim \leqslant p$, then $dim\, Q \leqslant p$. Now the lemma follows from Proposition II, sec. 1.1.

Q.E.D.

Corollary: If O is an open subset of $\mathbb{R}^n$, then $dim\, O \leqslant 7^n$.

Proof of Proposition IV: Observe that M admits a basis consisting of open subsets O_α homeomorphic to open subsets of $\mathbb{R}^n$. Hence, by the corollary above, $dim\, O_\alpha \leqslant 7^n$. Now Lemma III (applied with $O = M$) yields the proposition.

Q.E.D.

Proof of Theorem I: Let $\mathscr{O}$ be any open covering of M. According to Proposition IV there exists a locally finite refinement of finite order. Thus we may assume that $\mathscr{O}$ is locally finite and of finite order p. Moreover, we may assume that $\mathscr{O}$ is indexed by $\mathbb{N}$, $\mathscr{O} = \{O_j \mid j \in \mathbb{N}\}$.

Now we proceed by induction on p. If $p = 1$, there is nothing to prove. Assume that the theorem holds for coverings of order $p - 1$ and that $\mathscr{O}$ has order p. For each distinct set $\nu_1 < \cdots < \nu_{p+1}$ of $(p + 1)$ indices let

$$O_{\nu_1 \cdots \nu_{p+1}} = \bigcap_{k=1}^{p+1} O_{\nu_k}.$$

Since $\mathscr{O}$ has order p these sets are disjoint. Denote them by V_{1i} $(i = 1, 2, \ldots)$ and set

$$V_1 = \bigcup V_{1i}.$$

Next choose open sets U_j so that $\bar{U}_j \subset O_j$ and $\bigcup_j U_j = M$. Let A denote the union of all sets of the form $\bar{U}_{\nu_1} \cap \cdots \cap \bar{U}_{\nu_{p+1}}$ $(\nu_1 < \cdots < \nu_{p+1})$. Then A is closed because the O_j are locally finite.

Now the U_j provide a locally finite covering of $M - A$ of order $p - 1$. Since $M = (M - A) \cup V_1$, the theorem follows by induction.

Q.E.D.

Corollary: A topological manifold M admits a finite atlas.

Proof: Let $\{(U_\alpha, u_\alpha) \mid \alpha \in \mathscr{I}\}$ be any countable atlas for M such that the sets $u_\alpha(U_\alpha)$ are disjoint. Let $\{V_{ij} \mid i \leqslant \dim M + 1, j \in \mathbb{N}\}$ be the refinement of Theorem I.

Choose $\alpha(i, j)$ such that $V_{ij} \subset U_{\alpha(i,j)}$. Now let

$$W_i = \bigcup_{j \in \mathbb{N}} V_{ij}$$

and define $w_i \colon W_i \to \mathbb{R}^n$ by

$$w_i(x) = u_{\alpha(i,j)}(x), \qquad x \in V_{ij}.$$

Since the V_{ij} are disjoint for fixed i, w_i is a homeomorphism of W_i onto

$$\bigcup_{j \in \mathbb{N}} u_{\alpha(i,j)}(V_{ij}) \subset \mathbb{R}^n$$

and hence $\{(W_i, w_i)\}$ is a finite atlas.

Q.E.D.

§2. Smooth manifolds

1.3. Smooth atlases. Let M be a topological manifold and let $\{(U_\alpha, u_\alpha) \mid \alpha \in \mathscr{I}\}$ be an atlas for M. Consider two neighbourhoods U_α, U_β such that $U_{\alpha\beta} = U_\alpha \cap U_\beta \neq \varnothing$. Then a homeomorphism

$$u_{\alpha\beta}\colon u_\beta(U_{\alpha\beta}) \to u_\alpha(U_{\alpha\beta})$$

is defined by $u_{\alpha\beta} = u_\alpha \circ u_\beta^{-1}$. This map is called the *identification* map for U_α and U_β. By definition $u_{\gamma\beta} \circ u_{\beta\alpha} = u_{\gamma\alpha}$ in $u_\alpha(U_{\alpha\beta\gamma})$, and $u_{\alpha\alpha}(x) = x$, $x \in u_\alpha(U_\alpha)$. These relations imply that the inverse of $u_{\alpha\beta}$ is $u_{\beta\alpha}$.

The atlas $\{(U_\alpha, u_\alpha)\}$ is called *smooth* if all its identification maps are smooth (as mappings between open subsets of real vector spaces).

Two smooth atlases are *equivalent* if their *union* is again a smooth atlas; i.e., $\{(U_\alpha, u_\alpha)\}$ and $\{(V_i, v_i)\}$ are equivalent if all the maps

$$v_i \circ u_\alpha^{-1}\colon u_\alpha(U_\alpha \cap V_i) \to v_i(U_\alpha \cap V_i)$$

and their inverses are smooth. A *smooth structure* on M is an equivalence class of smooth atlases on M. A topological manifold endowed with a smooth structure is called a *smooth manifold.* An argument similar to that of the corollary to Theorem I shows that every smooth manifold admits a finite smooth atlas.

Henceforth we shall use the word "manifold" in the sense of a *smooth* manifold. An atlas for a manifold will mean a member of its smooth structure and the term *chart* will refer to a member of an atlas.

1.4. Examples of manifolds. 1. *Spheres*: Let E be an n-dimensional Euclidean space with inner product $\langle\ ,\ \rangle$. The unit sphere S^{n-1} is $\{x \in E \mid \langle x, x\rangle = 1\}$. S^{n-1} is a Hausdorff space with a countable basis in the relative topology. Let $a \in S^{n-1}$ and $U_+ = S^{n-1} - \{a\}$, $U_- = S^{n-1} - \{-a\}$. Define maps $u_+\colon U_+ \to a^\perp$, $u_-\colon U_- \to a^\perp$ by

$$u_+(x) = \frac{x - \langle x, a\rangle a}{1 - \langle x, a\rangle}, \qquad u_-(x) = \frac{x - \langle x, a\rangle a}{1 + \langle x, a\rangle}.$$

Then $\{(U_i, u_i) \mid i = +, -\}$ is a smooth atlas for S^{n-1}. Moreover, the atlas obtained, in this way, from a second point $b \in S^{n-1}$ is equivalent to this one. Thus the smooth manifold structure of S^{n-1} is independent of the choice of a.

2. *Projective Spaces*: Consider the equivalence relation on S^n whose equivalence classes are the pairs $\{x, -x\}$, $x \in S^n$, and introduce the quotient topology on the set of equivalence classes. We call the result *real projective n-space*, $\mathbb{R}P^n$.

To construct a smooth atlas on $\mathbb{R}P^n$, consider the projection $\pi: S^n \to \mathbb{R}P^n$ given by $\pi(x) = \{x, -x\}$. If O is an open set in S^n such that $x \in O$ implies that $-x \notin O$, then $\pi(O)$ is open in $\mathbb{R}P^n$ and $\pi: O \to \pi(O)$ is a homeomorphism. Now let $\{(U_\alpha, u_\alpha)\}$ be an atlas for S^n such that, if $x \in U_\alpha$, $-x \notin U_\alpha$. Then $\{(\pi(U_\alpha), u_\alpha \circ \pi^{-1})\}$ is a smooth atlas for $\mathbb{R}P^n$.

3. *Tori*: Denote the elements of $\mathbb{R}^n$ by $x = (\xi^1, \ldots, \xi^n)$, $\xi^i \in \mathbb{R}$. Define an equivalence relation in $\mathbb{R}^n$ by $x' \sim x$ if and only if $\xi^{i'} - \xi^i \in \mathbb{Z}$, $i = 1, \ldots, n$. Let the set of equivalence classes, with the quotient topology, be denoted by T^n and let $\pi: \mathbb{R}^n \to T^n$ be the canonical projection.

Consider the smooth atlas for $\mathbb{R}^n$ given by $\{(U_a, u_a)\}$, $a \in \mathbb{R}^n$, where

$$U_a = \{x \in \mathbb{R}^n \mid |\xi^i - \alpha^i| < \tfrac{1}{4}, i - 1, \ldots, n\}, \qquad u_a(x) = x.$$

Then $\{(\pi(U_a), u_a \circ \pi^{-1})\}$ is a smooth atlas for T^n.

4. *One-point compactifications*: Let E be a finite-dimensional Euclidean space. Then its one-point compactification E_∞ (cf. [4, p. 246]) can be given a smooth atlas (U_0, u_0), (U_∞, u_∞) as follows. Let x_∞ denote the point at ∞ and set

$$U_0 = E, \qquad u_0 = \iota_E$$

$$U_\infty = E_\infty - \{0\}, \qquad u_\infty(x) = \begin{cases} \dfrac{x}{|x|^2}, & x \neq x_\infty \\ 0, & x = x_\infty . \end{cases}$$

$(|x|^2 = \langle x, x \rangle)$

In the case when E is $\mathbb{C}$, regarded as a two-manifold, we also obtain a smooth atlas when u_∞ is replaced by $\bar{u}_\infty$, where

$$\bar{u}_\infty(z) = \begin{cases} z^{-1} = |z|^{-2}\bar{z}, & z \neq z_\infty \\ 0, & z = z_\infty \end{cases}$$

5. *Open subsets*: Let O be an open subset of a manifold M. If $\{(U_\alpha, u_\alpha)\}$ is a smooth atlas for M, then $\{(O \cap U_\alpha, u_\alpha|_{O \cap U_\alpha})\}$ is a smooth atlas for O. Equivalent atlases on M yield equivalent atlases on O. Thus the smooth structure of M induces a smooth structure on O. Any open subset of a manifold, with the induced smooth structure, is called an *open submanifold*.

1.5. Smooth maps. Let M, N be manifolds and assume that $\varphi: M \to N$ is a continuous map. Let $\{(U_\alpha, u_\alpha)\}$ and $\{(V_i, v_i)\}$ be atlases for M and N, respectively. Then φ determines continuous maps

$$\varphi_{i\alpha}: u_\alpha(U_\alpha \cap \varphi^{-1}(V_i)) \to v_i(V_i)$$

by $\varphi_{i\alpha} = v_i \circ \varphi \circ u_\alpha^{-1}$.

We say that $\varphi: M \to N$ is *smooth* if the maps $\varphi_{i\alpha}$ (as mappings of open subsets of vector spaces) are smooth. (This definition is independent of the choice of atlases for M and N). Moreover, if $\varphi: M \to N$ and $\psi: N \to P$ are smooth maps, then $\psi \circ \varphi: M \to P$ is smooth. The set of smooth maps $M \to N$ is denoted by $\mathscr{S}(M; N)$.

Proposition V: (1) If $\varphi: M \to N$ is smooth and $O \subset M$ is open, then the restriction of φ to O is smooth.

(2) If $\varphi: M \to N$ is a set map such that the restriction of φ to each element of an open covering of M is smooth, then φ is smooth.

A smooth map $\varphi: M \to N$ is called a *diffeomorphism* if it has a smooth inverse $\varphi^{-1}: N \to M$. Every diffeomorphism is a homeomorphism. Two manifolds M and N are *diffeomorphic* if there exists a diffeomorphism $\varphi: M \to N$. The fundamental equivalence relation for smooth manifolds is that of being diffeomorphic.

Examples: 1. Let M and N be manifolds and let $b \in N$. The *constant map* $\varphi: M \to N$ given by $\varphi(x) = b$, $x \in M$, is smooth.

2. Let E be a Euclidean space of finite dimension and let B be the open ball of radius r (about 0). The map $\varphi: B \to E$ given by $\varphi(x) = (r^2 - |x|^2)^{-1} x$, where $|x|^2 = \langle x, x \rangle$, is a diffeomorphism.

3. Let S^{n-1} be the unit sphere in a Euclidean space E. Then the inclusion map $i: S^{n-1} \to E$ is smooth (use Proposition V).

4. The projection map $\pi: S^n \to \mathbb{R}P^n$ is smooth.

5. Let O be any open submanifold of a manifold M. Then the inclusion map $i: O \to M$ is smooth.

6. The canonical projection of Example 3, sec. 1.4 is smooth.

7. Let M be a topological manifold and suppose that $\{(U_\alpha, u_\alpha)\}$ and $\{(V_i, v_i)\}$ are smooth atlases on M. Denote the corresponding smooth manifolds by $M_{(1)}$ and $M_{(2)}$. The identity map $\iota: M_{(1)} \to M_{(2)}$ is a diffeomorphism if and only if the two atlases are equivalent: i.e., if and only if $M_{(1)} = M_{(2)}$.

8. Given real numbers α, β with $\alpha < \beta$, there exists a smooth nondecreasing function $g: \mathbb{R} \to [0, 1]$ such that

$$g(t) = \begin{cases} 0, & t \leqslant \alpha \\ 1, & t \geqslant \beta. \end{cases}$$

In fact define $f: \mathbb{R} \to \mathbb{R}$ by

$$f(t) = \begin{cases} 0, & t \leqslant 0 \\ \exp(-t^{-2}), & t > 0. \end{cases}$$

Then f is smooth and a suitable g is given by

$$g(t) = \frac{f(t-\alpha)}{f(\beta - t) + f(t - \alpha)}$$

9. *The Cayley map*: Let E be a real or complex vector space of finite dimension. Then

$$O = \{\sigma \in L_E \mid \det(\iota + \sigma) \neq 0\}$$

is an open submanifold of the vector space L_E containing $0 \in L_E$. We will show that the Cayley map $f: O \to L_E$ given by

$$f(\sigma) = (\iota - \sigma)(\iota + \sigma)^{-1} \qquad \sigma \in O$$

is an involution of O. In fact,

$$f(\sigma) + \iota = (\iota - \sigma)(\iota + \sigma)^{-1} + (\iota + \sigma)(\iota + \sigma)^{-1} = 2(\iota + \sigma)^{-1},$$

whence $f: O \to O$. Next, observe that

$$\sigma(f(\sigma) + \iota) = 2\sigma(\iota + \sigma)^{-1} = \iota - f(\sigma)$$

and so

$$\sigma = (\iota - f(\sigma))(\iota + f(\sigma))^{-1} = f(f(\sigma));$$

this shows that $f^2 = \iota_O$. Since f is clearly smooth, it is a diffeomorphism.

10. *One-point compactifications of vector spaces*: We shall show that, if E is an n-dimensional Euclidean vector space, its one-point compactification E_∞ (cf. Example 4, sec. 1.4) is diffeomorphic to S^n (cf. Example 1, sec. 1.4).

Let S^n be defined as in sec. 1.4 with charts (U_-, u_-), (U_+, u_+), constructured from a point $a \in S^n$ and identify E with the orthogonal complement of a. Consider the map $\varphi\colon E_\infty \to S^n$, defined by

$$\varphi(x) = \begin{cases} u_+^{-1}(x), & x \in E \\ a, & x = x_\infty . \end{cases}$$

To examine the smoothness properties of φ, we use the atlas (U_0, u_0), (U_∞, u_∞) for E_∞ described in sec. 1.4. It follows immediately that

$$u_+ \circ \varphi \circ u_0^{-1} = \iota_E ,$$

while the relation

$$u_- \circ \varphi \circ u_\infty^{-1} = \iota_E ,$$

is obtained from a straightforward computation. These formulae show that both φ and φ^{-1} are smooth. Hence φ is a diffeomorphism.

In the case when $E = \mathbb{C}$, regarded as a two-dimensional Euclidean vector space, the corresponding two-sphere is customarily referred to as the Riemann sphere.

11. *The exponential map*: Let E be a real or complex vector space of dimension n and assume that a positive definite symmetric (resp. Hermitian) inner product is defined in E. Consider the space $S(E)$ of self-adjoint linear transformations of E. Then $S(E)$ is a real vector space of dimension $\frac{1}{2}n(n+1)$ (resp. n^2). A self-adjoint map $\varphi\colon E \to E$ is called *positive*, if

$$\langle \varphi(x), x \rangle > 0, \qquad x \neq 0.$$

The positive self-adjoint maps form an open subset of $S(E)$ which will be denoted by $S^+(E)$. It is easy to see that the exponential map restricts to a map

$$\exp\colon S(E) \to S^+(E).$$

It will be shown that the map so obtained is a diffeomorphism.

We consider first the complex case. Let

$$\mathbb{C}^+ = \{\lambda \in \mathbb{C} \mid \operatorname{Re}(\lambda) > 0\}$$

and define a map $\log\colon \mathbb{C}^+ \to \mathbb{C}$ by

$$\log \lambda = \int_\gamma z^{-1}\, dz,$$

where γ is the line segment from 1 to λ.

Now suppose that $\sigma \in S^+(E)$ so that the eigenvalues of σ are real and positive, and choose a circle $C \subset \mathbb{C}^+$ such that each eigenvalue of σ lies inside C. Regard $\lambda \mapsto (\log \lambda)(\lambda\iota - \sigma)^{-1}$ as a smooth $S(E)$-valued function in $\mathbb{C}^+$ and put

$$\log \sigma = \frac{1}{2\pi i} \int_C \log \lambda(\lambda\iota - \sigma)^{-1}\, d\lambda, \qquad \sigma \in S^+(E).$$

($\log \sigma$ is independent of the choice of C, see below.) Since the same C may be used in the construction of $\log \sigma$ for all σ belonging to some neighbourhood in $S^+(E)$, we conclude that log is smooth. To show that exp is a diffeomorphism it is then sufficient to show that log inverts it.

But, if $\sigma \in S^+(E)$, then σ has real eigenvalues $\lambda_\nu > 0$ corresponding to eigenvectors x_ν which, for $\nu = 1, 2, \ldots, n$, form a basis of E. Now the relations $\sigma(x_\nu) = \lambda_\nu x_\nu$ imply

$$(\lambda\iota - \sigma)^{-1} x_\nu = (\lambda - \lambda_\nu)^{-1} x_\nu\,, \qquad \lambda \neq \lambda_\nu\,,$$

and hence

$$(\log \sigma)\, x_\nu = \left(\frac{1}{2\pi i} \int_C \frac{\log \lambda}{\lambda - \lambda_\nu}\, d\lambda\right) x_\nu = (\log \lambda_\nu)\, x_\nu\,,$$

by Cauchy's theorem. It follows that $\log \sigma$ is independent of C and that

$$[(\exp \circ \log)(\sigma)](x_\nu) = \exp(\log \lambda_\nu)\, x_\nu = \lambda_\nu x_\nu = \sigma(x_\nu),$$

for $\nu = 1, \ldots, n$ and $\sigma \in S^+(E)$. Hence

$$\exp \circ \log = \iota_{S^+(E)}\,.$$

A similar computation shows that

$$\log \circ \exp = \iota_{S(E)}$$

and therefore completes the proof.

Now suppose that E is a real Euclidean space. Consider the Hermitian space $E^{\mathbb{C}}$ (cf. sec. 0.1). If $\varphi\colon E \to E$ is a self-adjoint map, then so is the map

$$\iota_{\mathbb{C}} \otimes \varphi\colon E^{\mathbb{C}} \to E^{\mathbb{C}}.$$

Hence we have an inclusion map $j\colon S(E) \to S(E^{\mathbb{C}})$. It restricts to a map $j^+\colon S^+(E) \to S^+(E^{\mathbb{C}})$. It follows from the definitions that the diagram

$$\begin{array}{ccc} S(E) & \xrightarrow{\;j\;} & S(E^{\mathbb{C}}) \\ {\scriptstyle \exp}\downarrow & & \downarrow{\scriptstyle \exp_{\mathbb{C}}} \\ S^+(E) & \xrightarrow{\;j^+\;} & S^+(E^{\mathbb{C}}) \end{array}$$

commutes. Hence so does the diagram

$$\begin{array}{ccc} S(E) & \xrightarrow{j} & S(E^{\mathbb{C}}) \\ {\scriptstyle \exp'(\alpha)}\downarrow & & \downarrow{\scriptstyle \exp_{\mathbb{C}}'(j(\alpha))}, \\ S(E) & \xrightarrow{j} & S(E^{\mathbb{C}}) \end{array} \qquad \alpha \in S(E).$$

It follows that the map

$$\exp'(\alpha): S(E) \to S(E)$$

is injective and hence a linear isomorphism. On the other hand, the first diagram shows that the map

$$\exp: S(E) \to S^+(E)$$

is injective. Finally, if $\varphi \in S^+(E)$, we can choose an orthonormal basis $x_1, \ldots, x_n$ of E for which

$$\varphi x_\nu = \lambda_\nu x_\nu \qquad \text{with} \quad \lambda_\nu > 0.$$

Define $\psi \in S(E)$ by

$$\psi x_\nu = \log \lambda_\nu x_\nu .$$

Then $\varphi = \exp \psi$ and so exp is surjective. Hence $\exp: S(E) \to S^+(E)$ is a diffeomorphism.

12. The power maps $P_k: S^+(E) \to S^+(E)$, defined by

$$P_k(\sigma) = \sigma^k \qquad (k \text{ is a nonzero integer}),$$

where E is a real or complex finite dimensional vector space, are diffeomorphisms. In fact, let $\mu_k: S(E) \to S(E)$ be the diffeomorphism given by $\mu_k(\sigma) = k\sigma$. Then

$$P_k = \exp \circ \mu_k \circ \exp^{-1}$$

is a diffeomorphism.

1.6. Construction of smooth manifolds. Proposition I has the following analogue:

Proposition VI: Let M be an n-manifold and $\{U_\alpha\}$ be an arbitrary open covering. There is a countable atlas $\{(V_i, v_i, \mathbb{R}^n)\}$ of M such that

(1) The covering $\{V_i\}$ refines $\{U_\alpha\}$.
(2) $\overline{V}_i$ is compact, $i \in \mathbb{N}$.

Proposition VII: Let M be a set which is the union of a countable collection $\{W_i\}$ of subsets such that

(1) For each $i \in \mathbb{N}$, there is a bijection $\varphi_i: W_i \to M_i$, where M_i is an n-manifold (n independent of i).

(2) For every pair i, j, the subsets $\varphi_i(W_{ij}) \subset M_i$ and $\varphi_j(W_{ij}) \subset M_j$ are open and the map

$$\varphi_{ji} = \varphi_j \circ \varphi_i^{-1} : \varphi_i(W_{ij}) \to \varphi_j(W_{ij})$$

is a diffeomorphism.

(3) For distinct points $a_i \in W_i$ and $a_j \in W_j$, there are disjoint subsets U_i, U_j such that $a_i \in U_i \subset W_i$, $a_j \in U_j \subset W_j$ and $\varphi_i(U_i)$, $\varphi_j(U_j)$ are open.

Then there is a unique smooth manifold structure on M such that the W_i are open and the φ_i are diffeomorphisms.

1.7. Products of manifolds. Let M and N be manifolds and consider the topological product $M \times N$. If $\{(U_\alpha, u_\alpha) \mid \alpha \in \mathscr{I}\}$ and $\{(V_i, v_i) \mid i \in \mathscr{J}\}$ are atlases for M and N, respectively, then $\{(U_\alpha \times V_i, u_\alpha \times v_i) \mid \alpha \in \mathscr{I}, i \in \mathscr{J}\}$ is an atlas for $M \times N$. It is easy to see that equivalent atlases on M and N induce equivalent atlases on $M \times N$. Hence a smooth structure on $M \times N$ is induced by the smooth structures of M and N.

The smoothness of the following maps follows from the definitions:

(1) the projection maps $\pi_M: M \times N \to M$, $\pi_N: M \times N \to N$, given by

$$\pi_M(x, y) = x, \qquad \pi_N(x, y) = y;$$

(2) the diagonal map $\Delta: M \to M \times M$, defined by

$$\Delta(x) = (x, x), \qquad x \in M;$$

(3) the interchange map $M \times N \to N \times M$ given by

$$(x, y) \mapsto (y, x);$$

(4) the "product" map $\chi: P \to M \times N$, given by

$$\chi(z) = (\varphi(z), \psi(z)), \qquad z \in P,$$

where $\varphi: P \to M$, $\psi: P \to N$ are smooth.

1.8. Smooth functions and partitions of unity. A *smooth function* on a manifold M is a smooth map $f: M \to \mathbb{R}$. If f and g are two such functions, then smooth functions $\lambda f + \mu g$ and fg are defined by

$$(\lambda f + \mu g)(x) = \lambda f(x) + \mu g(x), \qquad \lambda, \mu \in \mathbb{R}$$
$$(fg)(x) = f(x)\, g(x), \qquad x \in M.$$

These operations make the set of smooth functions on M into an algebra over $\mathbb{R}$, which we denote by $\mathscr{S}(M)$. The unit element of $\mathscr{S}(M)$ is the constant function $M \mapsto 1$.

If $\mathscr{M}$ and $\mathscr{N}$ are $\mathscr{S}(M)$-modules, we denote their tensor product (over $\mathscr{S}(M)$) by

$$\mathscr{M} \otimes_M \mathscr{N}.$$

The module of $\mathscr{S}(M)$-linear maps of $\mathscr{M}$ into $\mathscr{N}$ will be denoted by

$$\mathrm{Hom}_M(\mathscr{M}; \mathscr{N}).$$

Now suppose that $\varphi: M \to N$ is a smooth map. φ determines an algebra homomorphism

$$\varphi^*: \mathscr{S}(M) \leftarrow \mathscr{S}(N)$$

given by

$$\varphi^* f = f \circ \varphi, \qquad f \in \mathscr{S}(N).$$

If φ is surjective, φ^* is injective. If $\psi: N \to Q$ is a second smooth map, then

$$(\psi \circ \varphi)^* = \varphi^* \circ \psi^*.$$

Definition: The *carrier* (or *support*) of a smooth function f on M is the closure of the set $\{x \in M \mid f(x) \neq 0\}$. We denote this set by $\mathrm{carr}\, f$.

If O is an open subset of M and f is a smooth function on O whose carrier is closed in M, then f extends to the smooth function g on M, given by

$$g(x) = \begin{cases} f(x), & x \in O \\ 0, & x \in M - \mathrm{carr}\, f. \end{cases}$$

In particular, if $f \in \mathscr{S}(M)$ has carrier in O, and $h \in \mathscr{S}(O)$, a smooth function $f \cdot h \in \mathscr{S}(M)$ is given by

$$(f \cdot h)(x) = f(x)\, h(x), \quad x \in O \qquad \text{and} \qquad (f \cdot h)(x) = 0, \quad x \notin \mathrm{carr}\, f.$$

Next, suppose that $\{U_\alpha\}$ is a locally finite family of open sets of M, and let $f_\alpha \in \mathscr{S}(M)$ satisfy $\operatorname{carr} f_\alpha \subset U_\alpha$. Then for each $a \in M$ there is a neighbourhood $V(a)$ which meets only finitely many of the U_α. Thus *in this neighbourhood* $\sum_\alpha f_\alpha$ is a *finite* sum. It follows that a smooth function f on M is defined by

$$f(x) = \sum_\alpha f_\alpha(x), \qquad x \in M.$$

We write $f = \sum_\alpha f_\alpha$.

Proposition VIII: Let K, O be subsets of M such that K is closed, O is open and $K \subset O$. There exists a smooth function f such that

(1) $\operatorname{carr} f$ is contained in O
(2) $0 \leqslant f(x) \leqslant 1$, $x \in M$
(3) $f(x) = 1$, $x \in K$.

Lemma IV: Let E be a Euclidean space and $\alpha, \beta \in \mathbb{R}$ be such that $0 < \alpha < \beta$. There exists a smooth function $h: E \to [0, 1] \subset \mathbb{R}$ such that $h(x) = 1$, for $|x|^2 \leqslant \alpha$, $h(x) = 0$, for $|x|^2 \geqslant \beta$.

Proof: Define h by $h(x) = 1 - g(|x|^2)$, where $g: \mathbb{R} \to \mathbb{R}$ is the function of Example 8, sec. 1.5).

Q.E.D.

Proof of the proposition: Choose open sets $U_\alpha \subset M$ and compact sets $K_\alpha \subset U_\alpha$, subject to the following conditions (cf. sec. 1.1)

(1) $\{U_\alpha\}$, $M - K$ is a locally finite open cover of M.
(2) Each U_α is diffeomorphic to $\mathbb{R}^n$ and $\overline{\cup U_\alpha} \subset O$.
(3) $\cup K_\alpha = K$.

It follows at once (via Lemma IV) that there are smooth functions h_α in U_α such that $\operatorname{carr} h_\alpha$ is compact and

$$h_\alpha(x) = 1, \quad x \in K_\alpha.$$

In particular $\operatorname{carr} h_\alpha$ is closed in M.

Next, extend the h_α to smooth functions f_α in M with $\operatorname{carr} f_\alpha = \operatorname{carr} h_\alpha \subset U_\alpha$. Then we can form $\sum_\alpha f_\alpha \in \mathscr{S}(M)$. Evidently,

$$\operatorname{carr} \sum_\alpha f_\alpha \subset \overline{\bigcup U_\alpha} \subset O$$

and

$$\left(\sum_\alpha f_\alpha\right)(x) \geqslant 1, \qquad x \in K.$$

Finally, choose a smooth map $g\colon \mathbb{R} \to [0, 1]$ so that $g(0) = 0$ and $g(t) = 1$, $t \geqslant 1$ (cf. Example 8, sec. 1.5). Then the function

$$f = g \circ \left(\sum_\alpha f_\alpha\right)$$

satisfies the desired conditions.

Q.E.D.

Definition: *A partition of unity, subordinate to a locally finite open covering* $\{U_\alpha\}$ *of* M is a family $\{f_\alpha\}$ of smooth functions on M satisfying

(1) $0 \leqslant f_\alpha(x) \leqslant 1$, $x \in M$
(2) $\operatorname{carr} f_\alpha \subset U_\alpha$
(3) $\sum f_\alpha = 1$.

Theorem II: Every locally finite open covering of a manifold admits a subordinate partition of unity, $\{f_\alpha\}$.

Proof: Let $\{U_\alpha\}$ be such a covering of M and let $\{V_\alpha\}$ be a second open covering such that $\overline{V}_\alpha \subset U_\alpha$. In view of Proposition VIII, there are nonnegative smooth functions g_α on M which have carriers in U_α and take the value 1 at points of $\overline{V}_\alpha$. Thus $g = \sum g_\alpha$ is smooth and positive. Set $f_\alpha = g_\alpha/g$.

Q.E.D.

Corollary: If $\{U_\alpha \mid \alpha \in \mathscr{I}\}$ is any open covering of M, there is a partition of unity $\{f_i \mid i \in \mathscr{J}\}$ and a map $i \mapsto \alpha(i)$ of $\mathscr{J}$ into $\mathscr{I}$ such that $\operatorname{carr} f_i \subset U_{\alpha(i)}$, $i \in \mathscr{J}$.

1.9. Function germs. Let a be a fixed point of M. Two members f, g of $\mathscr{S}(M)$ will be called *a-equivalent*, $f \underset{a}{\sim} g$, if and only if there is a neighbourhood U of a such that $f(x) = g(x)$, $x \in U$. The equivalence classes so obtained are called *function germs* at a. We write $[f]_a$ for the germ represented by $f \in \mathscr{S}(M)$ and $\mathscr{S}_a(M)$ for the set of function germs at a. By setting $[f]_a + [g]_a = [f + g]_a$ and $[f]_a\,[g]_a = [fg]_a$, we make $\mathscr{S}_a(M)$ an algebra.

The map $\mathscr{S}(M) \to \mathscr{S}_a(M)$ given by $f \mapsto [f]_a$ is a surjective homomorphism.

4. If $\varphi \sim \psi: M \to N$ via a homotopy H, and $\varphi_1 \sim \psi_1: N \to Q$ via a homotopy G, then

$$\varphi_1 \circ \varphi \sim \psi_1 \circ \psi: M \to Q$$

via the homotopy K given by

$$K(t, x) = G(t, H(t, x)).$$

1.11. Smooth paths. A smooth *path* on M is a smooth map $\varphi: \mathbb{R} \to M$. A manifold is called smoothly path-connected if, for every two points $a, b \in M$, there exists a smooth path φ such that $\varphi(0) = a$ and $\varphi(1) = b$.

Proposition IX: If a, b are points of a connected manifold M, there is a smooth path φ on M such that

$$\varphi(t) = \begin{cases} a, & t \leqslant 0 \\ b, & t \geqslant 1. \end{cases}$$

In particular, M is smoothly path-connected.

Proof: φ exists if and only if the inclusion maps

$$j_a: \{\text{point}\} \to a \in M \qquad \text{and} \qquad j_b: \{\text{point}\} \to b \in M$$

are homotopic (cf. Lemma V). Since homotopy is an equivalence relation an equivalence relation is induced on the points of M:

$$a \sim b \text{ if and only if } a \text{ can be joined to } b \text{ by some } \varphi.$$

If $M = \mathbb{R}^n$, the proposition is obviously true (use Example 1, sec. 1.10). Thus in general, if $(U, u, \mathbb{R}^n)$ is a chart in M, then all the points of U are equivalent. Hence the equivalence classes are all open and M is their disjoint union. Since M is connected, there is only one class; i.e., every $a, b \in M$ are equivalent.

Q.E.D.

1.12. Diffeomorphisms of smooth manifolds. In this section we prove

Theorem III: Let C be a closed subset of a manifold M such that $M - C$ is nonvoid and connected. Let a, b be arbitrary points of $M - C$. Then there is a diffeomorphism $\varphi: M \to M$ homotopic to ι_M and such that $\varphi(a) = b$ and $\varphi(x) = x$, $x \in C$.

To this end we give the following lemma and its consequence.

Lemma VI: There is a smooth function f on $\mathbb{R}$ such that

(1) $\operatorname{carr} f \subset [-3, 3]$
(2) $0 \leqslant f(t) \leqslant 1$, $t \in \mathbb{R}$ and $f(0) = 1$,
(3) $|f'(t)| < 1$, $t \in \mathbb{R}$.

Proof: Define f by

$$f(t) = \begin{cases} \exp\left(-\dfrac{t^2}{9 - t^2}\right), & t \in (-3, 3) \\ 0, & \text{otherwise.} \end{cases}$$

Q.E.D.

Corollary: There exists a diffeomorphism φ of $\mathbb{R}^n$ such that

(1) $\varphi(0, ..., 0) = (1, 0, ..., 0)$
(2) $\varphi(x) = x$, for every $x = (\xi^1, ..., \xi^n)$ such that $\max_i |\xi^i| > 3$.

Proof: Define φ by

$$\varphi(\xi^1, ..., \xi^n) = \left(\xi^1 + \prod_{i=1}^n f(\xi^i), \xi^2, ..., \xi^n\right)$$

where $f: \mathbb{R} \to \mathbb{R}$ is the function of Lemma VI. Then the Jacobian of φ is given by

$$\det \varphi'(x) = 1 + f'(\xi^1) \prod_{i=2}^n f(\xi^i).$$

Det $\varphi'(x) > 0$, as follows from conditions (2), (3) on f. Thus φ is a local diffeomorphism. To see that it is in fact a global diffeomorphism it is only necessary to note that it induces a bijection on each of the lines

$$\xi^2 = \xi_0^2, \quad \xi^3 = \xi_0^3, \quad ..., \quad \xi^n = \xi_0^n.$$

That φ satisfies conditions (1) and (2) is immediate from the properties of f.

Q.E.D.

Proof of Theorem III: Let $\sim$ be the equivalence relation on $M - C$ defined by

$x_1 \sim x_2$ if and only if there is a diffeomorphism $\varphi: M \to M$, homotopic to ι_M, such that $\varphi(x_1) = x_2$ and $\varphi(x) = x$, $x \in C$.

We shall show that the equivalence classes are open. In fact, if $a \in M - C$, let $(U, u, \mathbb{R}^n)$ be a chart of M such that $a \in U \subset M - C$. If $b \in U$ is arbitrary, we can compose u with an affine transformation of $\mathbb{R}^n$, if necessary, and assume that $u(a) = 0$, $u(b) = (1, 0, ..., 0)$. Applying the corollary to Lemma VI, we obtain a diffeomorphism $\varphi_0: U \to U$ such that $\varphi_0(a) = b$ and φ_0 is the identity outside a compact set K such that $b \in K \subset U$. Then $\varphi: M \to M$ defined by

$$\varphi(x) = \begin{cases} \varphi_0(x), & x \in U \\ x, & x \notin U \end{cases}$$

is a diffeomorphism which establishes the equivalence of a and b; hence all points of U are equivalent to a.

Since the equivalence classes are open and $M - C$ is connected, all points of $M - C$ are equivalent, as required.

Q.E.D.

Corollary: Let M be a connected manifold of dimension $n \geqslant 2$ and $\{a_1, ..., a_k\}$, $\{b_1, ..., b_k\}$ be two finite subsets of M. Then there is a diffeomorphism $\varphi: M \to M$, homotopic to ι_M, such that $\varphi(a_i) = b_i$ $(i = 1, ..., k)$.

Proof: If $k = 1$, the result follows from the theorem with $C = \varnothing$. Suppose that the result has been proved for $k - 1$; i.e. a diffeomorphism φ_0 of M, homotopic to ι_M, has been found such that $\varphi_0(a_i) = b_i$, $i = 1, ..., k - 1$. Noting that $M - \{b_1, ..., b_{k-1}\}$ is nonvoid and connected, we obtain, from Theorem III, a diffeomorphism φ_1 of M, homotopic to ι_M, such that $\varphi_1(\varphi_0(a_k)) = b_k$ and $\varphi_1(b_i) = b_i$ for $i = 1, 2, ..., k - 1$. Set $\varphi = \varphi_1 \circ \varphi_0$ (cf. Example 4, sec. 1.10).

Q.E.D.

§3. Smooth fibre bundles

1.13. Local product property. Let $\pi: E \to B$ be a smooth map between manifolds. The map π will be said to have the *local product property* with respect to a manifold F if there is an open covering $\{U_\alpha\}$ of B and a family $\{\psi_\alpha\}$ of diffeomorphisms

$$\psi_\alpha: U_\alpha \times F \to \pi^{-1}(U_\alpha),$$

such that

$$\pi\psi_\alpha(x, y) = x, \qquad x \in U_\alpha, \quad y \in F.$$

The system $\{(U_\alpha, \psi_\alpha)\}$ will be called a *local decomposition of* π.

Clearly any mapping with the local product property is surjective and open.

Definition: A *smooth fibre bundle* is a four-tuple (E, π, B, F) where $\pi: E \to B$ is a smooth map which has the local product property with respect to F. A local decomposition for π is called a *coordinate representation* for the fibre bundle.

We call E the *total* or *bundle* space, B the *base* space, and F the *typical fibre*. For each $x \in B$, the set $F_x = \pi^{-1}(x)$ will be called the *fibre* over x. Every fibre is a closed subset of E, and E is the disjoint union of the fibres.

A smooth *cross-section* of a fibre bundle (E, π, B, F) is a smooth map $\sigma: B \to E$ such that $\pi \circ \sigma = \iota_B$.

If $\{(U_\alpha, \psi_\alpha)\}$ is a coordinate representation for the bundle, we obtain bijections $\psi_{\alpha,x}: F \to F_x$, $x \in U_\alpha$, defined by

$$\psi_{\alpha,x}(y) = \psi_\alpha(x, y), \qquad y \in F.$$

In particular, if $x \in U_{\alpha\beta}$, we obtain maps $\psi_{\beta,x}^{-1} \circ \psi_{\alpha,x}: F \to F$. These are diffeomorphisms. In fact, since ψ_α and ψ_β define diffeomorphisms of $U_{\alpha\beta} \times F$ onto $\pi^{-1}(U_{\alpha\beta})$, they determine a diffeomorphism $\psi_{\beta\alpha} = \psi_\beta^{-1} \circ \psi_\alpha$ of $U_{\alpha\beta} \times F$ onto itself. But

$$\psi_{\beta\alpha}(x, y) = (x, \psi_{\beta,x}^{-1}\psi_{\alpha,x}(y)), \qquad x \in U_{\alpha\beta}, \quad y \in F,$$

and hence $\psi_{\beta,x}^{-1} \circ \psi_{\alpha,x}$ is a diffeomorphism of F.

Suppose now that (E', π', B', F') is a second fibre bundle. Then a smooth map $\varphi: E \to E'$ is called *fibre preserving* if, whenever $\pi z_1 = \pi z_2$, $(z_1, z_2 \in E)$, then $\pi'\varphi(z_1) = \pi'\varphi(z_2)$. Any fibre preserving map φ determines a set map $\varphi_B: B \to B'$ by the requirement that the following diagram commute:

$$\begin{array}{ccc} E & \xrightarrow{\varphi} & E' \\ \pi\downarrow & & \downarrow\pi' \\ B & \xrightarrow{\varphi_B} & B' \end{array} .$$

We now show that φ_B is always smooth. In fact, if $\{(U_\alpha, \psi_\alpha)\}$ is a local decomposition for π and $y \in F$ is fixed, then

$$\varphi_B(x) = (\pi' \circ \varphi \circ \psi_\alpha)(x, y), \qquad x \in U_\alpha .$$

Hence φ_B is smooth on each member U_α of a covering of B.

Let (E'', π'', B'', F'') be a third fibre bundle and assume that $\varphi: E \to E'$, $\varphi': E' \to E''$ are fibre preserving. Then $\varphi' \circ \varphi: E \to E''$ is fibre preserving and $(\varphi' \circ \varphi)_B = \varphi'_{B'} \circ \varphi_B$.

Proposition X: Let B, F be manifolds and let E be a set. Assume that a surjective set map $\pi: E \to B$ is given with the following properties:

(1) There is an open covering $\{U_\alpha\}$ of B and a family $\{\psi_\alpha\}$ of bijections

$$\psi_\alpha: U_\alpha \times F \to \pi^{-1}U_\alpha .$$

(2) For every $x \in U_\alpha$, $y \in F$, $\pi\psi_\alpha(x, y) = x$.

(3) The maps $\psi_{\beta\alpha}: U_{\alpha\beta} \times F \to U_{\alpha\beta} \times F$ defined by $\psi_{\beta\alpha}(x, y) = (\psi_\beta^{-1} \circ \psi_\alpha)(x, y)$ are diffeomorphisms.

Then there is exactly one manifold structure on E for which (E, π, B, F) is a fibre bundle with coordinate representation $\{(U_\alpha, \psi_\alpha)\}$.

Proof: We may assume that $\{\alpha\}$ is countable and thus apply Proposition VII, sec. 1.6, with $W_\alpha = \pi^{-1}U_\alpha$, $\varphi_\alpha = \psi_\alpha^{-1}$, and $M_\alpha = U_\alpha \times F$ to obtain a unique manifold structure on E such that the ψ_α are diffeomorphisms.

Hypothesis (2) then says that the restriction of π to $\pi^{-1}U_\alpha$ is $\pi_\alpha \circ \psi_\alpha^{-1}$, where $\pi_\alpha: U_\alpha \times F \to U_\alpha$ denotes the projection onto the first factor. Since (cf. sec. 1.7) π_α is smooth, π is smooth on $\pi^{-1}U_\alpha$. Hence π is smooth on E and then, by definition, $\{(U_\alpha, \psi_\alpha)\}$ is a local decomposition for π. Hence (E, π, B, F) is a fibre bundle with coordinate representation $\{(U_\alpha, \psi_\alpha)\}$.

Q.E.D.

Proposition XI: Every smooth fibre bundle has a finite coordinate representation.

Proof: Let $\{(U_\alpha, \psi_\alpha)\}$ be any coordinate representation for (E, π, B, F) Choose a refinement $\{V_{ij} \mid i = 1, \dots, p; j \in \mathbb{N}\}$ of $\{U_\alpha\}$ such that $V_{ij} \cap V_{ik} = \varnothing$ for $j \neq k$ (cf. Theorem I, scc. 1.2). Let $V_i = \bigcup_j V_{ij}$ and define $\psi_i \colon V_i \times F \to \pi^{-1}V_i$ by

$$\psi_i(x, y) = \psi_{ij}(x, y) \qquad \text{if} \quad x \in V_{ij}, \quad y \in F,$$

where ψ_{ij} is the restriction of some ψ_α.

Q.E.D.

Problems

1. The fact that a topological space is locally Euclidean (each point has a neighbourhood homeomorphic to an open subset of $\mathbb{R}^n$) implies neither that the manifold is second countable nor that it is Hausdorff. Construct one-dimensional examples to prove this.

2. Let M and N be manifolds and $\varphi: M \to N$ be a map such that $\varphi^* g \in \mathscr{S}(M)$ whenever $g \in \mathscr{S}(N)$. Show that φ is smooth.

3. Construct a smooth injection of the two-dimensional torus $T^2 = \mathbb{R}^2/\mathbb{Z}^2$ into $\mathbb{R}^3$.

4. Show that the n-torus T^n is diffeomorphic to the product of n circles S^1.

5. Let M and N be smooth manifolds and suppose that ρ is a metric on N.

(i) Prove that if $\varphi: M \to N$ is a continuous map and $\epsilon > 0$ is given, then there is a smooth map $\psi: M \to N$ such that $\rho(\varphi x, \psi x) < \epsilon$, $x \in M$.

(ii) Two continuous maps $\varphi, \psi: M \to N$ are called *continuously homotopic*, if there is a continuous map $H: I \times M \to N$ such that $H(0, x) = \varphi x$ and $H(1, x) = \psi x$.

Prove that every continuous map is continuously homotopic to a smooth map. Prove that two smooth maps are smoothly homotopic if and only if they are continuously homotopic.

6. Let M and N be compact smooth manifolds. Assume that $\alpha: \mathscr{S}(M) \leftarrow \mathscr{S}(N)$ is a homomorphism. Show that there is a unique smooth map $\varphi: M \to N$ such that $\varphi^* = \alpha$. Conclude that if α is an isomorphism, then φ is a diffeomorphism.

7. Classify the one-dimensional topological manifolds.

8. Construct a nontrivial fibre bundle over S^1 with fibre

(i) $\mathbb{R}$ (Möbius strip).

(ii) S^1 (Klein bottle).

9. Let τ be a diffeomorphism of a compact manifold, M, such that (i) $\tau^k = \iota$, for some k, and (ii) for each $x \in M$, $x, \tau x, \ldots, \tau^{k-1}x$ are distinct. Define an equivalence relation in M by setting $x \sim y$ if $\tau^p x = y$ for some p. Show that the corresponding quotient space is a smooth manifold N and that (M, π, N, F) is a smooth bundle, where F is the set $\{0, 1, \ldots, k - 1\}$.

10. Regard S^{2n+1} as the unit sphere of an $(n + 1)$-dimensional Hermitian space E. Define an equivalence relation on S^{2n+1} by setting $x \sim y$ if $y = e^{i\theta}x$ for some $\theta \in \mathbb{R}$.

(i) Show that the equivalence classes with the quotient topology form a smooth $2n$-manifold. It is called the *complex projective space* $\mathbb{C}P^n$.

(ii) Show that the projection $\pi: S^{2n+1} \to \mathbb{C}P^n$ is the projection of a smooth fibre bundle $(S^{2n+1}, \pi, \mathbb{C}P^n, S^1)$. It is called the *Hopf fibering*.

(iii) Show that $\mathbb{C}P^1$ is the Riemann sphere S^2 and that π is given by

$$\pi(z_1, z_2) = \begin{cases} z_1 z_2^{-1}, & z_2 \neq 0 \\ z_\infty, & z_2 = 0 \end{cases} \qquad z_1, z_2 \in \mathbb{C}.$$

Construct an explicit coordinate representation (U, φ), (V, ψ) for this Hopf fibering, where $U = S^2 - \{0\}$ and $V = S^2 - \{z_\infty\}$, so that

$$\psi^{-1}\varphi(z, t) = \left(z, \frac{z}{|z|}\, t\right) \qquad z \in U \cap V, \quad t \in S^1.$$

11. Replace $\mathbb{C}$ by $\mathbb{H}$ in problem 10 and define the quaternionic projective space $\mathbb{H}P^n$. Obtain the Hopf fiberings $(S^{4n+3}, \pi, \mathbb{H}P^n, S^3)$. Discuss the case $n = 1$.

12. Imitate the definition of topological and smooth manifolds to define real and complex analytic manifolds. Do they admit analytic partitions of unity?

13. **Grassmann manifolds.** Let $\mathbb{R}^n$ have a positive definite inner product. For every k-dimensional subspace $E \subset \mathbb{R}^n$, let $\rho_E: \mathbb{R}^n \to E$ and $\rho_E^\perp: \mathbb{R}^n \to E^\perp$ be the orthogonal projections. Consider the set $\mathscr{G}_{\mathbb{R}}(n; k)$ of all k-dimensional subspaces of $\mathbb{R}^n$. For $E \in \mathscr{G}_{\mathbb{R}}(n; k)$ set

$$U_E = \{F \in \mathscr{G}_{\mathbb{R}}(n; k) \mid F \cap E^\perp = \{0\}\}$$

and define $u_E : U_E \to L(E; E^\perp)$ by

$$u_E(F) = \rho_{E^\perp} \circ (\rho_{FE})^{-1}$$

($\rho_{FE} : F \xrightarrow{\cong} E$ is the restriction of ρ_E).

(i) Make $\mathscr{G}_{\mathbb{R}}(n; k)$ into a smooth manifold with atlas $\{(U_E, u_E)\}$.

(ii) Show that $\dim \mathscr{G}_{\mathbb{R}}(n; k) = k(n - k)$. Define a natural diffeomorphism between $\mathscr{G}_{\mathbb{R}}(n; k)$ and $\mathscr{G}_{\mathbb{R}}(n; n - k)$.

(iii) Show that $\mathscr{G}_{\mathbb{R}}(n + 1; 1)$ is diffeomorphic to $\mathbb{R}P^n$.

(iv) Do (i) and (iii) with $\mathbb{R}$ replaced by $\mathbb{C}$. Find $\dim \mathscr{G}_{\mathbb{C}}(n; k)$.

14. Let E, F and H be real vector spaces of dimensions m, n and k with $k \leqslant m, n$. Let $S(E; H)$ (resp. $I(H; F)$, $GL(H)$) denote the set of linear surjections (resp. injections, bijections) and let $L(E; F; k)$ denote the set of linear maps $E \to F$ of rank k.

(i) Show that composition defines a set map

$$\pi: S(E; H) \times I(H; F) \to L(E; F; k).$$

(ii) Show that $S(E; H)$, $I(H; F)$, and $GL(H)$ are open subsets of the spaces $L(E; H)$, $L(H; F)$, and L_H. Conclude that they are smooth manifolds.

(iii) Construct a unique smooth structure in $L(E; F; k)$ so that $(S(E; H) \times I(H; F), \pi, L(E; F; k), GL(H))$ is a smooth bundle. Find the dimension of $L(E; F; k)$.

Chapter II

Vector Bundles

§1. Basic concepts

2.1. Definitions. A *vector bundle* is a quadruple $\xi = (E, \pi, B, F)$ where

(1) (E, π, B, F) is a smooth fibre bundle (cf. sec. 1.13)
(2) F, and the fibres $F_x = \pi^{-1}(x)$, $x \in B$, are real linear spaces
(3) there is a coordinate representation $\{(U_\alpha, \psi_\alpha)\}$ such that the maps

$$\psi_{\alpha,x}: F \to F_x$$

are linear isomorphisms.

The dimension of F is called the *rank* of ξ. A coordinate representation for the bundle which satisfies (3) is called a *coordinate representation* for the *vector* bundle ξ. We shall often denote a bundle ξ by its total space E.

If $\{(U_\alpha, \psi_\alpha)\}$ is a coordinate representation for ξ, then the maps $g_{\alpha\beta}: U_{\alpha\beta} \to GL(F)$ given by

$$g_{\alpha\beta}(x) = \psi_{\alpha,x}^{-1} \circ \psi_{\beta,x}$$

are smooth. They are called the *coordinate transformations* for ξ corresponding to $\{(U_\alpha, u_\alpha)\}$. ($GL(F)$ is an open submanifold of $L(F; F)$.)

A neighbourhood U in B is called a *trivializing neighbourhood* for ξ if there is a diffeomorphism

$$\psi_U: U \times F \to \pi^{-1}U$$

such that $\pi\psi_U(x, y) = x$ $(x \in U, y \in F)$ and such that the induced maps

$$\psi_{U,x}: F \to F_x$$

are linear isomorphisms. ψ_U is called a *trivializing map* for ξ.

A *subbundle* ξ' of a vector bundle ξ is a vector bundle with the same base such that each of its fibres F_x' is a linear subspace of F_x, and for which the induced inclusion map $i: E' \to E$ of total spaces is smooth.

2.2. Bundle maps. If $\xi = (E, \pi, B, F)$ and $\xi' = (E', \pi', B', F')$ are vector bundles, a *bundle map* (also called a *homomorphism of vector bundles*) $\varphi: \xi \to \xi'$ is a smooth fibre-preserving map $\varphi: E \to E'$ such that the restrictions

$$\varphi_x: F_x \to F'_{\psi(x)}, \qquad x \in B,$$

are linear ($\psi: B \to B'$ denotes the smooth map induced by φ, cf. sec. 1.13).

If $\varphi': \xi' \to \xi''$ is a second bundle map, then so is $\varphi' \circ \varphi$. Let ψ, ψ', and ψ'' denote the smooth maps of base manifolds induced by φ, φ', and $\varphi' \circ \varphi$. Then

$$\psi'' = \psi' \circ \psi.$$

A bundle map $\varphi: \xi \to \eta$ is called an *isomorphism* if it is a diffeomorphism. The inverse of a bundle isomorphism is clearly again a bundle isomorphism. Inverse bundle isomorphisms induce inverse diffeomorphisms between the base manifolds. Two vector bundles ξ and ξ' are called isomorphic, $\xi \cong \xi'$, if there is a bundle isomorphism $\varphi: \xi \xrightarrow{\cong} \xi'$.

A strong bundle map between two vector bundles with the same base is a bundle map which induces the identity in the base.

Now let $\varphi: \xi \to \xi'$ be an arbitrary bundle map inducing $\psi: B \to B'$ and choose coordinate representations $\{(U_\alpha, \psi_\alpha)\}$ and $\{(V_i, \chi_i)\}$ for ξ and ξ', respectively. Then smooth maps

$$\varphi_{i\alpha}: \psi^{-1}(V_i) \cap U_\alpha \to L(F; F')$$

are defined by

$$\varphi_{i\alpha}(x) = \chi_{ix'}^{-1} \circ \varphi_x \circ \psi_{\alpha,x}, \qquad x' = \psi(x).$$

They are called the *mapping transformations* for φ corresponding to the given coordinate representations.

Proposition I: Let $\varphi: \xi \to \xi'$ be a homomorphism of vector bundles inducing $\psi: B \to B'$ between the base manifolds. Then φ is an isomorphism if and only if

(1) $\psi: B \to B'$ is a diffeomorphism,
(2) each $\varphi_x: F_x \to F'_{\psi(x)}$ $(x \in B)$ is a linear isomorphism.

Proof: If φ is an isomorphism, then (1) and (2) are obvious. Conversely, assume (1) and (2) hold. Then φ is bijective and φ^{-1} restricts to the linear isomorphisms $\varphi_x^{-1}: F_x \xleftarrow{\cong} F'_{\psi(x)}$. It remains to prove that φ^{-1} is smooth.

With the aid of trivializing neighbourhoods for ξ and ξ' we can reduce to the case (E, E' are the total manifolds for ξ, ξ')

$$B = B', \qquad E = B \times F, \qquad E' = B \times F'$$

and ψ is the identity map. Then $x \mapsto \varphi_x$ defines a smooth map

$$\Phi\colon B \to L(F; F')$$

and φ^{-1} is the smooth map given by

$$\varphi^{-1}(x, y') = (x, \Phi(x)^{-1}(y')), \qquad x \in B, \quad y' \in F'.$$

Q.E.D.

2.3. Examples. 1. The *trivial bundle of rank r over B*

$$\xi = (B \times F, \pi, B, F),$$

where $\pi(x, y) = x$ and F is an r-dimensional real vector space. This bundle is often denoted by $B \times F$ or by ϵ^r.

2. *Restriction*: Let $\xi = (E, \pi, B, F)$ be a vector bundle. The restriction, $\xi|_O$, of ξ to an open submanifold $O \subset B$ is the bundle

$$\xi|_O = (\pi^{-1}(O), \pi_O, O, F),$$

where π_O is the restriction of π to the open set $\pi^{-1}(O)$.

3. *Cartesian product*: Let $\xi^i = (E^i, \pi^i, B^i, F^i)$ be vector bundles ($i = 1, 2$). Their *Cartesian product* is the vector bundle

$$\xi^1 \times \xi^2 = (E^1 \times E^2, \pi^1 \times \pi^2, B^1 \times B^2, F^1 \oplus F^2)$$

whose fibre at (x_1, x_2) is the vector space

$$F^1_{x_1} \times F^2_{x_2} = F^1_{x_1} \oplus F^2_{x_2}.$$

If $\{(U_\alpha, \varphi_\alpha)\}$ and $\{(V_\nu, \psi_\nu)\}$ are coordinate representations for ξ^1 and ξ^2, then a coordinate representation $\{(U_\alpha \times V_\nu, \chi_{\alpha\nu})\}$ for $\xi^1 \times \xi^2$ is given by

$$\chi_{\alpha\nu}(x_1, x_2; y_1 \oplus y_2) = (\varphi_\alpha(x_1, y_1), \psi_\nu(x_2, y_2)).$$

The projections $\rho^1\colon E^1 \times E^2 \to E^1$, $\rho^2\colon E^1 \times E^2 \to E^2$ are bundle maps $\xi^1 \times \xi^2 \to \xi^1$ and $\xi^1 \times \xi^2 \to \xi^2$.

The Cartesian product $\xi^1 \times \xi^2$ has the following factorization property: If $\xi = (E, \pi, B, F)$ is a third vector bundle and $p^1\colon E \to E^1$,

$p^2: E \to E^2$ are bundle maps, then there exists a unique bundle map $p: E \to E^1 \times E^2$ such that

$$\rho^1 \circ p = p^1 \qquad \text{and} \qquad \rho^2 \circ p = p^2 .$$

2.4. Multilinear maps. Let $\xi^1, \xi^2, \ldots, \xi^p, \xi$ be vector bundles over the same base B. A *p-linear bundle map*

$$\Phi: (\xi^1, \ldots, \xi^p) \to \xi$$

is a collection of p-linear maps

$$\Phi_x: F_x^1 \times \cdots \times F_x^p \to F_x ,$$

indexed by B, which satisfies the following smoothness condition:

If $\{(U_\alpha, \psi_\alpha^1)\}, \ldots, \{(U_\alpha, \psi_\alpha^p)\}$, and $\{(U_\alpha, \psi_\alpha)\}$ are coordinate representations of $\xi^1, \ldots, \xi^p$ and ξ respectively (we may assume that the covering $\{U_\alpha\}$ of B is the same for all bundles), then the *mapping transformations*

$$\Phi_\alpha: U_\alpha \to L(F^1, \ldots, F^p; F)$$

defined by

$$\Phi_\alpha(x) = \psi_{\alpha,x}^{-1} \circ \Phi_x \circ (\psi_{\alpha,x}^1 \times \cdots \times \psi_{\alpha,x}^p),$$

are smooth.

This definition is independent of the choice of coordinate representations and coincides with the definition of a strong bundle map when $p = 1$. However, if $p > 1$, Φ may *not* be regarded as a set map on the Cartesian product of the total spaces of $\xi^1, \ldots, \xi^p$.

When it is convenient to do so we shall use the notation $\Phi(x; z_1, \ldots, z_p)$ for $\Phi_x(z_1, \ldots, z_p)$.

2.5. Construction of vector bundles. Proposition X in sec. 1.13 provides a useful tool for the construction of vector bundles over a given manifold B. In fact, consider a manifold B and an r-dimensional vector space F. Assume that to every point $x \in B$ there is assigned an r-dimensional vector space F_x.

Consider the disjoint union $E = \bigcup_{x \in B} F_x$ and the natural projection $\pi: E \to B$. Assume given an open covering $\{U_\alpha\}$ of B, together with linear isomorphisms

$$\psi_{\alpha,x}: F \xrightarrow{\cong} F_x , \qquad x \in U_\alpha$$

subject to the following condition:

Condition S: The maps $g_{\alpha\beta}: U_{\alpha\beta} \to GL(F)$ given by

$$g_{\alpha\beta}(x) = \psi_{\alpha,x}^{-1} \circ \psi_{\beta,x}$$

are smooth.

Define bijections $\psi_\alpha: U_\alpha \times F \to \pi^{-1}(U_\alpha)$ by setting

$$\psi_\alpha(x, y) = \psi_{\alpha,x}(y).$$

Then Condition S implies that the bijections $\psi_{\beta\alpha} = \psi_\beta^{-1} \circ \psi_\alpha$ of $U_{\alpha\beta} \times F$ are smooth. Now, by Proposition X, sec. 1.13, there is a unique smooth manifold structure on E which makes (E, π, B, F) into a bundle with coordinate representation $\{(U_\alpha, \psi_\alpha)\}$. It is clear from the construction that the bundle so obtained is a vector bundle. The fibre at $x \in B$ is the vector space F_x.

Example. *Pull-backs*: Let $\xi = (E, \pi, B, F)$ be a vector bundle and let $\sigma: M \to B$ be a smooth map. Assign to each $x \in M$ the vector space $F_{\sigma(x)}$. Let $\{(V_\alpha, \varphi_\alpha)\}$ be a coordinate representation for ξ and set $U_\alpha = \sigma^{-1}(V_\alpha)$. Define linear isomorphisms

$$\psi_{\alpha,x}: F \to F_{\sigma(x)}, \qquad x \in U_\alpha$$

by $\psi_{\alpha,x} = \varphi_{\alpha,\sigma(x)}$. Then the map $x \mapsto \psi_{\alpha,x}^{-1} \circ \psi_{\beta,x}$ $(x \in V_\alpha \cap V_\beta)$ can be written

$$x \mapsto g_{\alpha\beta}(\sigma(x)),$$

where $g_{\alpha\beta}$ are the coordinate transformations for ξ. Hence it is smooth.

Thus there is a vector bundle $\sigma^*\xi = (N, \rho, M, F)$, with $N = \bigcup_{x\in M} F_{\sigma(x)}$ and with coordinate representation $\{(U_\alpha, \psi_\alpha)\}$. $\sigma^*\xi$ is called the ***pull-back*** of ξ over σ.

The identity maps $F_{\sigma(x)} \to F_{\sigma(x)}$ define a bundle map

$$\tau: \sigma^*\xi \to \xi$$

which induces $\sigma: M \to B$. τ restricts to linear isomorphisms in each fibre.

If $\eta = (N', \rho', M, H)$ is a second vector bundle over M and $\varphi: \eta \to \xi$ is a bundle map inducing $\sigma: M \to B$, then we may restrict φ to linear maps

$$\varphi_x: H_x \to F_{\sigma(x)}, \qquad x \in M.$$

These define a *strong* bundle map $\hat{\varphi}: \eta \to \sigma^*\xi$, and the diagram

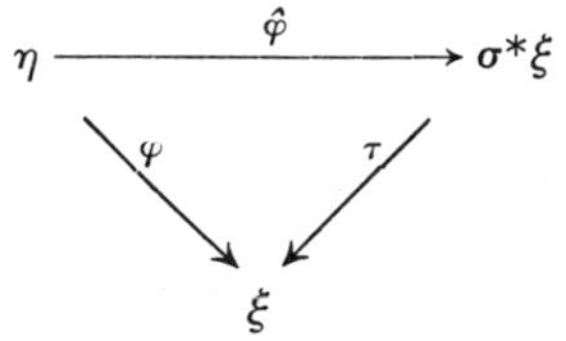

commutes. In particular, if each φ_x is a linear isomorphism, then $\hat{\varphi}$ is a strong bundle isomorphism; i.e., η is isomorphic to the pull-back of ξ over σ.

§2. Algebraic operations with vector bundles

In this article all vector bundles will have a fixed base B. Moreover, ξ, ξ^{ν}, η, ζ will denote fixed vector bundles with typical fibres F, F^{ν}, H, and K. In particular, we shall write $\xi = (E, \pi, B, F)$.

2.6. The module of strong bundle maps. Let $\varphi, \psi: \xi \to \eta$ be strong bundle maps and let $f \in \mathcal{S}(B)$. Then strong bundle maps

$$\varphi + \psi,\ f\varphi : \xi \to \eta$$

are given by

$$(\varphi + \psi)(z) = \varphi(z) + \psi(z), \qquad z \in E, \quad \pi(z) = x.$$
$$(f\varphi)(z) = f(x)\,\varphi(z),$$

The operations $(\varphi, \psi) \mapsto \varphi + \psi$ and $(f, \varphi) \mapsto f\varphi$ make the set of strong bundle maps into an $\mathcal{S}(B)$-module, which we denote by $\operatorname{Hom}(\xi; \eta)$.

Let $\varphi \in \operatorname{Hom}(\xi; \eta)$, $\psi \in \operatorname{Hom}(\eta; \zeta)$. Then the composite $\psi \circ \varphi$ is a strong bundle map,

$$\psi \circ \varphi: \xi \to \zeta.$$

The correspondence $(\varphi, \psi) \mapsto \psi \circ \varphi$ defines an $\mathcal{S}(B)$-bilinear map

$$\operatorname{Hom}(\xi; \eta) \times \operatorname{Hom}(\eta; \zeta) \to \operatorname{Hom}(\xi; \zeta).$$

Example 1: If $\xi = B \times F$, $\eta = B \times H$ are trivial bundles, then to each $\varphi \in \operatorname{Hom}(\xi; \eta)$ we can associate the mapping transformation $\hat{\varphi}: B \to L(F; H)$ defined by

$$\varphi(x, v) = (x, \hat{\varphi}(x)v), \qquad x \in B, \quad v \in F.$$

This defines an isomorphism of $\mathcal{S}(B)$-modules:

$$\operatorname{Hom}(\xi; \eta) \xrightarrow{\cong} \mathcal{S}(B; L(F; H)).$$

If Φ, Ψ: $(\xi^1, \ldots, \xi^p) \to \eta$ are p-linear maps and if $f \in \mathscr{S}(B)$, we define $\Phi + \Psi, f\Phi$: $(\xi^1, \ldots, \xi^p) \to \eta$ by

$$\begin{aligned} (\Phi + \Psi)_x &= \Phi_x + \Psi_x , \\ (f\Phi)_x &= f(x)\, \Phi_x , \end{aligned} \qquad x \in B.$$

The operations $(\Phi, \Psi) \mapsto \Phi + \Psi$, $(f, \Phi) \mapsto f\Phi$ make the set of p-linear maps $(\xi^1, \ldots, \xi^p) \to \eta$ into an $\mathscr{S}(B)$-module, which we denote by $\mathrm{Hom}(\xi^1, \ldots, \xi^p; \eta)$. If $\eta = B \times \mathbb{R}$, the elements of the module $\mathrm{Hom}(\xi^1, \ldots, \xi^p; \eta)$ are called *p-linear functions.*

Example 2: An $\mathscr{S}(B)$-bilinear map

$$\mathrm{Hom}(\xi^1, \ldots, \xi^p; \eta) \times \mathrm{Hom}(\eta; \xi) \to \mathrm{Hom}(\xi^1, \ldots, \xi^p; \xi)$$

is given by $(\Phi, \varphi) \mapsto \varphi \circ \Phi$, where

$$(\varphi \circ \Phi)_x = \varphi_x \circ \Phi_x , \qquad x \in B.$$

An element $\Phi \in \mathrm{Hom}\,(\xi, \ldots, \xi; \eta)$ is called *skew-symmetric* (*symmetric*) if, for every $x \in B$, the p-linear fibre maps

$$\Phi_x : F_x \times \cdots \times F_x \to H_x$$

are skew-symmetric (symmetric). Skew-symmetric bundle maps will be of particular importance in Chapter III. The skew-symmetric (symmetric) bundle maps are submodules of $\mathrm{Hom}(\xi, \ldots, \xi; \eta)$ which will be denoted by $A^p(\xi; \eta)$ $(S^p(\xi; \eta))$, $p \geqslant 1$. We extend this definition to the case $p = 0$ by setting

$$S^0(\xi; \eta) = A^0(\xi; \eta) = \mathrm{Hom}(B \times \mathbb{R}; \eta).$$

Now set

$$A(\xi; \eta) = \textstyle\bigoplus_{p=0}^{\infty} A^p(\xi; \eta), \qquad S(\xi; \eta) = \bigoplus_{p=0}^{\infty} S^p(\xi; \eta).$$

Since, evidently $A^p(\xi; \eta) = 0$, $p > \operatorname{rank} \xi \;(= r)$, we have

$$A(\xi; \eta) = \textstyle\bigoplus_{p=0}^{r} A^p(\xi; \eta).$$

The module of skew-symmetric p-linear functions in ξ is denoted by $A^p(\xi)$, and we write

$$A(\xi) = \textstyle\bigoplus_{p=0}^{r} A^p(\xi).$$

As in the case of $\mathrm{Hom}(\xi;\eta)$, if $\xi^1, \ldots, \xi^p, \eta$ are trivial bundles, then

$$\begin{aligned}\mathrm{Hom}(\xi^1, \ldots, \xi^p; \eta) &\cong \mathscr{S}(B; L(F^1, \ldots, F^p; H)),\\ A^p(\xi;\eta) &\cong \mathscr{S}(B; A^p(F; H)),\\ S^p(\xi;\eta) &\cong \mathscr{S}(B; S^p(F; H)),\end{aligned}$$

and

$$A(\xi) \cong \mathscr{S}(B; A(F)).$$

2.7. Dual bundles. Suppose that a bilinear function $\langle\ ,\ \rangle$ in (η, ξ) has been defined (cf. sec. 2.6). We say that $\langle\ ,\ \rangle$ is *nondegenerate*, or a *scalar product*, if the $\mathbb{R}$-bilinear functions

$$\langle\ ,\ \rangle_x : H_x \times F_x \to \mathbb{R}, \qquad x \in B,$$

are all nondegenerate. In this case ξ and η are called *dual* with respect to $\langle\ ,\ \rangle$.

Now it will be shown that every vector bundle ξ admits a dual bundle ξ^*. Let F_x^* be dual to the vector space F_x $(x \in B)$ with respect to a scalar product $\langle\ ,\ \rangle_x$. Let $\{(U_\alpha, \varphi_\alpha)\}$ be a coordinate representation for ξ and define linear isomorphisms

$$\psi_{\alpha,x} : F^* \to F_x^*$$

by setting $\psi_{\alpha,x} = (\varphi_{\alpha,x}^*)^{-1}$.

Then the construction principle of sec. 2.5 yields a vector bundle $\xi^* = (E^*, \pi^*, B, F^*)$ where

$$E^* = \bigcup_{x \in B} F_x^*$$

and $\pi^* : E^* \to B$ is the obvious projection.

The scalar products $\langle\ ,\ \rangle_x$ define a scalar product between ξ^* and ξ; thus ξ^* and ξ are dual bundles.

Evidently, $\operatorname{rank} \xi^* = \operatorname{rank} \xi$.

Next, let ξ^*, ξ and η^*, η be two pairs of dual vector bundles, with fibres F_x^*, F_x and H_x^*, H_x at $x \in B$. Suppose $\varphi : \xi \to \eta$ is a strong bundle map. Then a strong bundle map $\varphi^* : \xi^* \leftarrow \eta^*$ is given by

$$(\varphi^*)_x = (\varphi_x)^* : F_x^* \leftarrow H_x^*, \qquad x \in B.$$

φ^* is called the *dual* of φ.

Proposition II: If ξ, η are dual with respect to $\langle\ ,\ \rangle$, then an $\mathscr{S}(B)$-module isomorphism

$$\mathrm{Hom}(\zeta, \xi; B \times \mathbb{R}) \xrightarrow{\cong} \mathrm{Hom}(\zeta; \eta)$$

is defined by $\Phi \mapsto \varphi$, where

$$\langle \varphi(w), z \rangle_x = \Phi(x; w, z), \qquad w \in K_x, \quad z \in F_x, \quad x \in B.$$

Proof: The relation above defines unique linear maps $\varphi_x \colon K_x \to H_x$ for each $x \in B$. The collection of all these maps determines a set map φ from the total space of ζ to that of η. The smoothness of φ follows from that of Φ; thus φ is a strong bundle map.

The correspondence $\Phi \mapsto \varphi$ is an $\mathscr{S}(B)$-module homomorphism. On the other hand, a module homomorphism

$$\mathrm{Hom}(\zeta, \xi; B \times \mathbb{R}) \leftarrow \mathrm{Hom}(\zeta; \eta)$$

is defined by associating to each $\varphi \in \mathrm{Hom}(\zeta; \eta)$ the bilinear function Φ given by

$$\Phi(x; w, z) = \langle \varphi(w), z \rangle_x, \qquad w \in K_x, \quad z \in F_x, \quad x \in B.$$

These homomorphisms are inverse to each other, and so they are isomorphisms.

Q.E.D.

Corollary I: Any two vector bundles which are dual to a given vector bundle are strongly isomorphic.

Proof: Suppose ξ^1, ξ^2 are dual to ξ with respect to bilinear functions $\langle\ ,\ \rangle^1$ and $\langle\ ,\ \rangle^2$. Then the bilinear function $\langle\ ,\ \rangle^2$ determines a $\varphi \in \mathrm{Hom}(\xi^2; \xi^1)$ such that

$$\langle \varphi_x(w), z \rangle_x^1 = \langle w, z \rangle_x^2, \qquad w \in F_x^2, \quad z \in F_x, \quad x \in B.$$

It follows from standard linear algebra that φ_x is a linear isomorphism and so φ is a strong bundle isomorphism.

Q.E.D.

Corollary II: If ξ^*, ξ and η^*, η are two pairs of dual bundles, then

$$\mathrm{Hom}(\eta; \xi^*) \cong \mathrm{Hom}(\eta, \xi; B \times \mathbb{R}) \cong \mathrm{Hom}(\xi, \eta; B \times \mathbb{R}) \cong \mathrm{Hom}(\xi; \eta^*).$$

In particular, setting $\eta = \eta^* = B \times \mathbb{R}$ (with the standard scalar product $\langle(x, \lambda), (x, \mu)\rangle = \lambda\mu$) yields

$$\operatorname{Hom}(B \times \mathbb{R}; \xi^*) \cong \operatorname{Hom}(\xi; B \times \mathbb{R}).$$

2.8. Whitney sum. A vector bundle ξ is called the *Whitney sum* of the bundles $\xi^\nu (\nu = 1, \ldots, p)$ if there are defined strong bundle maps

$$i^\nu: \xi^\nu \to \xi \qquad \text{and} \qquad \rho^\nu: \xi \to \xi^\nu$$

such that

$$\rho^\nu \circ i^\mu = \begin{cases} 0, & \nu \neq \mu \\ \iota, & \nu = \mu \end{cases}$$

and

$$\sum_{\nu=1}^{p} i^\nu \circ \rho^\nu = \iota_\xi .$$

In particular, the fibre F_x in ξ over a point $x \in B$ is then the direct sum of the fibres F_x^ν. In this case ξ is denoted by $\xi^1 \oplus \cdots \oplus \xi^p$.

Next, suppose that $\varphi^\nu: \xi^\nu \to \eta$ are strong bundle maps. Then a strong bundle map $\varphi: \xi \to \eta$ is given by

$$\varphi = \sum_\nu \varphi^\nu \circ \rho^\nu .$$

The correspondence $(\varphi^1, \ldots, \varphi^p) \mapsto \varphi$ defines a module isomorphism

$$\bigoplus_\nu \operatorname{Hom}(\xi^\nu; \eta) \xrightarrow{\cong} \operatorname{Hom}(\xi^1 \oplus \cdots \oplus \xi^p; \eta).$$

In particular, $\xi^1 \oplus \cdots \oplus \xi^p$ is determined up to strong isomorphism.

Now we shall show that the Whitney sum of vector bundles always exists. We shall restrict ourselves to the case $p = 2$, the generalization being obvious. Assign to each $x \in B$ the vector space $F_x^1 \oplus F_x^2$. Let $\{(U_\alpha, \varphi_\alpha^1)\}$ and $\{(U_\alpha, \varphi_\alpha^2)\}$ be coordinate representations for ξ^1, ξ^2 and assign to $x \in U_\alpha$ the linear isomorphism

$$\psi_{\alpha,x} = \varphi_{\alpha,x}^1 \oplus \varphi_{\alpha,x}^2: F^1 \oplus F^2 \to F_x^1 \oplus F_x^2 .$$

Then the construction principle of sec. 2.5 yields a vector bundle

$$\tilde{\xi} = (\tilde{E}, \pi, B, F^1 \oplus F^2),$$

where

$$\tilde{E} = \bigcup_{x \in B} F_x^1 \oplus F_x^2$$

and π is the obvious projection.

The inclusions $F_x^1, F_x^2 \to F_x^1 \oplus F_x^2$ define strong bundle maps

$$i^1: \xi^1 \to \tilde{\xi}, \qquad i^2: \xi^2 \to \tilde{\xi}.$$

The projections $F_x^1 \oplus F_x^2 \to F_x^1$, $F_x^1 \oplus F_x^2 \to F_x^2$ define strong bundle maps

$$\rho^1: \tilde{\xi} \to \xi^1, \qquad \rho^2: \tilde{\xi} \to \xi^2.$$

These maps satisfy the required conditions (check this at each $x \in B$ separately) and so $\tilde{\xi}$ is the Whitney sum of ξ^1 and ξ^2. Clearly $\operatorname{rank}(\xi^1 \oplus \xi^2) = \operatorname{rank} \xi^1 + \operatorname{rank} \xi^2$.

2.9. Tensor product. A vector bundle η is called a *tensor product* of the bundles ξ^ν ($\nu = 1, ..., p$) and is denoted by $\xi^1 \otimes \cdots \otimes \xi^p$ if there is defined a p-linear bundle map

$$\otimes^p: (\xi^1, ..., \xi^p) \to \eta$$

subject to the following factorization property: For each vector bundle ζ over B, and to every p-linear bundle map $\Phi \in \operatorname{Hom}(\xi^1, ..., \xi^p; \zeta)$ corresponds a unique strong bundle map $\varphi: \eta \to \zeta$ such that

$$\varphi \circ \otimes^p = \Phi.$$

A tensor product of vector bundles ξ^ν always exists. It is constructed in exactly the same way as the Whitney sum, with $F_x^1 \otimes F_x^2$ replacing $F_x^1 \oplus F_x^2$. In particular, $\operatorname{rank}(\xi^1 \otimes \xi^2) = (\operatorname{rank} \xi^1)(\operatorname{rank} \xi^2)$.

Let $\xi^1 \otimes \cdots \otimes \xi^p$ be a tensor product for vector bundles ξ^ν. Then the correspondence $\varphi \mapsto \varphi \circ \otimes^p$ defines a module isomorphism

$$\operatorname{Hom}(\xi^1 \otimes \cdots \otimes \xi^p; \zeta) \xrightarrow{\cong} \operatorname{Hom}(\xi^1, ..., \xi^p; \zeta).$$

In particular, $\xi^1 \otimes \cdots \otimes \xi^p$ is determined up to strong isomorphism.

If $\xi^1 = \cdots = \xi^p = \xi$, we denote the tensor product by $\otimes^p \xi$. We extend the definition by setting

$$\otimes^0 \xi = B \times \mathbb{R}.$$

Finally, let ξ, ξ^* and η, η^* be two pairs of dual vector bundles. Then the bilinear function in $(\xi^* \otimes \eta^*, \xi \otimes \eta)$ given by

$$\langle z^* \otimes w^*, z \otimes w \rangle_x = \langle z^*, z \rangle_x \langle w^*, w \rangle_x, \qquad x \in B,$$

is a scalar product. Hence we can write

$$(\xi \otimes \eta)^* \cong \xi^* \otimes \eta^*.$$

2.10. The bundle $L(\xi; \eta)$. Recall that ξ, η denote vector bundles over B with typical fibres F, H. Let $\{(U_\alpha, \varphi_\alpha)\}$ and $\{(U_\alpha, \psi_\alpha)\}$ denote coordinate representations for these bundles. Now assign to each $x \in B$ the vector space $L(F_x; H_x)$. Further to each $x \in U_\alpha$ assign the linear isomorphism

$$\hat{\psi}_{\alpha,x}: L(F; H) \xrightarrow{\cong} L(F_x; H_x)$$

given by

$$\hat{\psi}_{\alpha,x}(\sigma) = \psi_{\alpha,x} \circ \sigma \circ \varphi_{\alpha,x}^{-1}.$$

Then the construction principle of sec. 2.5 yields a vector bundle

$$L(\xi; \eta) = (\hat{E}, \hat{\pi}, B, L(F; H)),$$

where

$$\hat{E} = \bigcup_{x \in B} L(F_x; H_x).$$

Its fibre at x is the space $L(F_x; H_x)$.

If $\xi = \eta$, we sometimes denote $L(\xi; \xi)$ by L_ξ.

If ξ^* is a vector bundle dual to ξ, then the canonical isomorphisms

$$F_x^* \otimes H_x \xrightarrow{\cong} L(F_x; H_x), \qquad x \in B$$

define a strong bundle isomorphism

$$\xi^* \otimes \eta \xrightarrow{\cong} L(\xi; \eta).$$

More generally, we may construct the vector bundle $L(\xi^1, ..., \xi^p; \eta)$ whose fibre at $x \in B$ is the space of p-linear maps $F_x^1 \times \cdots \times F_x^p \to H_x$. The canonical linear isomorphisms

$$L(F_x^1, ..., F_x^p; H_x) \cong L(F_x^1 \otimes \cdots \otimes F_x^p; H_x) \cong (F_x^1)^* \otimes \cdots \otimes (F_x^p)^* \otimes H_x$$

define strong bundle isomorphisms

$$L(\xi^1, ..., \xi^p; \eta) \cong L(\xi^1 \otimes \cdots \otimes \xi^p; \eta) \cong (\xi^1)^* \otimes \cdots \otimes (\xi^p)^* \otimes \eta.$$

The bilinear bundle map

$$\epsilon: (L(\xi; \eta), \xi) \to \eta,$$

given by

$$\epsilon(x; \alpha_x, z_x) = \alpha_x(z_x), \qquad \alpha_x \in L(F_x; H_x), \quad z_x \in F_x, \quad x \in B,$$

is called the *evaluation map*. The bilinear bundle map

$$\Phi: L(\xi; \eta) \times L(\eta; \zeta) \to L(\xi; \zeta),$$

defined by

$$\Phi(x; \alpha_x, \beta_x) = \beta_x \circ \alpha_x, \qquad \alpha_x \in L(F_x; H_x), \quad \beta_x \in L(H_x, K_x), \quad x \in B$$

is called the *composition map*.

2.11. Exterior algebra. A *pth exterior power of* ξ is a vector bundle $\wedge^p \xi$, together with a p-linear skew-symmetric bundle map

$$\wedge^p : (\xi, \ldots, \xi) \to \wedge^p \xi,$$

with the following factorization property: Every p-linear skew-symmetric map $\Phi: (\xi, \ldots, \xi) \to \eta$ can be written uniquely in the form

$$\Phi = \varphi \circ \wedge^p,$$

where $\varphi: \wedge^p \xi \to \eta$ is a strong bundle map.

The map $\varphi \mapsto \varphi \circ \wedge^p$ defines a module isomorphism

$$\text{Hom}(\wedge^p \xi; \eta) \xrightarrow{\cong} A^p(\xi; \eta).$$

An argument similar to the one used in sec. 2.9 shows that the bundle $\wedge^p \xi$ exists and is uniquely determined by ξ up to a strong isomorphism. $\wedge^p \xi$ has fibre $\wedge^p F_x$ at $x \in B$ and the map $\wedge^p$ is given by

$$\wedge^p(x; z_1, \ldots, z_p) = z_1 \wedge \cdots \wedge z_p, \qquad z_\nu \in F_x, \quad x \in B.$$

In particular, $\wedge^p \xi = B \times \{0\}$, $p >$ rank ξ; and $\text{rank}(\wedge^p \xi) = \binom{r}{p}$, $1 \leqslant p \leqslant r$, where r is the rank of ξ.

Now suppose ξ and ξ^* are a pair of dual vector bundles. Then (for each $x \in B$) the scalar product $\langle\ ,\ \rangle_x$ between F_x and F_x^* induces a scalar product, $\langle\ ,\ \rangle_x$ between $\wedge^p F_x$ and $\wedge^p F_x^*$, given by

$$\langle z^{*1} \wedge \cdots \wedge z^{*p}, z_1 \wedge \cdots \wedge z_p \rangle_x = \det(\langle z^{*\nu}, z_\mu \rangle_x), \qquad z^{*\nu} \in F_x^*, \quad z_\mu \in F_x.$$

These scalar products, in turn, define a scalar product between $\wedge^p \xi$ and $\wedge^p \xi^*$. Thus these bundles are dual, and we can write

$$(\wedge^p \xi)^* = \wedge^p \xi^*.$$

We now extend the definition of $\wedge^p \xi$ to the case $p = 0$ by setting $\wedge^0 \xi = B \times \mathbb{R}$. We define the *exterior algebra bundle* $\wedge \xi$ to be the Whitney sum

$$\wedge \xi = \bigoplus_{p=0}^{r} \wedge^p \xi, \qquad r = \text{rank } \xi.$$

$\wedge \xi$ has fibre $\wedge F_x$ at $x \in B$, and its rank is 2^r.

A strong bundle isomorphism $\wedge\xi \otimes \wedge\eta \cong \wedge(\xi \oplus \eta)$ is given by $z \otimes w \mapsto z \wedge w$, $\quad z \in \wedge F_x, \quad w \in \wedge H_x, \quad x \in B$.

If $\varphi: \xi \to \eta$ is a strong bundle map, then the induced linear mappings

$$\varphi_x: F_x \to H_x, \qquad x \in B,$$

extend to algebra homomorphisms

$$\wedge\varphi_x: \wedge F_x \to \wedge H_x .$$

These define a strong bundle map, written $\wedge\varphi: \wedge\xi \to \wedge\eta$.

2.12. Symmetric algebra. In precisely the same manner as in the preceding sections we obtain for every $p \geqslant 1$ a unique vector bundle $\vee^p\xi$ together with a p-linear symmetric bundle map

$$\vee^p: (\xi, ..., \xi) \to \vee^p\xi,$$

with the following property: Every symmetric p-linear map

$$\Psi: (\xi, ..., \xi) \to \eta$$

can be written uniquely in the form $\Psi = \psi \circ \vee^p$, where $\psi: \vee^p\xi \to \eta$ is a strong bundle map. $\vee^p\xi$ is called a *pth symmetric power of* ξ. The map $\psi \mapsto \psi \circ \vee^p$ defines a module isomorphism

$$\mathrm{Hom}(\vee^p\xi; \eta) \xrightarrow{\cong} S^p(\xi; \eta).$$

The bundle $\vee^p\xi$ has fibre $\vee^pF_x$ at $x \in B$ and the map $\vee^p: (\xi, ..., \xi) \to \vee^p\xi$ is given by

$$\vee^p(x; z_1, ..., z_p) = z_1 \vee \cdots \vee z_p, \qquad z_\nu \in F_x, \quad \nu = 1, ..., p, \quad x \in B.$$

Finally, assume that ξ and ξ^* is a pair of dual vector bundles. Then (for each $x \in B$) the scalar product $\langle\ ,\ \rangle_x$ between F_x and F_x^* induces a scalar product $\langle\ ,\ \rangle_x$ between $\vee^pF_x$ and $\vee^pF_x^*$ by

$$\langle z^{*1} \vee \cdots \vee z^{*p}, z_1 \vee \cdots \vee z_p\rangle = \mathrm{perm}(\langle z^{*\nu}, z_\mu\rangle_x), \qquad z^{*\nu} \in F_x^*, \quad z_\mu \in F_x .$$

These scalar products define a scalar product between the bundles $\vee^p\xi$ and $\vee^p\xi^*$. Thus these bundles are dual, and we can write

$$(\vee^p\xi)^* = \vee^p\xi^*.$$

§3. Cross-sections

In this article $\xi = (E, \pi, B, F)$ denotes a fixed vector bundle.

2.13. Definitions. A *cross-section* σ in a vector bundle ξ is a smooth map $\sigma: B \to E$ such that $\pi \circ \sigma = \iota$. Every vector bundle ξ admits the *zero cross-section* o defined by

$$o(x) = 0_x \in F_x, \qquad x \in B.$$

The *carrier* (or *support*) of a cross-section σ is the set

$$\operatorname{carr} \sigma = \operatorname{closure}\{x \in B \mid \sigma(x) \neq 0_x\}.$$

Let $\sigma: U \to E$ be a cross-section defined in an open subset U of B (i.e. σ is a cross-section in the bundle $\xi|_U$). Let $f \in \mathscr{S}(B)$ satisfy $\operatorname{carr} f \subset U$. Then a cross-section $f \cdot \sigma$ in ξ is given by

$$(f \cdot \sigma)(x) = \begin{cases} f(x)\,\sigma(x), & x \in U \\ 0_x, & x \notin \operatorname{carr} f. \end{cases}$$

More generally, let $\{U_\nu\}$ be a locally finite open cover of B, and for each ν let σ_ν be a cross-section defined over U_ν. Let $\{f_\nu\}$ be a partition of unity subordinate to the open cover $\{U_\nu\}$. Then a cross-section $\sum_\nu f_\nu \cdot \sigma_\nu$, in ξ is given by

$$\Big(\sum_\nu f_\nu \cdot \sigma_\nu\Big)(x) = \sum_\nu f_\nu(x)\,\sigma_\nu(x).$$

If σ is a cross-section in ξ and $\sigma_\nu: U_\nu \to E$ is the restriction of σ to U_ν, then

$$\sigma = \sum_\nu f_\nu \cdot \sigma_\nu .$$

Examples: 1. Let $a \in E$ and let $b = \pi(a)$. Then there exists a cross-section σ such that $\sigma(b) = a$.

In fact, choose a trivializing neighbourhood U of b. Then $\xi|_U \cong U \times F$ and so there is a cross-section $\tau: U \to E$ such that $\tau(b) = a$. Choose $f \in \mathscr{S}(B)$ with $f(b) = 1$ and $\operatorname{carr} f \subset U$; then set $\sigma = f \cdot \tau$.

2. A vector bundle of rank 1 is trivial if and only if it admits a cross-section σ such that

$$\sigma(x) \neq 0_x, \qquad x \in B.$$

Indeed, if $\xi = B \times \mathbb{R}$ is trivial define σ by $\sigma(x) = (x, 1)$. Conversely, if σ exists, define a strong isomorphism $B \times \mathbb{R} \to \xi$ by

$$(x, t) \mapsto t\sigma(x), \qquad x \in B, \quad t \in \mathbb{R}.$$

2.14. The module of cross-sections. If σ, τ are cross-sections in ξ and $f \in \mathscr{S}(B)$, cross-sections $\sigma + \tau$, $f\sigma$ in ξ are given by

$$(\sigma + \tau)(x) = \sigma(x) + \tau(x), \qquad (f\sigma)(x) = f(x)\,\sigma(x), \qquad x \in B.$$

The operations $(\sigma, \tau) \mapsto \sigma + \tau$ and $(f, \sigma) \mapsto f\sigma$ make the set of cross-sections in ξ into an $\mathscr{S}(B)$-module, which will be denoted by Sec ξ.

Examples: 1. If $\xi = B \times F$ is a trivial bundle, then every cross-section σ determines a smooth map $\varphi: B \to F$ given by

$$\sigma(x) = (x, \varphi(x)).$$

This defines a canonical isomorphism

$$\operatorname{Sec} \xi \xrightarrow{\cong} \mathscr{S}(B; F)$$

of $\mathscr{S}(B)$-modules. In particular, if $e_1, \ldots, e_r$ is a basis of F, then the cross-sections σ_i given by

$$\sigma_i(x) = (x, e_i), \qquad i = 1, \ldots, r, \quad x \in B,$$

are a basis for Sec ξ: thus it is a *free* $\mathscr{S}(B)$-module.

2. $\operatorname{Sec}(\xi^1 \oplus \xi^2) \cong \operatorname{Sec} \xi^1 \oplus \operatorname{Sec} \xi^2$. In fact, the homomorphisms

$$\operatorname{Sec}(\xi^1 \oplus \xi^2) \to \operatorname{Sec} \xi^1 \oplus \operatorname{Sec} \xi^2, \qquad \operatorname{Sec} \xi^1 \oplus \operatorname{Sec} \xi^2 \to \operatorname{Sec}(\xi^1 \oplus \xi^2),$$

given by

$$\sigma \mapsto (\rho^1 \circ \sigma, \rho^2 \circ \sigma), \qquad (\sigma_1, \sigma_2) \mapsto i^1 \circ \sigma_1 + i^2 \circ \sigma_2,$$

respectively, are easily seen to be inverse (cf. sec. 2.8).

3. Let η be a second vector bundle over B, with typical fibre H. We shall define a canonical isomorphism

$$\omega: \operatorname{Sec} L(\xi; \eta) \xrightarrow{\cong} \operatorname{Hom}(\xi; \eta)$$

of $\mathscr{S}(B)$-modules.

Indeed, fix $\sigma \in \operatorname{Sec} L(\xi; \eta)$. Then, for $x \in B$, $\sigma(x) \in L(F_x; H_x)$. Let $\varphi: \xi \to \eta$ be the fibre-preserving set map whose restriction to each F_x is the linear map $\sigma(x)$. φ is smooth, as follows directly from the smoothness of σ, with the aid of coordinate representations for ξ and for η. Hence φ is a strong bundle map.

Now define ω to be the correspondence

$$\omega: \sigma \mapsto \varphi,$$

just given. ω is clearly an $\mathscr{S}(B)$-module homomorphism. To show that it is an isomorphism we construct the inverse. Let $\varphi \in \operatorname{Hom}(\xi; \eta)$ and define $\sigma \in \operatorname{Sec} L(\xi; \eta)$ by

$$\sigma(x) = \varphi_x \in L(F_x; H_x), \qquad x \in B.$$

Then the correspondence $\varphi \mapsto \sigma$ defines a module homomorphism inverse to ω.

4. A canonical isomorphism

$$\omega: \operatorname{Sec} L(\xi^1, ..., \xi^p; \eta) \to \operatorname{Hom}(\xi^1, ..., \xi^p; \eta)$$

is defined by

$$\omega(\sigma)(z_1, ..., z_p) = \sigma(x; z_1, ..., z_p), \qquad x \in B, \quad z_\nu \in F_x^\nu, \quad \nu = 1, ..., p$$

(same argument as in Example 3).

5. A canonical isomorphism

$$\omega: \operatorname{Sec} L(\wedge^p \xi; \eta) \xrightarrow{\cong} A^p(\xi; \eta)$$

is defined by

$$\omega(\sigma)(z_1, ..., z_p) = \sigma(x)(z_1 \wedge \cdots \wedge z_p), \qquad x \in B, \quad z_\nu \in F_x, \quad \nu = 1, ..., p.$$

6. A canonical isomorphism

$$\omega: \operatorname{Sec} L(\vee^p \xi; \eta) \xrightarrow{\cong} S^p(\xi; \eta)$$

is defined by

$$\omega(\sigma)(z_1, ..., z_p) = \sigma(x)(z_1 \vee \cdots \vee z_p), \qquad x \in B, \quad z_\nu \in F_x, \quad \nu = 1, ..., p.$$

2.15. Mappings of cross-sections. In this section we describe three types of mappings (induced by bundle maps) between modules of cross-sections.

First, let $\eta = (E', \pi', B', H)$ be a second vector bundle and let $\varphi: \xi \to \eta$ be a bundle map inducing $\psi: B \to B'$. Let ξ^*, η^* be bundles dual to ξ and η. Then φ induces a map

$$\varphi^*: \operatorname{Sec} \xi^* \leftarrow \operatorname{Sec} \eta^*$$

defined as follows:

Let

$$\varphi_x^* : F_x^* \leftarrow H_{\psi(x)}^*, \qquad x \in B$$

be the dual of the linear map $\varphi_x: F_x \to H_{\psi(x)}$. Then, for $\tau \in \operatorname{Sec} \eta^*$, define a map $\sigma: B \to E^*$ (the total space of ξ^*) by

$$\sigma(x) = \varphi_x^*(\tau(\psi x)), \qquad x \in B.$$

Use the trivializing maps to show that σ is smooth, and hence a cross-section in ξ^*.

We denote σ by $\varphi^*(\tau)$. The map $\varphi^*: \operatorname{Sec} \eta^* \to \operatorname{Sec} \xi^*$, so defined satisfies

$$\varphi^*(\tau_1 + \tau_2) = \varphi^*(\tau_1) + \varphi^*(\tau_2),$$
$$\varphi^*(f\tau) = \psi^*f \cdot \varphi^*(\tau), \qquad f \in \mathscr{S}(B'), \quad \tau, \tau_1, \tau_2 \in \operatorname{Sec} \eta^*.$$

The *second* type of mapping occurs when we consider bundle maps $\varphi: \xi \to \eta$ which restrict to linear *isomorphisms* φ_x in each fibre. In this case a map

$$\varphi^\#: \operatorname{Sec} \xi \leftarrow \operatorname{Sec} \eta$$

is given by

$$[\varphi^\#(\tau)](x) = \varphi_x^{-1}(\tau(\psi x)), \qquad x \in B, \quad \tau \in \operatorname{Sec} \eta.$$

It satisfies the relations

$$\varphi^\#(\tau_1 + \tau_2) = \varphi^\#(\tau_1) + \varphi^\#(\tau_2)$$

and

$$\varphi^\#(f\tau) = \psi^*(f) \cdot \varphi^\#(\tau), \qquad \tau, \tau_1, \tau_2 \in \operatorname{Sec} \eta, \quad f \in \mathscr{S}(B').$$

Thirdly consider the case that ξ and η are bundles over the same base B, and let $\varphi: \xi \to \eta$ be a *strong* bundle map. Define

$$\varphi_*: \text{Sec}\, \xi \to \text{Sec}\, \eta$$

by

$$(\varphi_*\sigma)(x) = \varphi(\sigma(x)) = \varphi_x(\sigma(x)), \qquad x \in B, \quad \sigma \in \text{Sec}\, \xi.$$

Then φ_* is a homomorphism of $\mathscr{S}(B)$-modules.

If $\psi: \eta \to \zeta$ is a second strong bundle map, then

$$(\psi \circ \varphi)_* = \psi_* \circ \varphi_* .$$

Moreover,

$$(\iota_\xi)_* = \iota_{\text{Sec}\, \xi} .$$

§4. Vector bundles with extra structure

2.16. Orientable vector bundles. Let $\xi = (E, \pi, B, F)$ be a vector bundle of rank r, with dual bundle ξ^*. Then (cf. sec. 2.11) $\wedge^r \xi^*$ is a vector bundle of rank 1. We say that ξ is *orientable* if there exists a $\Delta \in \operatorname{Sec} \wedge^r \xi^*$ such that

$$\Delta(x) \neq 0, \qquad x \in B.$$

Such a cross-section is called a *determinant function* in ξ. Clearly $\Delta(x)$ is a determinant function in the vector space F_x.

According to Example 2, sec. 2.13, a vector bundle ξ of rank r is orientable if and only if $\wedge^r \xi^*$ is trivial.

If Δ_1, Δ_2 are both determinant functions in ξ, then there is a unique $f \in \mathscr{S}(B)$ such that $\Delta_1 = f\Delta_2$. Moreover, $f(x) \neq 0$ $(x \in B)$. Thus an equivalence relation on determinant functions is given by

$$\Delta_1 \sim \Delta_2 \Leftrightarrow f(x) > 0, \qquad x \in B.$$

Each class is called an *orientation* of ξ.

If a given class containing Δ is chosen, the vector bundle is said to be *oriented*, Δ is said to *represent the orientation* and is called *positive*.

Suppose then that ξ is oriented and Δ represents the orientation. The determinant functions $\Delta(x)$ orient the fibres F_x $(x \in B)$ in the sense of linear algebra [cf. 5, p. 127]. This orientation is independent of the choice of Δ. A sequence $z_1, \ldots, z_r$ of vectors in F_x will be called *positive* (with respect to the orientation) if

$$\Delta(x; z_1, \ldots, z_r) > 0.$$

Now assume B is connected, and ξ is orientable. Then the functions f, above, are either everywhere positive or everywhere negative; thus ξ admits precisely two orientations.

Proposition III: A vector bundle $\xi = (E, \pi, B, F)$ is orientable if and only if it admits a coordinate representation $\{(U_\alpha, \varphi_\alpha)\}$ whose coordinate transformations $g_{\alpha\beta}(x) = \varphi_{\alpha,x}^{-1} \circ \varphi_{\beta,x}$ have positive determinant.

Proof: Assume that ξ is orientable and let Δ be a determinant function in ξ. Let $\{(U_\alpha, \psi_\alpha)\}$ be a coordinate representation for ξ such that the U_α are connected. Choose a fixed determinant function Δ_F in F.

Since the U_α are connected, for each α, the linear maps

$$\psi_{\alpha,x}: F \to F_x, \qquad x \in U_\alpha$$

either all preserve, or all reverse the orientations. Let ρ be an orientation-reversing isomorphism of F and define a coordinate representation $(U_\alpha, \varphi_\alpha)$ for ξ by

$$\varphi_\alpha(x, y) = \begin{cases} \psi_\alpha(x, y), & \text{if } \psi_{\alpha,x} \text{ preserves orientations} \\ \psi_\alpha(x, \rho(y)), & \text{if } \psi_{\alpha,x} \text{ reverses orientations.} \end{cases}$$

Then each $\varphi_{\alpha,x}$ preserves orientations. Hence so does $\varphi_{\alpha,x}^{-1} \circ \varphi_{\beta,x}$; i.e. $\det(\varphi_{\alpha,x}^{-1} \circ \varphi_{\beta,x}) > 0$.

Conversely, assume that ξ admits a coordinate representation $\{(U_\alpha, \varphi_\alpha)\}$ such that

$$\det(\varphi_{\alpha,x}^{-1} \circ \varphi_{\beta,x}) > 0, \qquad x \in U_\alpha \cap U_\beta .$$

Let Δ_F be a determinant function in F and define $\Delta_\alpha \in A^r(\xi|_{U_\alpha})$ ($r = \operatorname{rank} \xi$) by

$$\Delta_\alpha(x; z_1, \ldots, z_r) = \Delta_F(\varphi_{\alpha,x}^{-1}(z_1), \ldots, \varphi_{\alpha,x}^{-1}(z_r)).$$

A simple computation shows that

$$\Delta_\alpha(x; z_1, \ldots, z_r) = \det(\varphi_{\alpha,x}^{-1} \circ \varphi_{\beta,x})\, \Delta_\beta(x; z_1, \ldots, z_r), \qquad x \in U_\alpha \cap U_\beta, \quad z_i \in F_x .$$

Now, assume that the cover $\{U_\alpha\}$ of B is locally finite, and let $\{p_\alpha\}$ be a subordinate partition of unity. Define $\Delta \in \operatorname{Sec} \wedge^r \xi^*$ by

$$\Delta(x; z_1, \ldots, z_r) = \sum_\alpha p_\alpha(x)\, \Delta_\alpha(x; z_1, \ldots, z_r), \qquad x \in B, \quad z_i \in F_x .$$

Since $\sum_\alpha p_\alpha(x) = 1$, $p_\alpha(x) \geqslant 0$ and $\det(\varphi_{\alpha,x}^{-1} \circ \varphi_{\beta,x}) > 0$, it follows that

$$\Delta(x) \neq 0, \qquad x \in B,$$

and so ξ is orientable.

Q.E.D.

Corollary: Let ξ be an oriented vector bundle and choose a fixed orientation in the typical fibre F. Then there exists a coordinate representation $\{(U_\alpha, \varphi_\alpha)\}$ for ξ such that the isomorphisms

$$\varphi_{\alpha,x}: F \xrightarrow{\cong} F_x$$

preserve the orientations.

If $\varphi: \xi \to \eta$ is a bundle map inducing linear isomorphisms on the fibres, then an orientation in η induces an orientation in ξ. In fact, let Δ_η be a cross-section in $\wedge^r \eta^*$ ($r = \operatorname{rank} \eta$) such that $\Delta_\eta(y) \neq 0$ ($y \in B_\eta$) and set

$$\Delta_\xi = \varphi^* \Delta_\eta .$$

Then $\Delta_\xi(x) \neq 0$ ($x \in B_\xi$) and so Δ_ξ orients ξ.

Let ξ, η be vector bundles over B oriented by Δ_ξ and Δ_η. Then $\xi \oplus \eta$ is oriented by the determinant function Δ given by

$$\Delta(x) = \Delta_\xi(x) \wedge \Delta_\eta(x), \qquad x \in B$$

(cf. sec. 2.11). The orientation represented by Δ depends only on the orientations represented by Δ_ξ and Δ_η, and is called the *induced orientation.*

2.17. Riemannian and pseudo-Riemannian vector bundles. Let $\xi = (E, \pi, B, F)$ be a vector bundle. A *pseudo-Riemannian metric* in ξ is an element $g \in S^2(\xi)$ such that, for each $x \in B$, the symmetric bilinear form $g(x)$ in F_x is nondegenerate. The pair (ξ, g) is called a *pseudo-Riemannian vector bundle.*

If the bilinear forms $g(x)$ are positive definite for every $x \in B$, then g is called a *Riemannian metric*, and (ξ, g) is called a Riemannian vector bundle. A cross-section σ in a pseudo-Riemannian vector bundle is called *normed* if

$$g(x; \sigma(x), \sigma(x)) = 1, \qquad x \in B.$$

A pseudo-Riemannian metric g in ξ defines a duality between ξ and itself. Hence (cf. sec. 2.7) if ξ^* is any vector bundle dual to ξ, g determines a strong isomorphism

$$\tau: \xi \xrightarrow{\cong} \xi^*$$

by the equation

$$\langle \tau(z), w \rangle_x = g(x; z, w), \qquad x \in B, \quad z, w \in F_x .$$

Examples: **1.** Let ξ, η be vector bundles over B with typical fibres F, H. Assume g_ξ, g_η are Riemannian metrics in ξ and η. Then a Riemannian metric g in $\xi \oplus \eta$ is given by ($x \in B$)

$$g(x; z, w) = \begin{cases} g_\xi(x; z, w), & z, w \in F_x \\ 0, & z \in F_x, \quad w \in H_x \\ 0, & z \in H_x, \quad w \in F_x \\ g_\eta(x; z, w), & z, w \in H_x . \end{cases}$$

2. Let ξ, η, g_ξ, g_η be as above. Then a pseudo-Riemannian metric g in $\xi \oplus \eta$ is given by replacing g_η by $-g_\eta$ in the above definition.

3. Let ξ, η, g_ξ, g_η be as above. Then a Riemannian metric g in $\xi \otimes \eta$ is given by

$$g(x; z \otimes w, z' \otimes w') = g_\xi(x; z, z')\, g_\eta(x; w, w').$$

4. A Riemannian metric in $\wedge^p \xi$ is given by

$$g(x; z_1 \wedge \cdots \wedge z_p, z_1' \wedge \cdots \wedge z_p') = \det g(x; z_i, z_j').$$

Proposition IV: Every vector bundle ξ admits a Riemannian metric.

Proof: If $\xi = B \times F$ is trivial and $\langle\, , \rangle$ is a Euclidean metric in F, then

$$g(x; y_1, y_2) = \langle y_1, y_2 \rangle, \qquad x \in B, \quad y_1, y_2 \in F$$

defines a Riemannian metric in ξ.

Now, let ξ be arbitrary, and let $\{(U_\alpha, \varphi_\alpha)\}$ be a coordinate representation for ξ such that $\{U_\alpha\}$ is a locally finite open cover of B. Let $\{p_\alpha\}$ be a subordinate partition of unity.

Since the restriction ξ_α of ξ to U_α is trivial, there is a Riemannian metric g_α in ξ_α. Define g by $\sum_\alpha p_\alpha g_\alpha$. Then $g(x)$ is a Euclidean metric in F_x; hence g is a Riemannian metric in ξ.

Q.E.D.

Corollary: If ξ, ξ^* are dual vector bundles, then $\xi^* \cong \xi$.

Let $\xi = (E_\xi, \pi_\xi, B, F)$ and $\eta = (E_\eta, \pi_\eta, B', H)$ be Riemannian bundles and let $\varphi: \xi \to \eta$ be a bundle map. φ is called *isometric* or an *isometry* if the linear maps φ_x are isomorphisms which preserve the inner product.

Proposition V: Let $\xi = (E, \pi, B, F)$ be a vector bundle with Riemannian metric g. Let $\langle\, , \rangle$ be a fixed Euclidean inner product in F. Then there is a coordinate representation $\{(U_\alpha, \varphi_\alpha)\}$ for ξ such that the maps $\varphi_{\alpha,x}: F \to F_x$ are isometries.

Proof: It is sufficient to consider the case that $\xi = B \times F$ is trivial. Denote $g(x)$ by $\langle\, , \rangle_x$ and let F_x denote the vector space F endowed with the inner product $\langle\, , \rangle_x$. Let $e_1, \ldots, e_r$ be an orthonormal basis of F (with respect to $\langle\, , \rangle$).

Now let $\tau_1(x), \ldots, \tau_r(x)$ be the orthonormal basis of F_x obtained from $e_1, \ldots, e_r$ by the Gram–Schmidt process:

$$\tau_i(x) = \langle \omega_i(x), \omega_i(x) \rangle_x^{-1/2} \omega_i(x),$$

where

$$\omega_i(x) = e_i - \sum_{j=1}^{i-1} \langle e_i, \tau_j(x) \rangle_x \tau_j(x).$$

It follows from this formula that the $\tau_i : B \to F$ are smooth.

Hence a coordinate representation for ξ is given by (B, ψ) where $\psi: B \times F \to E$ is defined by

$$\psi(x, y) = \left(x, \sum_i \langle e_i, y \rangle \tau_i(x)\right).$$

Moreover each $\psi_x : E \to F_x$ is an isometry.

Q.E.D.

Definition: A coordinate representation satisfying the condition of Proposition V will be called a *Riemannian coordinate representation.*

Proposition VI: If (ξ, g), (η, h) are Riemannian vector bundles over the same base B and $\varphi \in \mathrm{Hom}(\xi; \eta)$ is an isomorphism, then there exists an isometric isomorphism

$$\psi: \xi \xrightarrow{\cong} \eta.$$

Proof: Since h induces a Riemannian metric $\tilde{h}$ on ξ with respect to which φ is an isometry we may assume that $\eta = \xi$.

Define $\alpha \in \mathrm{Hom}(\xi; \xi)$ by

$$h(x; \alpha_x(z), w) = g(x; z, w), \qquad x \in B, \quad z, w \in F_x.$$

Since $h(x)$ and $g(x)$ are inner products, each $\alpha_x \in S^+(F_x)$. In view of Example 12, sec. 1.5, there is a unique $\psi_x \in S^+(F_x)$ which satisfies $\psi_x^2 = \alpha_x$ and which depends smoothly on α_x. The induced bundle map $\psi: \xi \to \xi$ is a strong isometric isomorphism.

Q.E.D.

2.18. Subbundles. Proposition VII: If η is a subbundle of ξ (sec. 2.1), there is a second subbundle ζ of ξ such that ξ is the Whitney sum of η and ζ.

Lemma I: Suppose that $\eta = (E_\eta, \pi_\eta, B, H)$ is a subbundle of a Riemannian vector bundle $(\xi = (E, \pi, B, F), g)$ and that $\langle\, , \rangle$ is an inner product for F. Then there exists a Riemannian coordinate representation $\{(U_a, \varphi_a) \mid a \in B\}$ for ξ such that if ψ_a is the restriction of φ_a to $U_a \times H$ (without loss of generality we may assume $H \subset F$), then $\{(U_a, \psi_a)\}$ is a coordinate representation of η.

Proof: According to Example 1, sec. 2.14, we can find, for each $a \in B$, a neighbourhood V_a and a basis $\sigma_1, \ldots, \sigma_s$ $(s = \operatorname{rank} \eta)$ of $\operatorname{Sec}(\eta\,|_{V_a})$. In particular $\sigma_1(a), \ldots, \sigma_s(a)$ is a linearly independent set of vectors in F_a and so there are $\sigma_{s+1}, \ldots, \sigma_r \in \operatorname{Sec} \xi$ such that $\sigma_1(a), \ldots, \sigma_r(a)$ is a basis for F_a.

In view of the continuity of the map

$$x \mapsto \sigma_1(x) \wedge \cdots \wedge \sigma_r(x) \in \wedge^r F_x, \qquad x \in V_a, \quad r = \operatorname{rank} \xi,$$

there exists a neighbourhood U_a of a such that $\sigma_1(x), \ldots, \sigma_r(x)$ forms a basis of F_x, $x \in U_a$. Apply the Gram–Schmidt process (proof of Proposition V above) to obtain new cross-sections $\tau_1, \ldots, \tau_r$ in $\operatorname{Sec}(\xi\,|_{U_a})$ such that $\tau_1(x), \ldots, \tau_r(x)$ is an orthonormal basis of F_x (with respect to $g(x)$). Since $\sigma_1(x), \ldots, \sigma_s(x)$ is a basis of H_x $(x \in U_a)$, it follows from the construction that $\tau_1(x), \ldots, \tau_s(x)$ is a basis of H_x.

Now choose an orthonormal basis $e_1, \ldots, e_r$ of F such that $e_1, \ldots, e_s$ is a basis for H. Define maps $\varphi_a\colon U_a \times F \to \pi^{-1} U_a$ by

$$\varphi_a\left(x, \sum \lambda_i e_i\right) = \sum \lambda_i \tau_i(x), \qquad x \in U_a.$$

Then $\{(U_a, \varphi_a) \mid a \in B\}$ is the required coordinate representation of ξ.

Q.E.D.

Proof of Proposition VII: Assign to ξ a Riemannian metric g. Choose an inner product $\langle\, , \rangle$ for F, and let $\{(U_a, \varphi_a) \mid a \in B\}$ be a coordinate representation for ξ satisfying the conditions of Lemma I.

To construct the subbundle ζ we use the construction principle of sec. 2.5. Assign to $x \in B$ the vector space $H_x^\perp$,

$$H_x^\perp = \{z \in F_x \mid g(x; z, w) = 0 \quad \text{when} \quad w \in H_x\}.$$

Since $\varphi_{a,x}: F \to F_x$ $(x \in U_a)$ is an isometry which carries H to H_x, it restricts to a linear isomorphism

$$\psi_{a,x}\colon H^\perp \xrightarrow{\cong} H_x^\perp.$$

The induced maps $U_a \cap U_b \to L(H^\perp; H^\perp)$ given by

$$x \mapsto \psi_{b,x}^{-1} \circ \psi_{a,x}$$

satisfy Condition S, sec. 2.5. Thus we obtain a vector bundle

$$\zeta = (E_\zeta, \pi_\zeta, B, H^\perp)$$

where $E_\zeta = \cup_x H_x^\perp$. Evidently ζ is a subbundle of ξ.

The inclusions $i: \eta \to \xi$, $j: \zeta \to \xi$ extend to a strong bundle map

$$\eta \oplus \zeta \to \xi.$$

Since $F_x = H_x \oplus H_x^\perp$, this map restricts to isomorphisms in each fibre; hence it is an isomorphism.

Q.E.D.

2.19. Oriented Riemannian bundles. Assume that g is a Riemannian metric in an oriented vector bundle $\xi = (E, \pi, B, F)$ of rank r. Let ξ^* be any dual bundle. Then the induced isomorphism $\xi \cong \xi^*$ induces a Riemannian metric in ξ^*, and hence in $\wedge^r \xi^*$. There is a unique normed cross-section $\Delta \in \operatorname{Sec} \wedge^r \xi^*$ which is positive with respect to the orientation of ξ. It is called the positive normed determinant function in ξ. For each $x \in B$, $\Delta(x)$ is the positive normed determinant function in F_x.

Proposition VIII: If (ξ, g, Δ) is an oriented Riemannian vector bundle and $(F, \langle\ ,\ \rangle, \Delta_F)$ is an oriented Euclidean vector space, there exists a coordinate representation $\{(U_\alpha, \varphi_\alpha)\}$ for ξ such that the maps $\varphi_{\alpha,x} : F \to F_x$ are orientation preserving isometries.

Proof: Apply the proof of Proposition III to a Riemannian coordinate representation, being careful to choose an *isometric* reflection, ρ, in F.

Q.E.D.

2.20. The bundle $\tilde{\mathscr{B}}$. Let $\xi = (E, \pi, B, F)$ be a Riemannian vector bundle of rank r and consider the rank 1 bundle $\wedge^r \xi$. Let S_x denote the unit sphere of the one-dimensional Euclidean space $\wedge^r F_x$ $(x \in B)$. We shall construct a smooth bundle $\tilde{\mathscr{B}} = (\tilde{B}, \rho, B, S^0)$ such that $\rho^{-1}(x) = S_x$.

Let $\tilde{B} = \cup_{x \in B} S_x$ and let $\rho: \tilde{B} \to B$ denote the obvious projection. Choose a Riemannian coordinate representation $\{(U_\alpha, \psi_\alpha)\}$ for ξ. Then each linear map

$$\hat{\psi}_{\alpha,x} = \wedge^r \psi_{\alpha,x}: \wedge^r F \to \wedge^r F_x$$

is an isometry and so $\{(U_\alpha, \hat{\psi}_\alpha)\}$ is a Riemannian coordinate representation for $\wedge^r \xi$. The corresponding coordinate transformations are given by

$$\hat{\psi}_{\beta\alpha,x} = \det \psi_{\beta\alpha,x} \cdot \iota = \epsilon_{\beta\alpha,x} \cdot \iota,$$

where $\epsilon_{\beta\alpha,x} = \pm 1$, and the function $x \mapsto \epsilon_{\beta\alpha,x}$ is smooth.

Thus $\hat{\psi}_\alpha$ restricts to a bijection

$$\varphi_\alpha \colon U_\alpha \times S^0 \to \rho^{-1}(U_\alpha).$$

In particular, the map

$$\varphi_{\beta\alpha} = \varphi_\beta^{-1} \circ \varphi_\alpha : U_\alpha \cap U_\beta \times S^0 \to U_\alpha \cap U_\beta \times S^0$$

is given by

$$\varphi_{\beta\alpha}(x, t) = (x, \epsilon_{\beta\alpha,x} t), \qquad x \in U_\alpha \cap U_\beta, \quad t \in S^0,$$

and hence it is smooth. Now apply Proposition X, sec. 1.13, to obtain a smooth structure on $\tilde{B}$ such that $(\tilde{B}, \rho, B, S^0)$ is a smooth fibre bundle. Since S^0 consists of two points, $\tilde{B}$ is called the *double cover of B induced by ξ*. The smooth involution ω of $\tilde{B}$ which interchanges the two points in each S_x is called the *covering transformation of $\tilde{B}$*.

Proposition IX: If B is connected, then $\tilde{B}$ is connected if and only if ξ is *not* orientable.

Proof: ρ preserves open and closed sets; hence it maps each component of $\tilde{B}$ *onto* the connected manifold B. Since $\rho^{-1}(x)$ consists of two points $(x \in B)$, there are two possibilities: either (i) $\tilde{B}$ is connected or (ii) $\tilde{B}$ has two components $\tilde{B}_1$, $\tilde{B}_2$, and ρ restricts to diffeomorphisms

$$\rho_i \colon \tilde{B}_i \xrightarrow{\cong} B.$$

If $\tilde{B}$ is not connected $\rho_1^{-1} \colon B \to \tilde{B}_1$ may be interpreted as a cross-section with no zeros in $\wedge^r \xi$; hence ξ is orientable.

Conversely, suppose ξ orientable. Choose orientations in ξ and in F, and choose ψ_α so that each $\psi_{\alpha,x}$ is orientation preserving (Proposition VIII, sec. 2.19). Then (in the notation above)

$$\varphi_{\alpha,x} = \varphi_{\beta,x} : S^0 \to S_x, \qquad x \in U_\alpha \cap U_\beta.$$

Thus the φ_α define a diffeomorphism

$$\varphi \colon B \times S^0 \xrightarrow{\cong} \tilde{B}.$$

In particular, $\tilde{B}$ is not connected.

Q.E.D.

Proposition X: Let $\tilde{\xi} = (\tilde{E}, \tilde{\pi}, \tilde{B}, F)$ be the pull-back of ξ to $\tilde{B}$. Then the bundle $\tilde{\xi}$ is orientable.

Proof: Set $\tilde{U}_\alpha = \rho^{-1} U_\alpha$ and let $\{(\tilde{U}_\alpha, \tilde{\psi}_\alpha)\}$ be the coordinate representation of $\tilde{\xi}$ induced by the coordinate representation $\{(U_\alpha, \psi_\alpha)\}$ for ξ. Choose a determinant function Δ_F in F. Then the cross-section

$$\Omega(x, t) = t \cdot \Delta_F, \qquad x \in U_\alpha, \quad t \in S^0,$$

orients the trivial vector bundle $(U_\alpha \times S^0) \times F$. Thus the bundle isomorphism

$$\lambda_\alpha: (U_\alpha \times S^0) \times F \xrightarrow[\varphi_\alpha \times \iota]{\cong} \tilde{U}_\alpha \times F \xrightarrow[\tilde{\psi}_\alpha]{\cong} \tilde{\pi}^{-1}\tilde{U}_\alpha$$

induces an orientation in the restriction $\tilde{\xi}_\alpha$ of $\tilde{\xi}$ to $\tilde{U}_\alpha$.

Now a simple computation shows that the maps

$$\lambda_\beta^{-1} \circ \lambda_\alpha: [(U_\alpha \cap U_\beta) \times S^0] \times F \xrightarrow{\cong} [(U_\alpha \cap U_\beta) \times S^0] \times F$$

are orientation preserving. It follows that the restrictions of $\tilde{\xi}_\alpha$ and $\tilde{\xi}_\beta$ to $\tilde{U}_\alpha \cap \tilde{U}_\beta$ have the same orientation. Hence the orientations in the $\tilde{\xi}_\alpha$ define a global orientation of $\tilde{\xi}$.

Q.E.D.

2.21. The bundle Sk_ξ. Let $(\xi = (E, \pi, B, F), g)$ be a Riemannian vector bundle. Consider, for every $x \in B$, the subspace

$$\mathrm{Sk}_{F_x} \subset L_{F_x},$$

whose elements are the *skew* transformations (with respect to the inner product $g(x)$). Let

$$\hat{E} = \bigcup_{x \in B} \mathrm{Sk}_{F_x} \subset L_\xi$$

and let $\hat{\pi}$ denote the restriction of the bundle projection of L_ξ to $\hat{E}$. Finally, let $\langle\ ,\ \rangle$ be an inner product in F and let $\mathrm{Sk}_F \subset L_F$ be the space of linear transformations of F which are skew with respect to $\langle\ ,\ \rangle$. We shall show that

$$\mathrm{Sk}_\xi = (\hat{E}, \hat{\pi}, B, \mathrm{Sk}_F)$$

is a vector bundle.

In fact, let $\{(U_\nu, \varphi_\nu)\}$ be a Riemannian coordinate representation for ξ. The corresponding coordinate representation $\{(U_\nu, \chi_\nu)\}$ for L_ξ is given by

$$\chi_{\nu,x}(\alpha) = \varphi_{\nu,x}\, \alpha\, \varphi_{\nu,x}^{-1}, \qquad \alpha \in L_F, \quad x \in U_\nu.$$

Since the $\varphi_{\nu,x}$ are isometries, $\chi_{\nu,x}$ restricts to a linear isomorphism

$$\psi_{\nu,x}: \mathrm{Sk}_F \xrightarrow{\cong} \mathrm{Sk}_{F_x}$$

Apply the construction principle of sec. 2.5 to obtain the desired bundle.

Evidently Sk_ξ is a subbundle of L_ξ.

A strong bundle isomorphism

$$\beta: \wedge^2\xi \xrightarrow{\cong} \mathrm{Sk}_\xi$$

is defined by

$$\beta_x(z \wedge w) = \alpha_x, \qquad z, w \in F_x, \quad x \in B,$$

where α_x is the skew linear map in F_x given by

$$\alpha_x(y) = \langle z, y\rangle_x w - \langle w, y\rangle_x z, \qquad y \in F_x, \qquad \langle,\rangle_x = g(x).$$

2.22. Complex vector bundles. A *complex vector bundle* is a quadruple $\xi = (E, \pi, B, F)$ where

(1) (E, π, B, F) is a smooth fibre bundle.

(2) F, and the fibres F_x $(x \in B)$ are complex linear spaces.

(3) There is a coordinate representation $\{(U_\alpha, \psi_\alpha)\}$ for ξ such that the maps

$$\psi_{\alpha,x}: F \to F_x$$

are complex linear isomorphisms.

The complex dimension of F is called the rank of the complex bundle ξ.

Let $\xi = (E, \pi, B, F)$ be a complex vector bundle of rank r. Let $F_{\mathbb{R}}$ be the $2r$-dimensional real vector space underlving F, and let

$$i: F_{\mathbb{R}} \to F_{\mathbb{R}}$$

be multiplication by $i \in \mathbb{C}$. Let $\xi_{\mathbb{R}} = (E, \pi, B, F_{\mathbb{R}})$ be the real vector bundle obtained by forgetting the complex structure and let $i_\xi \in L(\xi_{\mathbb{R}}; \xi_{\mathbb{R}})$ be the strong bundle isomorphism which restricts to multiplication by i in each $(F_x)_{\mathbb{R}} = (F_{\mathbb{R}})_x$. Then, if $\{(U_\alpha, \psi_\alpha)\}$ is a coordinate representation for ξ, we have

$$\psi_{\alpha,x} \circ i = i_\xi(x) \circ \psi_{\alpha,x}, \qquad x \in U_\alpha.$$

i_ξ is called the *complex structure of* ξ.

Proposition XI: Let $\eta = (E, \pi, B, F_{\mathbb{R}})$ be a real vector bundle of rank $2r$. Let $\gamma \in L_\eta$ be a strong bundle map such that $\gamma^2 = -\iota$. Let F be a complex space with underlying real space $F_{\mathbb{R}}$. Then η is the underlying real vector bundle of a complex bundle $\xi = (E, \pi\ B, F)$ with complex structure γ.

Proof: We must find a coordinate representation $\{(U_\alpha\ \varphi_\alpha)\}$ for η such that

$$\varphi_{\alpha,x} \circ i = \gamma_x \circ \varphi_{\alpha,x}, \qquad x \in U_\alpha .$$

Let $a \in B$ be arbitrary and choose a basis for $(F_{\mathbb{R}})_a$ of the form $z_1, ..., z_r, \gamma_a z_1, ..., \gamma_a z_r$. By Example 1, sec. 2.13, there are $\sigma_\nu \in \operatorname{Sec}\eta$ such that $\sigma_\nu(a) = z_\nu$ ($\nu = 1, ..., r$). By the continuity of the map

$$x \mapsto \sigma_1(x) \wedge \cdots \wedge \sigma_r(x) \wedge (\gamma_*\sigma_1)(x) \wedge \cdots \wedge (\gamma_*\sigma_r)(x),$$

there is a neighbourhood U of a such that

$$\tau_\nu = \sigma_\nu\,|_U, \qquad \bar{\tau}_\nu = (\gamma_*\sigma_\nu)|_U, \qquad \nu = 1, ..., r,$$

form a basis for $\operatorname{Sec}(\eta\,|_U)$.

Let $e_1, ..., e_r, i(e_1), ..., i(e_r)$ be a basis for $F_{\mathbb{R}}$ and define $\varphi: U \times F_{\mathbb{R}} \to \pi^{-1}U$ by

$$\varphi(x, e_\nu) = \tau_\nu(x), \qquad \varphi(x, i(e_\nu)) = \bar{\tau}_\nu(x), \qquad \nu = 1, ..., r.$$

It is easily checked that (U, φ) is a trivializing chart of ξ and that $\varphi_x \circ i = \gamma_x \circ \varphi_x$, $x \in U$.

Since U is a neighbourhood of an arbitrary point $a \in B$, the proposition is proved.

Q.E.D.

The results of §1, §2, and §3 are essentially unchanged if we replace $\mathbb{R}$ by $\mathbb{C}$ and real vector bundles by complex vector bundles. In particular we have the notion of complex bundle maps (the fibres being complex linear), the module of complex p-linear mappings, the complex tensor product, the complex exterior algebra bundle, and complex triviality.

Suppose $\xi = (E, \pi, B, F)$ is a complex vector bundle. A *Hermitian metric*, g, in ξ is an element

$$g \in \operatorname{Hom}(\xi_{\mathbb{R}}, \xi_{\mathbb{R}}; B \times \mathbb{C})$$

such that $g(x)$ is a Hermitian inner product in the complex vector space F_x for each $x \in B$.

In exactly the same way that the analogous theorems for Riemannian bundles were proved we obtain

Proposition XII: A Hermitian metric can be introduced in every complex vector bundle. If (ξ, g) is a Hermitian complex vector bundle, there exists a coordinate representation $\{(U_a, \varphi_a)\}$ of ξ such that the mappings

$$\varphi_{\alpha,x}: F \to F_x$$

are Hermitian isometries (F being given a fixed Hermitian metric).

§5. Structure theorems

2.23. The main theorem. This section is devoted to proving

Theorem I: For every vector bundle ξ there exists a vector bundle η, over the same base, such that $\xi \oplus \eta$ is trivial.

We write, as usual, $\xi = (E, \pi, B, F)$.

Lemma II: Sec ξ is a finitely generated $\mathscr{S}(B)$-module.

Proof: It follows from Theorem I, sec. 1.2 (as in Proposition XI, sec. 1.13) that ξ admits a finite coordinate representation $\{(U_\mu, \psi_\mu)\}$, $\mu = 1, ..., p$. Since the restrictions ξ_μ of ξ to U_μ are trivial, the $\mathscr{S}(U_\mu)$-modules Sec ξ_μ are free on bases $\{\sigma_{\mu i}\}_{i=1,\dots,r}$ ($r = \text{rank}\ \xi$), (cf. Example 1, sec. 2.14).

Let $f_\mu (\mu = 1, ..., p)$ denote a partition of unity for B subordinate to the covering $\{U_\mu\}$. Define cross-sections $\tau_{\mu i}$ in ξ by

$$\tau_{\mu i} = f_\mu \cdot \sigma_{\mu i}, \qquad \mu = 1, ..., p; \quad i = 1, ..., r.$$

We shall show that the $\tau_{\mu i}$ generate Sec ξ.

Since $\text{carr}\, f_\mu \subset U_\mu$, Proposition VIII, sec. 1.8, yields functions h_μ on B such that

$$\text{carr}\, h_\mu \subset U_\mu \qquad \text{and} \qquad h_\mu(x) = 1, \qquad x \in \text{carr}\, f_\mu.$$

Thus

$$h_\mu f_\mu = f_\mu .$$

Now let $\omega \in \text{Sec}\ \xi$ and denote by ω_μ the restriction of ω to U_μ. Write

$$\omega_\mu = \sum_{i=1}^{r} g_{\mu i}\sigma_{\mu i}, \qquad g_{\mu i} \in \mathscr{S}(U_\mu).$$

Define $p_{\mu i} \in \mathscr{S}(B)$ by $p_{\mu i} = h_\mu g_{\mu i}$. Then

$$\omega = \sum_\mu f_\mu \omega_\mu = \sum_\mu h_\mu f_\mu \omega_\mu = \sum_{\mu,i} (h_\mu g_{\mu i} f_\mu)\sigma_{\mu i} = \sum_{\mu,i} p_{\mu i}\tau_{\mu i} .$$

Q.E.D.

Lemma III: Let $\zeta = (\hat{E}, \hat{\pi}, B, \hat{F})$ be a second vector bundle over the same base and let $\psi: \zeta \to \xi$ be a strong bundle map such that each $\psi_x: \hat{F}_x \to F_x$ is surjective. Then there is a strong bundle map $\varphi: \xi \to \zeta$ such that $\psi \circ \varphi = \iota$.

Proof: Give ξ and η Riemannian metrics. Then each ψ_x determines an adjoint map $\psi_x^*: F_x \to \hat{F}_x$. Since ψ_x is surjective, $\psi_x \circ \psi_x^*$ is an isomorphism of F_x onto itself. Define $\varphi: E \to \hat{E}$ by

$$\varphi_x = \psi_x^*(\psi_x \circ \psi_x^*)^{-1}, \qquad x \in B$$

and use Riemannian coordinate representations for ξ and η to show that φ is smooth. Thus φ is a strong bundle map and $\psi \circ \varphi = \iota$.

Q.E.D.

Proof of the theorem: According to Lemma II there is a finite system of generators $\sigma_1, \ldots, \sigma_m$ for Sec ξ. Consider the trivial bundle $B \times \mathbb{R}^m$ and the strong bundle map $\psi: B \times \mathbb{R}^m \to \xi$ given by

$$\psi\left(x, \sum_{i=1}^{m} \lambda_i e_i\right) = \sum_{i=1}^{m} \lambda_i \sigma_i(x), \qquad x \in B,$$

where $e_1, \ldots, e_m$ is a basis for $\mathbb{R}^m$. Then each linear map $\psi_x: \mathbb{R}^m \to F_x$ is surjective.

In fact, if $z \in F_x$, choose $\sigma \in \text{Sec}\, \xi$ so that $\sigma(x) = z$ (cf. Example 1, sec. 2.13). Since the σ_i generate Sec ξ, we can write

$$\sigma = \sum_{i=1}^{m} f_i \sigma_i, \qquad f_i \in \mathscr{S}(B),$$

whence

$$z = \sigma(x) = \sum_{i=1}^{m} f_i(x)\, \sigma_i(x) = \psi_x\left(\sum_{i=1}^{m} f_i(x)\, e_i\right).$$

Now Lemma III yields a strong bundle map $\varphi: \xi \to B \times \mathbb{R}^m$ satisfying $\psi \circ \varphi = \iota$. Since every map $\varphi_x: F_x \to \mathbb{R}^m$ is injective, φ makes ξ into a subbundle of $B \times \mathbb{R}^m$. Thus Proposition VII, sec. 2.18, gives a second subbundle η of $B \times \mathbb{R}^m$ such that

$$\xi \oplus \eta = B \times \mathbb{R}^m.$$

Q.E.D.

Corollary: The $\mathscr{S}(B)$-module Sec ξ is finitely generated and projective.

Proof: By the theorem, we can write

$$\operatorname{Sec}(\xi \oplus \eta) = \operatorname{Sec}(B \times \mathbb{R}^m),$$

whence

$$\operatorname{Sec} \xi \oplus \operatorname{Sec} \eta = \operatorname{Sec}(B \times \mathbb{R}^m).$$

It follows that Sec ξ is a direct summand of a finitely generated free module.

Q.E.D.

2.24. Applications. In this section ξ, η, ξ^ν denote fixed vector bundles over B. Let $\varphi: \xi \to \eta$ be a strong bundle map and consider the map

$$\varphi_*: \operatorname{Sec} \xi \to \operatorname{Sec} \eta$$

(cf. sec. 2.15).

Proposition XIII: The map $\varphi \mapsto \varphi_*$ defines an isomorphism

$$*: \operatorname{Hom}(\xi; \eta) \xrightarrow{\cong} \operatorname{Hom}_B(\operatorname{Sec} \xi; \operatorname{Sec} \eta)$$

of $\mathscr{S}(B)$-modules, (cf. sec. 1.8 for the notation).

Proof: Clearly, $*$ is a homomorphism of $\mathscr{S}(B)$-modules. To show that it is an isomorphism consider first the case of trivial bundles, $\xi = B \times F$, $\eta = B \times H$. Let $a_1, \ldots, a_r$ and $b_1, \ldots, b_s$ be bases respectively for F and H. Then the constant cross-sections

$$\sigma_i: B \to a_i, \qquad \tau_j: B \to b_j$$

are, respectively, bases for Sec ξ and Sec η. Thus the elements $\varphi_{ik} \in \operatorname{Hom}_B(\operatorname{Sec} \xi; \operatorname{Sec} \eta)$ given by

$$\varphi_{ik}(\sigma_j) = \delta_{ij}\tau_k$$

form a basis for $\operatorname{Hom}_B(\operatorname{Sec} \xi; \operatorname{Sec} \eta)$.

On the other hand, a basis $\{\omega_{ik}\}$ of $L(F; H)$ is given by

$$\omega_{ik}(a_j) = \delta_{ij}b_k.$$

Hence the constant cross-sections $\psi_{ik}: B \to \omega_{ik}$ form a basis for

$$\mathscr{S}(B; L(F; H)) = \operatorname{Sec}(B \times L(F; H)) = \operatorname{Hom}(B \times F; B \times H).$$

Since $(\psi_{ik})_* = \varphi_{ik}$, it follows that $*$ is an isomorphism.

In the general case choose vector bundles $\hat{\xi}$ and $\hat{\eta}$ such that $\xi \oplus \hat{\xi}$ and $\eta \oplus \hat{\eta}$ are trivial. Then

$$*: \operatorname{Hom}(\xi \oplus \hat{\xi}; \eta \oplus \hat{\eta}) \xrightarrow{\cong} \operatorname{Hom}_B(\operatorname{Sec}(\xi \oplus \hat{\xi}); \operatorname{Sec}(\eta \oplus \hat{\eta}))$$

is an isomorphism. Distributing both sides over $\oplus$, we find that this isomorphism is the direct sum of four maps, each of which must then be an isomorphism. But one of these maps is

$$*: \operatorname{Hom}(\xi; \eta) \to \operatorname{Hom}_B(\operatorname{Sec} \xi; \operatorname{Sec} \eta).$$

Q.E.D.

This result may be extended to the multilinear case by extending the definition of $*$. In fact, if $\varphi \in \operatorname{Hom}(\xi^1, ..., \xi^p; \xi^0)$, we define $\varphi_* \in \operatorname{Hom}_B(\operatorname{Sec} \xi^1, ..., \operatorname{Sec} \xi^p; \operatorname{Sec} \xi^0)$ by

$$[\varphi_*(\sigma_1, ..., \sigma_p)](x) = \varphi_x(\sigma_1(x), ..., \sigma_p(x)),$$

where $\sigma_\nu \in \operatorname{Sec} \xi^\nu$ $(\nu = 1, ..., p)$. Then

$$*: \operatorname{Hom}(\xi^1, ..., \xi^p; \xi^0) \to \operatorname{Hom}_B(\operatorname{Sec} \xi^1, ..., \operatorname{Sec} \xi^p; \operatorname{Sec} \xi^0)$$

is an $\mathscr{S}(B)$-isomorphism. (The proof is similar to that given in the special case above.)

Corollary I: If $\xi^1 = \cdots = \xi^p = \xi$, $\xi^0 = \eta$ then $*$ restricts to an isomorphism

$$A^p(\xi; \eta) \xrightarrow{\cong} A_B^p(\operatorname{Sec} \xi; \operatorname{Sec} \eta).$$

Corollary II: (The localization isomorphism) The map

$$\epsilon: \operatorname{Sec} L(\xi^1, ..., \xi^p; \eta) \to \operatorname{Hom}_B(\operatorname{Sec} \xi^1, ..., \operatorname{Sec} \xi^p; \operatorname{Sec} \eta),$$

given by

$$([\epsilon(\Phi)](\sigma_1, ..., \sigma_p))(x) = \Phi(x; \sigma_1(x), ..., \sigma_p(x)),$$

is an isomorphism of $\mathscr{S}(B)$-modules.

Proof: It follows from the definitions that the diagram

$$\begin{array}{ccc} \operatorname{Sec} L(\xi^1, \ldots, \xi^p; \eta) & \xrightarrow{\epsilon} & \operatorname{Hom}_B(\operatorname{Sec} \xi^1, \ldots, \operatorname{Sec} \xi^p; \operatorname{Sec} \eta) \\ & {}_{\omega}\searrow^{\cong} \quad {}^{\cong}\nearrow_{*} & \\ & \operatorname{Hom}(\xi^1, \ldots, \xi^p; \eta) & \end{array}$$

commutes. Here, ω is the isomorphism of Example 4, sec. 2.14.

Q.E.D.

Remark: If ξ, ξ^* are dual, the isomorphism

$$\epsilon: \operatorname{Sec} \xi^* \xrightarrow{\cong} \operatorname{Hom}_B(\operatorname{Sec} \xi; \mathscr{S}(B))$$

is given by

$$[\epsilon(\sigma^*)](\sigma) = \langle \sigma^*, \sigma \rangle.$$

The following propositions are proved by the same argument as that of Proposition VIII.

Proposition XIV: The map

$$\mathcal{O}: \operatorname{Sec} \xi \otimes_B \operatorname{Sec} \eta \to \operatorname{Sec}(\xi \otimes \eta),$$

defined by

$$[\mathcal{O}(\sigma \otimes \tau)](x) = \sigma(x) \otimes \tau(x),$$

is an isomorphism of $\mathscr{S}(B)$-modules.

Corollary: An isomorphism

$$\alpha: \operatorname{Sec} \xi^* \otimes_B \operatorname{Sec} \eta \to \operatorname{Hom}_B(\operatorname{Sec} \xi; \operatorname{Sec} \eta)$$

is given by

$$\alpha(\sigma^* \otimes \tau): \sigma \mapsto \langle \sigma^*, \sigma \rangle \tau.$$

Proof of the Corollary: It follows from the definitions that the diagram

$$\begin{array}{ccccc} \operatorname{Sec} \xi^* \otimes_B \operatorname{Sec} \eta & \underset{\mathcal{O}}{\xrightarrow{\cong}} & \operatorname{Sec}(\xi^* \otimes \eta) & \xrightarrow{\cong} & \operatorname{Sec} L(\xi; \eta) \\ & {}_{\alpha}\searrow & & \swarrow^{\cong}_{\epsilon} & \\ & & \operatorname{Hom}_B(\operatorname{Sec} \xi; \operatorname{Sec} \eta) & & \end{array}$$

commutes. (Note that $\xi^* \otimes \eta \cong L(\xi; \eta)$, cf. sec. 2.10). Hence α is an isomorphism.

Q.E.D.

Proposition XV: An isomorphism

$$\Phi: \wedge_B^p \operatorname{Sec} \xi \xrightarrow{\cong} \operatorname{Sec} \wedge^p \xi$$

is defined by

$$\Phi(\sigma_1 \wedge \cdots \wedge \sigma_p): x \mapsto \sigma_1(x) \wedge \cdots \wedge \sigma_p(x).$$

Applying the isomorphism α to the case $\eta = \xi$, we see that there is a distinguished element $t \in \operatorname{Sec} \xi^* \otimes_B \operatorname{Sec} \xi$ which satisfies

$$\alpha(t) = \iota.$$

t is called the *unit tensor* for the pair ξ^*, ξ. Since $t \in \operatorname{Sec} \xi^* \otimes_B \operatorname{Sec} \xi$, there are finitely many cross-sections $\sigma_i^* \in \operatorname{Sec} \xi^*$ and $\sigma_i \in \operatorname{Sec} \xi$ such that

$$t = \sum_i \sigma_i^* \otimes_B \sigma_i .$$

Thus

$$\sigma = \sum_i \langle \sigma_i^*, \sigma \rangle \sigma_i , \qquad \sigma \in \operatorname{Sec} \xi.$$

The isomorphisms established above, together with those established earlier give the following isomorphisms:

$$\begin{aligned} A^p(\xi; \eta) &\cong \operatorname{Sec} L(\wedge^p \xi; \eta) \cong A_B^p(\operatorname{Sec} \xi; \operatorname{Sec} \eta) \\ &\cong \operatorname{Hom}_B(\wedge^p \operatorname{Sec} \xi; \operatorname{Sec} \eta) \cong \operatorname{Sec}(\wedge^p \xi^* \otimes \eta) \cong \wedge^p \operatorname{Sec} \xi^* \otimes_B \operatorname{Sec} \eta. \end{aligned}$$

In particular, we have module isomorphisms

$$A(\xi) \cong \wedge_B \operatorname{Sec} \xi^* \cong \operatorname{Sec} \wedge \xi^*$$

(obtained by setting $\eta = B \times \mathbb{R}$). The algebra structure induced in $A(\xi)$ by this isomorphism and the algebra structure in $\wedge_B \operatorname{Sec} \xi^*$ is given explicitly by

$$(\Phi \wedge \Psi)(x; z_1, \ldots, z_{p+q}) = \frac{1}{p!q!} \sum_\sigma \epsilon_\sigma \Phi(x; z_{\sigma(1)}, \ldots, z_{\sigma(p)}) \Psi(x; z_{\sigma(p+1)}, \ldots, z_{\sigma(p+q)})$$

$$x \in B, \quad z_i \in F_x, \quad \Phi \in A^p(\xi), \quad \Psi \in A^q(\xi),$$

where σ runs through the symmetric group of order $p + q$ and ϵ_σ is the sign of σ.

2.25. Cross-sections and multilinear bundle maps. Let $\xi_1, \ldots, \xi_p, \eta$ be vector bundles over B. Then, as was done above for $p = 1$, we can construct isomorphisms

$$\begin{aligned}\mathrm{Hom}(\xi_1, \ldots, \xi_p; \eta) &\cong \mathrm{Sec}(\xi_1^* \otimes \cdots \otimes \xi_p^* \otimes \eta)\\ &\cong (\mathrm{Sec}\, \xi_1)^* \otimes_B \cdots \otimes_B (\mathrm{Sec}\, \xi_p)^* \otimes_B \mathrm{Sec}\, \eta;\end{aligned}$$

i.e., we can represent multilinear bundle maps as cross-sections. For instance, giving a bilinear function in (ξ, η) is the same as constructing a cross-section in the bundle $\xi^* \otimes \eta^*$.

In general, theorems about these modules proved using analytical or geometrical techniques are most efficiently established using the form $\mathrm{Sec}(\xi_1^* \otimes \cdots \otimes \xi_p^*; \eta)$ or $\mathrm{Hom}(\xi_1, \ldots, \xi_p; \eta)$, while theorems which are established using algebraic methods are best proved using the forms $\mathrm{Sec}\, \xi_1^* \otimes_B \cdots \otimes_B \mathrm{Sec}\, \xi_p^* \otimes_B \mathrm{Sec}\, \eta$. In any case, we shall use these forms interchangeably and without further reference from now on. We shall, moreover, identify the various isomorphic modules above, also *without further reference.*

2.26. Pull backs. Let $\hat{\xi} = (\hat{E}, \hat{\pi}, ,\hat{B}, F)$ and $\xi = (E, \pi, B, F)$ be vector bundles and let

$$\begin{array}{ccc}\hat{E} & \xrightarrow{\varphi} & E\\ {\scriptstyle\hat{\pi}}\downarrow & & \downarrow{\scriptstyle\pi}\\ \hat{B} & \xrightarrow{\psi} & B\end{array}$$

be a bundle map restricting to linear isomorphisms in the fibres. Make $\mathscr{S}(\hat{B})$ into an $\mathscr{S}(B)$-module by setting

$$f \cdot g = \psi^* f \cdot g, \qquad f \in \mathscr{S}(B), \quad g \in \mathscr{S}(\hat{B})$$

and make $\mathscr{S}(\hat{B}) \otimes_B \mathrm{Sec}\, \xi$ into an $\mathscr{S}(\hat{B})$-module by left multiplication. Then an $\mathscr{S}(\hat{B})$-homomorphism

$$\alpha_\varphi \colon \mathscr{S}(\hat{B}) \otimes_B \mathrm{Sec}\, \xi \to \mathrm{Sec}\, \hat{\xi}$$

is given by

$$\alpha_\varphi(g \otimes \sigma) = g \cdot \varphi^\# \sigma, \qquad g \in \mathscr{S}(\hat{B}), \quad \sigma \in \mathrm{Sec}\, \xi.$$

Proposition XVI: α_φ is an isomorphism.

Proof: If ξ is trivial, let $\sigma_1, \ldots, \sigma_m$ be a basis for the $\mathscr{S}(B)$-module Sec ξ. Then $\varphi^{\#}\sigma_1, \ldots, \varphi^{\#}\sigma_m$ is a basis for Sec $\hat{\xi}$ and $1 \otimes \sigma_1, \ldots, 1 \otimes \sigma_m$ is a basis for the $\mathscr{S}(\hat{B})$-module $\mathscr{S}(\hat{B}) \otimes_B \operatorname{Sec} \xi$. Hence α_φ carries basis to basis and so it is an isomorphism.

In general, choose a vector bundle η over B such that $\xi \oplus \eta$ is trivial (cf. Theorem I, sec. 2.23) and consider the pull-back $\hat{\eta}$ of η via ψ (cf. sec. 2.5)

$$\begin{array}{ccc} \hat{\eta} & \xrightarrow{\varphi_1} & \eta \\ \downarrow & & \downarrow \\ \hat{B} & \xrightarrow{\psi} & B \end{array}$$

Then, since $\hat{\xi}$ is the pull-back of ξ via ψ (cf. sec. 2.5), it follows that $\hat{\xi} \oplus \hat{\eta}$ is the pull-back of $\xi \oplus \eta$. Hence, $\hat{\xi} \oplus \hat{\eta}$ is trivial. Hence, the map

$$\alpha_{\varphi \oplus \varphi_1} \colon \mathscr{S}(\hat{B}) \otimes_B \operatorname{Sec}(\xi \oplus \eta) \to \operatorname{Sec}(\hat{\xi} \oplus \hat{\eta})$$

is an isomorphism. On the other hand,

$$\alpha_{\varphi \oplus \varphi_1} = \alpha_\varphi \oplus \alpha_{\varphi_1}$$

and so α_φ is an isomorphism.

Q.E.D.

Problems

All vector bundles are real unless otherwise stated. The symbol $\cong$ denotes strong bundle isomorphism.

1. Quotient bundles. Suppose η is a subbundle of ξ with typical fibre $H \subset F$.

(i) Show that there is a unique bundle, ξ/η, whose fibre at x is F_x/H_x and such that the projection $\xi \to \xi/\eta$ is a bundle map. ξ/η is called the *quotient bundle* of ξ with respect to η.

(ii) A sequence $0 \to \eta \to \xi \to \rho \to 0$ is called *short exact*, if it restricts to a short exact sequence on each fibre. Show that a short exact sequence of strong bundle maps determines an isomorphism $\rho \xrightarrow{\cong} \xi/\eta$.

(iii) If η is a subbundle of ξ, show that $\xi \cong \eta \oplus \xi/\eta$.

2. Bundle maps. (i) Show that every bundle map is the composite of a strong bundle map and a bundle map which restricts to isomorphisms on the fibres.

(ii) A strong bundle map φ: $\xi \to \eta$ is said to have *constant rank*, if the rank of the linear maps φ_x is independent of x. Show that in this case $\ker \varphi = \bigcup_x \ker \varphi_x$ and $\operatorname{Im} \varphi = \bigcup_x \operatorname{Im} \varphi_x$ are subbundles of ξ and η, and that $\operatorname{Im} \varphi \cong \xi/\ker \varphi$.

(iii) Let d: $\xi \to \xi$ be a strong bundle map satisfying $d^2 = 0$ and set $H_x = \ker d_x/\operatorname{Im} d_x$. If $\dim H_x$ is independent of x, show that $\ker d$, $\operatorname{Im} d$, and $\ker d/\operatorname{Im} d$ are vector bundles.

(iv) Suppose φ: $\xi \to \eta$ is a strong bundle map. Construct a dual bundle map φ^*: $\xi^* \leftarrow \eta^*$. If φ has constant rank, show that so has φ^*. In this case prove that $\operatorname{Im} \varphi^* \oplus \ker \varphi \cong \xi$ and $\operatorname{Im} \varphi \oplus \ker \varphi^* \cong \eta$.

3. Given vector bundles ξ and η over the same base show that $\xi \oplus \eta$ is the pullback of $\xi \times \eta$ via the diagonal map.

4. External tensor product. Let ξ, η be vector bundles over M, N with typical fibres F, H.

(i) Construct a canonical bundle $\xi \boxtimes \eta$ over $M \times N$ whose fibre at (x, y) is $F_x \otimes H_y$. It is called the *external tensor product* of ξ and η.

(ii) Show that if $M = N$, then $\xi \otimes \eta$ is the pullback of $\xi \boxtimes \eta$ under the diagonal map.

5. Endomorphisms. Fix a strong bundle map $\varphi: \xi \to \xi$, where $\xi = (E, \pi, B, F)$.

(i) Let $c_p(\varphi_x)$ denote the pth characteristic coefficient of φ_x. Show that $x \mapsto c_p(\varphi_x)$ is a smooth function on B. Is this true for the coefficients of the minimum polynomial of φ_x?

(ii) Are there always smooth functions $\lambda_1, \ldots, \lambda_r$ on B such that $\lambda_1(x), \ldots, \lambda_r(x)$ are the eigenvalues of φ_x?

(iii) Assume that U and V are disjoint open subsets of $\mathbb{C}$ such that for each x, the eigenvalues of φ_x are contained in $U \cup V$. Construct a unique decomposition $\xi = \zeta \oplus \rho$ such that $\varphi = \psi \oplus \chi$ and the eigenvalues of each ψ_x (resp. χ_x) are contained in U (resp. V). Conclude that the number of eigenvalues of φ_x in U (counted with their multiplicities) is constant.

(iv) Each φ_x decomposes uniquely in the form $\varphi_x = \varphi_x^S \oplus \varphi_x^N$, where φ_x^S is semisimple and φ_x^N is nilpotent and $\varphi_x^S \circ \varphi_x^N = \varphi_x^N \circ \varphi_x^S$ (cf. [5, p. 415]). Do the φ_x^S (resp. φ_x^N) define strong bundle maps?

6. Pseudo-Riemannian bundles. Let g be a pseudo-Riemannian metric in $\xi = (E, \pi, B, F)$ and suppose B is connected.

(i) Show that the signature of g_x is independent of x.

(ii) Construct a pseudo-Riemannian coordinate representation for ξ.

(iii) Show that $\xi = \xi^+ \oplus \xi^-$, where $\xi^+ \perp \xi^-$ with respect to g and the restriction of g to ξ^+ (resp. ξ^-) is positive (resp. negative) definite.

7. Symplectic bundles. A *symplectic bundle* is a vector bundle ξ together with a skew-symmetric nondegenerate bilinear function g in ξ.

(i) Show that every symplectic bundle is of even rank and orientable.

(ii) If ξ is a complex bundle, make $\xi_{\mathbb{R}}$ into a symplectic bundle.

8. Projective spaces. Interpret $\mathbb{R}P^n$ as the space of straight lines through the origin in $\mathbb{R}^{n+1}$.

(i) Construct a rank 1 vector bundle ξ over $\mathbb{R}P^n$ whose fibre at a point l is the one-dimensional subspace $l \subset \mathbb{R}^{n+1}$.

(ii) Show that ξ is nontrivial.

(iii) Do (i) and (ii) with $\mathbb{R}$ replaced by $\mathbb{C}$ (cf. problem 10, Chap. I).

9. Classifying maps. (i) Construct a vector bundle $\xi_{n,k}$ over $\mathscr{G}_{\mathbb{R}}(n; k)$ (cf. problem 13, Chap. I) whose fibre at $E \subset \mathbb{R}^n$ is the vector space E.

(ii) Show that, if η is any vector bundle over a manifold B with rank k, then, for sufficiently large n, there is a smooth map $\psi: B \to \mathscr{G}_{\mathbb{R}}(n; k)$ such that $\psi^*\xi_{n,k} \cong \eta \cdot \psi$ is called the *classifying map for* η.

(iii) Repeat (i) and (ii) with $\mathbb{R}$ replaced by $\mathbb{C}$ (construct complex vector bundles and complex maps).

10. Homotopic maps. (i) Let ξ be a vector bundle over $B \times \mathbb{R}$. Define $\psi: B \times \mathbb{R} \to B \times \mathbb{R}$ by $\psi(x, t) = (x, 1)$. Prove that $\psi^*\xi \cong \xi$.

(ii) Let ξ be a vector bundle over B and let $\psi_0, \psi_1: \hat{B} \to B$ be homotopic maps. Show that $\psi_0^*\xi \cong \psi_1^*\xi$.

(iii) A connected manifold M is called contractible if ι_M is homotopic to a constant map. Show that every vector bundle over a contractible manifold is trivial.

11. Prove that every finitely generated projective module over $\mathscr{S}(B)$ is of the form Sec ξ, where ξ is a vector bundle over B.

12. The ring $V(B)$. The *isomorphism class* of a complex vector bundle ξ over B is the collection of all complex vector bundles which are strongly isomorphic to ξ. It is written $[\xi]$. Denote the set of isomorphism classes by Vect(B). Let $\mathscr{F}(B)$ be the free abelian group with Vect(B) as a basis. Consider the subgroup generated by the elements of the form $[\xi] + [\eta] - [\xi \oplus \eta]$ and denote the factor group by $V(B)$.

(i) Show that the composition $(\xi, \eta) \to \xi \otimes \eta$ defines a commutative ring structure in $V(B)$. Is there an identity? If B is compact, the ring so obtained is denoted by $K(B)$.

(ii) Let $\bar{\xi}, \bar{\eta}$ denote the images of ξ, η in $V(B)$. Show that $\bar{\xi} = \bar{\eta}$ if and only if $\xi \oplus \epsilon^p \cong \eta \oplus \epsilon^p$ for some p, where ϵ^p denotes the trivial complex bundle of rank p.

(iii) Show that a smooth map $\varphi: \hat{B} \to B$ induces a homomorphism $\varphi^*: V(\hat{B}) \leftarrow V(B)$ which depends only on the homotopy class of φ.

(iv) Show that the external tensor product (cf. problem 4) determines a homomorphism $V(M) \otimes V(N) \to V(M \times N)$. If $M = N$, show that this map, composed with Δ^*, is the ring multiplication.

13. Consider the set of all isomorphism classes of real vector bundles of rank 1 over B. Show that $\otimes$ makes this set into a commutative group in which each element has order 2.

Chapter III

Tangent Bundle and Differential Forms

§1. Tangent bundle

3.1. Tangent space. Let M be a smooth manifold, and let $\mathscr{S}(M)$ be the ring of smooth functions on M.

Definition: A *tangent vector* of M at a point $a \in M$ is a linear map $\xi: \mathscr{S}(M) \to \mathbb{R}$ such that

$$\xi(fg) = \xi(f)\, g(a) + f(a)\, \xi(g), \qquad f, g \in \mathscr{S}(M).$$

The tangent vectors form a real vector space, $T_a(M)$, under the linear operations

$$(\lambda\xi + \mu\eta)(f) = \lambda\xi(f) + \mu\eta(f), \qquad \lambda, \mu \in \mathbb{R}, \quad \xi, \eta \in T_a(M), \quad f \in \mathscr{S}(M)$$

$T_a(M)$ is called the *tangent space* of M at a. In sec. 3.3 it will be shown that $\dim T_a(M) = \dim M$.

Lemma I: Let $\xi \in T_a(M), f \in \mathscr{S}(M)$. Then the number $\xi(f)$ depends only on the germ of f at a.

Proof: It has to be shown that $\xi(f) = 0$ if f is zero in some neighbourhood U of a. Choose $g \in \mathscr{S}(M)$ so that $g(a) = 0$ and $gf = f$. Then

$$\xi(f) = \xi(gf) = \xi(g) f(a) + g(a)\, \xi(f) = 0.$$

Q.E.D.

Corollary: If f is constant in a neighbourhood of a, then $\xi(f) = 0$.

Proof: Set $f(a) = \lambda$ and let λ also denote the constant function $M \to \lambda$. Then

$$\xi(f) = \xi(\lambda) = \lambda\xi(1) = \lambda\xi(1 \cdot 1) = 2\lambda\xi(1) = 0.$$

Q.E.D.

Example: Let $t \mapsto x(t)$ $(t_0 < t < t_1)$ be a smooth path on M. For each $t \in (t_0, t_1)$ we obtain $\xi \in T_{x(t)}(M)$ by setting

$$\xi(f) = \lim_{s\to 0} \frac{1}{s} [f(x(t+s)) - f(x(t))] = \frac{d}{ds}(f \circ x)\Big|_{s=t}, \qquad f \in \mathscr{S}(M).$$

That ξ does in fact belong to $T_{x(t)}(M)$ follows from elementary calculus. It is called the *tangent vector* to the path at $x(t)$ and we will denote it by $\dot{x}(t)$.

3.2. The derivative of a smooth map. Let $\varphi: M \to N$ be a smooth map. Recall that φ induces a homomorphism

$$\varphi^*: \mathscr{S}(M) \leftarrow \mathscr{S}(N)$$

given by

$$(\varphi^* f)(x) = f(\varphi(x)), \qquad f \in \mathscr{S}(N), \quad x \in M.$$

Lemma II: Let $\xi \in T_a(M)$. Then $\xi \circ \varphi^* \in T_{\varphi(a)}(N)$, and the correspondence $\xi \mapsto \xi \circ \varphi^*$ defines a linear map from $T_a(M)$ to $T_{\varphi(a)}(N)$.

Proof: $\xi \circ \varphi^*$ is a linear map from $\mathscr{S}(N)$ to $\mathbb{R}$. Moreover,

$$(\xi \circ \varphi^*)(fg) = \xi(\varphi^* f \cdot \varphi^* g) = \xi(\varphi^* f) \cdot g(\varphi(a)) + f(\varphi(a)) \cdot \xi(\varphi^* g)$$

$(f, g \in \mathscr{S}(N))$ and so $\xi \circ \varphi^* \in T_{\varphi(a)}(N)$. Clearly $\xi \mapsto \xi \circ \varphi^*$ is linear.

Q.E.D.

Definition: Let $\varphi: M \to N$ be a smooth map and let $a \in M$. The linear map $T_a(M) \to T_{\varphi(a)}(N)$ defined by $\xi \mapsto \xi \circ \varphi^*$ is called *the derivative of φ at a*. It will be denoted by $(d\varphi)_a$,

$$((d\varphi)_a \xi)(g) = \xi(\varphi^* g), \qquad g \in \mathscr{S}(N), \quad \xi \in T_a(M).$$

If $\psi: N \to Q$ is a second smooth map, then

$$(d(\psi \circ \varphi))_a = (d\psi)_{\varphi(a)} \circ (d\varphi)_a, \qquad a \in M.$$

Moreover, for the identity map $\iota: M \to M$, we have

$$(d\iota)_a = \iota_{T_a(M)}, \qquad a \in M$$

In particular, if $\varphi: M \to N$ is a diffeomorphism, then

$$(d\varphi)_a: T_a(M) \to T_{\varphi(a)}(N) \qquad \text{and} \qquad (d\varphi^{-1})_{\varphi(a)}: T_{\varphi(a)}(N) \to T_a(M)$$

are inverse linear isomorphisms.

Examples: 1. Let $t \mapsto x(t)$ $(t_0 < t < t_1)$ be a smooth path on a manifold M. Let $\xi_t \in T_t(\mathbb{R})$ be the tangent vector given by

$$\xi_t(f) = f'(t), \qquad f \in \mathscr{S}(\mathbb{R}).$$

Then the tangent vector $\dot{x}(t) \in T_{x(t)}(M)$ (cf. example, sec. 3.1) is given by

$$\dot{x}(t) = (dx)_t\, \xi_t\,.$$

2. Let O be an open subset of a manifold M and let $j\colon O \to M$ denote the inclusion map. Then

$$(dj)_a\colon T_a(O) \to T_a(M)$$

is a linear isomorphism for each $a \subset O$.

We shall prove this by constructing an inverse map. Fix $a \in O$ and choose $p \in \mathscr{S}(M)$ so that $p = 1$ in a neighbourhood of a, and carr $p \subset O$. Define $\beta\colon T_a(M) \to T_a(O)$ by

$$(\beta(\xi))(f) = \xi(pf), \qquad f \in \mathscr{S}(O), \quad \xi \in T_a(M)$$

That β is well defined, and inverse to $(dj)_a$ follows easily from Lemma I, sec. 3.1.

Using this example, we obtain

3. Let $\varphi\colon M \to N$ be a smooth map which sends a neighbourhood U of a point $a \in M$ diffeomorphically onto a neighbourhood V of $\varphi(a)$ in N. Then

$$(d\varphi)_a\colon T_a(M) \to T_{\varphi(a)}(N)$$

is a linear isomorphism.

3.3. Open subsets of vector spaces. Let E be an n-dimensional real vector space. Let O be an open subset of E and let $a \in O$. We shall define a linear isomorphism

$$\lambda_a\colon E \xrightarrow{\cong} T_a(O).$$

First recall that if $\varphi\colon O \to F$ is a smooth map of O into a second vector space F, then the classical derivative of φ at a is the linear map $\varphi'(a)\colon E \to F$ given by

$$\varphi'(a; h) = \lim_{t \to 0} \frac{\varphi(a + th) - \varphi(t)}{t}, \qquad h \in E.$$

Moreover, in the special case $F = \mathbb{R}$ we have the product formula

$$(fg)'(a; h) = f'(a; h)\, g(a) + f(a)\, g'(a; h), \qquad f, g \in \mathscr{S}(O).$$

This shows that the linear map $\xi_h\colon \mathscr{S}(O) \to \mathbb{R}$ given by

$$\xi_h(f) = f'(a; h)$$

is a tangent vector of O at a.

Hence we have a canonical linear map $\lambda_a\colon E \to T_a(O)$ given by

$$\lambda_a\colon h \mapsto \xi_h\,, \qquad h \in E.$$

If $\varphi\colon O \to F$ is a smooth map into a second vector space, then the diagram

$$\begin{array}{ccc} E & \xrightarrow{\varphi'(a)} & F \\ {\scriptstyle\lambda_a}\downarrow & & \downarrow{\scriptstyle\lambda_{\varphi(a)}} \\ T_a(O) & \xrightarrow[(d\varphi)_a]{} & T_{\varphi(a)}(F) \end{array} \tag{3.1}$$

commutes, as follows easily from the ordinary chain rule.

Proposition I: The canonical linear map $\lambda_a\colon E \to T_a(O)$ is an isomorphism.

Lemma III: Let $e_i (i = 1, \ldots, n)$ be a basis for E and let $f \in \mathscr{S}(E)$. Then

$$f = f(a) + \sum_{i=1}^{n} h_i g_i\,,$$

where

(1) the functions $h_i \in \mathscr{S}(E)$ are given by

$$x - a = \sum_{i=1}^{n} h_i(x)\, e_i\,, \qquad x \in E,$$

(2) the functions $g_i \in \mathscr{S}(E)$ satisfy

$$g_i(a) = f'(a; e_i), \qquad i = 1, \ldots, n.$$

Proof: By the fundamental theorem of calculus we have

$$f(x) - f(a) = \int_0^1 \frac{d}{dt} f(a + t(x - a))\, dt$$
$$= \int_0^1 f'(a + t(x - a); x - a)\, dt$$
$$= \sum_{i=1}^n h_i(x) \int_0^1 f'(a + t(x - a); e_i)\, dt.$$

Thus the lemma follows, with

$$g_i(x) = \int_0^1 f'(a + t(x - a); e_i)\, dt, \qquad x \in E.$$

Q.E.D.

Proof of Proposition I: Consider first the case $O = E$. We show first that λ_a is surjective. Let $\xi \in T_a(E)$ and let $f \in \mathscr{S}(E)$. Write

$$f = f(a) + \sum_{i=1}^n h_i g_i,$$

where the h_i, g_i satisfy conditions (1) and (2) of Lemma III. By the corollary to Lemma I, sec. 3.1, ξ maps the constant function $f(a)$ into zero. Thus,

$$\xi(f) = \sum_{i=1}^n \xi(h_i)\, g_i(a) + \sum_{i=1}^n h_i(a)\, \xi(g_i)$$
$$= \sum_{i=1}^n \xi(h_i) f'(a; e_i) = \left(\sum_{i=1}^n \xi(h_i)\, \xi_{e_i}\right)(f).$$

Since the functions h_i are independent of f, we can write

$$\xi = \sum_{i=1}^n \xi(h_i)\, \xi_{e_i} = \lambda_a\left(\sum_{i=1}^n \xi(h_i)\, e_i\right).$$

Thus λ_a is surjective.

To show that λ_a is injective, let f be any *linear* function in E. Then for $h \in E$

$$\lambda_a(h)(f) = f(h).$$

Now suppose $\lambda_a(h) = 0$. Then

$$f(h) = 0, \qquad f \in E^*.$$

Hence $h = 0$, and so λ_a is injective.

Finally, let O be any open subset of E and let $j\colon O \to E$ be the inclusion map. Then $j'(a)\colon E \to E$ is the identity map and so formula (3.1) yields the commutative diagram

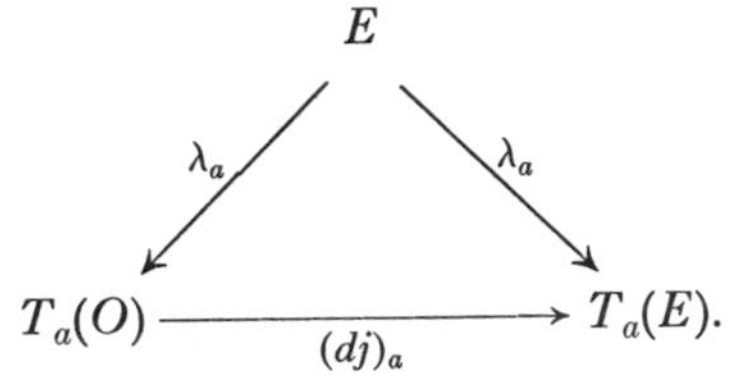

According to Example 2, sec. 3.2, $(dj)_a$ is a linear isomorphism. Hence so is the map $\lambda_a\colon E \to T_a(O)$.

Q.E.D.

Corollary: Let M be a smooth manifold and let $a \in M$. Then

$$\dim T_a(M) = \dim M.$$

Proof: Let $(U, u, \hat{U})$ be a chart for M such that $a \in U$. Using the result of Example 2, sec. 3.2, and the Proposition we find

$$\dim T_a(M) = \dim T_a(U) = \dim U = \dim M.$$

Q.E.D.

Proposition II: The derivative of a constant map is zero. Conversely, let $\varphi\colon M \to N$ be a smooth map such that $(d\varphi)_a = 0$, $a \in M$. Then, if M is connected, φ is a constant map.

Proof: Assume that φ is the constant map $M \to b \in N$. Then, for $g \in \mathscr{S}(N)$, $\varphi^* g$ is the constant function given by

$$(\varphi^* g)(x) = g(b), \qquad x \in M.$$

Hence (Corollary to Lemma I, sec. 3.1), for $\xi \in T_x(M)$, $x \in M$,

$$(d\varphi)_x(\xi)(g) = \xi(\varphi^* g) = 0.$$

It follows that each $(d\varphi)_x = 0$.

Conversely, assume that $\varphi: M \to N$ is a smooth map satisfying $(d\varphi)_a = 0$, $a \in M$; and let M be connected. Then, given two points $x_0 \in M$ and $x_1 \in M$, there exists a smooth curve $f: \mathbb{R} \to M$ such that $f(0) = x_0$ and $f(1) = x_1$ (cf. Proposition IX, sec. 1.11).

Consider the map $g = \varphi \circ f: \mathbb{R} \to N$. We have

$$(dg)_t = (d\varphi)_{f(t)} \circ (df)_t = 0, \qquad t \in \mathbb{R}.$$

Now using an atlas for N and applying formula (3.1) we see that g must be constant. In particular $g(0) = g(1)$ and so

$$\varphi(x_0) = g(0) = g(1) = \varphi(x_1).$$

Q.E.D.

3.4. Example. Let S^n be the unit sphere in an $(n + 1)$-dimensional Euclidean space E. We shall establish a canonical isomorphism between $T_a(S^n)$ and the subspace $a^\perp$, the orthogonal complement of a in E.

The injection $i: S^n \to E$ determines a linear map

$$(di)_a: T_a(S^n) \to T_a(E).$$

Combining this map with the linear isomorphism $\lambda_a^{-1}: T_a(E) \to E$, we obtain a linear map

$$j_a: T_a(S^n) \to E.$$

We show first that $\operatorname{Im} j_a \subset a^\perp$. Consider the function $f \in \mathscr{S}(E)$ given by

$$f(x) = \langle x, x \rangle, \qquad x \in E.$$

i^*f is the constant function, 1, on S^n. It follows that

$$\xi(i^*f) = 0, \qquad \xi \in T_a(S^n),$$

whence

$$f'(a; j_a(\xi)) = \xi(i^*f) = 0.$$

An elementary computation shows that

$$f'(a; h) = 2\langle a, h \rangle, \qquad h \in E.$$

Thus we obtain

$$2\langle a, j_a(\xi) \rangle = 0, \qquad \xi \in T_a(S^n);$$

i.e. $\operatorname{Im} j_a \subset a^\perp$.

Next we prove that j_a is injective. Consider the smooth map $\rho: E - \{0\} \to S^n$ given by

$$\rho(x) = \frac{x}{|x|}, \qquad x \in E - \{0\}.$$

Clearly, $\rho \circ i = \iota$ and hence

$$(d\rho)_a \circ (di)_a = \iota.$$

It follows that $(di)_a$ is injective. Hence so is j_a.

Finally, to show that j_a is a linear isomorphism onto $a^\perp$ observe that

$$\dim T_a(S^n) = n = \dim E - 1 = \dim a^\perp.$$

3.5. Tangent bundle. Let M be an n-manifold. Consider the disjoint union

$$T_M = \bigcup_{a \in M} T_a(M),$$

and let $\pi_M: T_M \to M$ be the projection,

$$\pi_M(\xi) = a, \qquad \xi \in T_a(M).$$

In this section we shall define a manifold structure on T_M so that

$$\tau_M = (T_M, \pi_M, M, \mathbb{R}^n)$$

is a vector bundle over M, whose fibre at a point $a \in M$ is the tangent space $T_a(M)$. τ_M is called the *tangent bundle of* M.

Let $(U_\alpha, u_\alpha, \hat{U}_\alpha)$ be a chart for M and let $j_\alpha: U_\alpha \to M$ be the inclusion map. For each $x \in U_\alpha$ there are linear isomorphisms

$$\lambda_{u_\alpha(x)}: \mathbb{R}^n \xrightarrow{\cong} T_{u_\alpha(x)}(\hat{U}_\alpha)$$

$$(du_\alpha)_x^{-1}: T_{u_\alpha(x)}(\hat{U}_\alpha) \xrightarrow{\cong} T_x(U_\alpha)$$

and

$$(dj_\alpha)_x: T_x(U_\alpha) \xrightarrow{\cong} T_x(M).$$

Composing them we obtain a linear isomorphism

$$\psi_{\alpha,x}: \mathbb{R}^n \xrightarrow{\cong} T_x(M).$$

Finally, let $\{(U_\alpha, u_\alpha)\}$ be an atlas for M. Define maps

$$\psi_\alpha: U_\alpha \times \mathbb{R}^n \to T_M$$

by

$$\psi_\alpha(x, h) = \psi_{\alpha,x}(h), \qquad x \in U_\alpha, \quad h \in \mathbb{R}^n.$$

If $U_\alpha \cap U_\beta \neq \varnothing$ and $u_{\beta\alpha} = u_\beta \circ u_\alpha^{-1}$, the map

$$\psi_{\beta\alpha} = \psi_\beta^{-1} \circ \psi_\alpha: U_\alpha \cap U_\beta \times \mathbb{R}^n \to U_\alpha \cap U_\beta \times \mathbb{R}^n$$

is given by

$$\psi_{\beta\alpha}(x, h) = (x, u'_{\beta\alpha}(u_\alpha(x); h)),$$

(cf. formula (3.1), sec. 3.3). Hence it is smooth.

Now it follows from sec. 2.5 that there is a unique vector bundle $\tau_M = (T_M, \pi_M, M, \mathbb{R}^n)$ for which $\{(U_\alpha, \psi_\alpha)\}$ is a coordinate representation. The fibre of this bundle at $x \in M$ is the tangent space $T_x(M)$.

Evidently this bundle structure is independent of the choice of atlas for M.

Example 1: If O is an open subset of a vector space E, then the tangent bundle τ_O is isomorphic to the product bundle $O \times E$. In fact, define a map $\lambda: O \times E \to T_O$ by setting

$$\lambda(a, h) = \lambda_a(h), \qquad a \in O, \quad h \in E,$$

where λ_a is the canonical linear map given in sec. 3.3. Then λ is a strong bundle isomorphism.

Next, suppose $\varphi: M \to N$ is a smooth map. Then a set map $d\varphi: T_M \to T_N$ is defined by

$$d\varphi(\xi) = (d\varphi)_x \xi, \qquad \xi \in T_x(M), \quad x \in M.$$

It is called the *derivative of* φ.

Proposition III: The derivative of a smooth map $\varphi: M \to N$ is a bundle map $d\varphi: \tau_M \to \tau_N$.

Proof: It follows from the definition that $d\varphi$ is fibre preserving and that the restriction of $d\varphi$ to each fibre is linear. To show that $d\varphi$ is smooth use atlases on M and N to reduce to the case $M = \mathbb{R}^n$, $N = \mathbb{R}^p$. In this case formula (3.1), sec. 3.3, shows that

$$d\varphi\colon \mathbb{R}^n \times \mathbb{R}^n \to \mathbb{R}^p \times \mathbb{R}^p$$

is given by

$$d\varphi(x; h) = (\varphi(x); \varphi'(x; h)).$$

Hence it is smooth.

Q.E.D.

Now let $\psi\colon N \to Q$ be a smooth map into a third manifold. Then

$$d(\psi \circ \varphi) = d\psi \circ d\varphi$$

as follows from the definition. Moreover, the derivative of the identity map $\iota\colon M \to M$ is the identity map of T_M,

$$d\iota_M = \iota_{T_M}.$$

It follows that if $\varphi\colon M \to N$ and $\psi\colon M \leftarrow N$ are inverse diffeomorphisms, then $d\varphi$ and $d\psi$ are inverse bundle isomorphisms.

Example 2: Let $U \subset M$ be an open subset and let $j\colon U \to M$ be the inclusion map. The derivative

$$dj\colon T_U \to T_M$$

can be regarded as a strong bundle map from τ_U to the restriction, $\tau_M|_U$, of τ_M to U (cf. Example 2, sec. 2.3). According to Example 2, sec. 3.2, the restriction of dj to each fibre is a linear isomorphism.

Thus (cf. Proposition I, sec. 2.2) dj defines a strong bundle isomorphism,

$$\tau_U \xrightarrow{\cong} \tau_M|_U.$$

We shall often identify these bundles under the above isomorphism. In the process, we identify $T_x(U)$ and $T_x(M)$ for each point $x \in U$.

3.6. Cotangent space and cotangent bundle. Let M be a manifold and let $a \in M$. A *cotangent vector* of M at a is a linear map

$$\omega_a\colon T_a(M) \to \mathbb{R};$$

i.e., it is an element of the dual space $T_a(M)^*$, which is called the *cotangent space of* M at a. Observe that

$$\dim T_a(M)^* = \dim M.$$

In the same way as in sec. 3.5 we can construct a vector bundle τ_M^* over M whose fibre at a is the cotangent space $T_a(M)^*$. τ_M^* is called the *cotangent bundle of* M. Clearly the tangent bundle τ_M and the cotangent bundle τ_M^* are dual.

3.7. Product manifolds. Let M and N be smooth manifolds and consider the product manifold $M \times N$. We shall construct a strong bundle isomorphism between the tangent bundle $\tau_{M\times N}$ and the Cartesian product $\tau_M \times \tau_N$.

Define a strong bundle map

$$\varphi\colon T_{M\times N} \to T_M \times T_N$$

by

$$\varphi(\zeta) = (d\pi_M(\zeta), d\pi_N(\zeta)), \qquad \zeta \in T_{M\times N},$$

where $\pi_M\colon M \times N \to M$ and $\pi_N\colon M \times N \to N$ are the canonical projections.

To show that φ is an isomorphism we need only prove that the linear maps

$$\varphi_{a,b}\colon T_{(a,b)}(M \times N) \to T_a(M) \oplus T_b(N)$$

are isomorphisms. Let $j_a\colon N \to M \times N$, $j_b\colon M \to M \times N$ be the *inclusion maps opposite* $a \in M$ *and* $b \in N$:

$$j_a(y) = (a, y) \qquad \text{and} \qquad j_b(x) = (x, b), \qquad x \in M, \quad y \in N.$$

Then

$$\pi_M \circ j_a = \gamma_a, \qquad \pi_N \circ j_a = \iota_N$$
$$\pi_M \circ j_b = \iota_M, \qquad \pi_N \circ j_b = \gamma_b,$$

where $\gamma_a\colon N \to a$ and $\gamma_b\colon M \to b$ are the constant maps. These relations yield

$$d\pi_M \circ dj_a = 0, \qquad d\pi_N \circ dj_a = \iota$$
$$d\pi_M \circ dj_b = \iota, \qquad d\pi_N \circ dj_b = 0.$$

Now define a linear map

$$\psi_{a,b}\colon T_a(M) \oplus T_b(N) \to T_{(a,b)}(M \times N)$$

by

$$\psi_{a,b}(\xi, \eta) = (dj_b)\xi + (dj_a)\eta, \qquad \xi \in T_a(M), \quad \eta \in T_b(N).$$

The equations above show that $\varphi_{a,b} \circ \psi_{a,b} = \iota$. Since these maps are linear maps between finite-dimensional vector spaces and since

$$\begin{aligned} \dim T_{(a,b)}(M \times N) &= \dim(M \times N) = \dim M + \dim N \\ &= \dim(T_a(M) \oplus T_b(N)), \end{aligned}$$

it follows that $\varphi_{a,b}$ and $\psi_{a,b}$ are inverse isomorphisms. In particular φ is a strong bundle isomorphism.

Henceforth we shall identify the bundles $\tau_{M\times N}$ and $\tau_M \times \tau_N$ via φ. In particular, we shall write

$$T_{(a,b)}(M \times N) = T_a(M) \oplus T_b(N).$$

Our remarks above show that we are identifying (ξ, η) with

$$(dj_b)\xi + (dj_a)\eta, \quad \xi \in T_a(M), \quad \eta \in T_b(N).$$

§2. Local properties of smooth maps

3.8. The main theorem. Let $\varphi: M \to N$ be a smooth map. Then φ is called a *local diffeomorphism* (resp. an *immersion*, a *submersion*) *at a point* $a \in M$ if the map

$$(d\varphi)_a: T_a(M) \to T_{\varphi(a)}(N)$$

is a linear isomorphism (resp. injective, surjective). If φ is a local diffeomorphism (resp. an immersion, a submersion) for all points $a \in M$, it is called a *local diffeomorphism* (resp. an *immersion*, a *submersion*) *of M into N*.

Theorem I: Let $\varphi: M \to N$ be a smooth map where $\dim M = n$ and $\dim N = r$. Let $a \in M$ be a given point. Then

(1) If φ is a local diffeomorphism at a, there are neighbourhoods U of a and V of b such that φ maps U diffeomorphically onto V.

(2) If $(d\varphi)_a$ is injective, there are neighbourhoods U of a, V of b, and W of 0 in $\mathbb{R}^{r-n}$, and a diffeomorphism

$$\psi: U \times W \xrightarrow{\cong} V$$

such that

$$\varphi(x) = \psi(x, 0), \qquad x \in U.$$

(3) If $(d\varphi)_a$ is surjective, there are neighbourhoods U of a, V of b, and W of 0 in $\mathbb{R}^{n-r}$, and a diffeomorphism

$$\psi: U \xrightarrow{\cong} V \times W$$

such that

$$\varphi(x) = \pi_V \psi(x), \qquad x \in U,$$

where $\pi_V : V \times W \to V$ is the projection.

Proof: By using charts we may reduce to the case $M = \mathbb{R}^n$, $N = \mathbb{R}^r$. In part (1), then, we are assuming that $\varphi'(a): \mathbb{R}^n \to \mathbb{R}^r$ is an isomorphism, and the conclusion is the inverse function theorem (cf. sec. 0.9).

For part (2), we choose a subspace E of $\mathbb{R}^r$ such that

$$\operatorname{Im} \varphi'(a) \oplus E = \mathbb{R}^r,$$

and consider the map $\psi\colon \mathbb{R}^n \times E \to \mathbb{R}^r$ given by

$$\psi(x, y) = \varphi(x) + y, \qquad x \in \mathbb{R}^n, \quad y \in E.$$

Then

$$\psi'(a, 0; h, k) = \varphi'(a; h) + k, \qquad h \in \mathbb{R}^n, \quad k \in E.$$

It follows that $\psi'(a, 0)$ is injective and thus an isomorphism ($r = \dim \operatorname{Im} \varphi'(a) + \dim E = n + \dim E$).

Thus part (1) implies the existence of neighbourhoods U of a, V of b, and W of 0 in E such that $\psi\colon U \times W \to V$ is a diffeomorphism. Clearly, $\psi(x, 0) = \varphi(x)$.

Finally, for part (3), we choose a subspace E of $\mathbb{R}^n$ such that

$$\ker \varphi'(a) \oplus E = \mathbb{R}^n.$$

Let $\rho\colon \mathbb{R}^n \to E$ be the projection induced by this decomposition, and define

$$\psi\colon \mathbb{R}^n \to \mathbb{R}^r \oplus E$$

by

$$\psi(x) = (\varphi(x), \rho(x)), \qquad x \in \mathbb{R}^n.$$

Then

$$\psi'(a; h) = (\varphi'(a; h), \rho(h)), \qquad a \in \mathbb{R}^n, \quad h \in \mathbb{R}^n.$$

It follows easily that $\psi'(a)$ is a linear isomorphism. Hence there are neighbourhoods U of a, V of b, and W of $0 \in E$ such that $\psi\colon U \to V \times W$ is a diffeomorphism.

Q.E.D.

Corollary: (1) If $(d\varphi)_a$ is a linear isomorphism there is a smooth map $\chi\colon V \to U$ such that

$$\varphi \circ \chi = \iota_V \qquad \text{and} \qquad \chi \circ \varphi_U = \iota_U,$$

where φ_U denotes the restriction of φ to U.

(2) If $(d\varphi)_a$ is injective, there is a smooth map $\chi\colon V \to U$ such that

$$\chi \circ \varphi_U = \iota_U.$$

(3) If $(d\varphi)_a$ is surjective, there is a smooth map $\chi\colon V \to U$ such that

$$\varphi \circ \chi = \iota_V.$$

Proposition IV: If $\varphi: M \to N$ is a smooth bijective map and if the maps

$$(d\varphi)_x: T_x(M) \to T_{\varphi(x)}(N)$$

are all injective, then φ is a diffeomorphism.

Proof: Let $\dim M = n$, $\dim N = r$. Since $(d\varphi)_x$ is injective, we have $r \geqslant n$. Now we show that $r = n$. In fact, according to Theorem I, part (2), for every $a \in M$ there are neighbourhoods $U(a)$ of a, V of $\varphi(a)$ and W of $0 \in \mathbb{R}^{r-n}$ together with a diffeomorphism

$$\psi_a: U(a) \times W \xrightarrow{\cong} V$$

such that the diagram

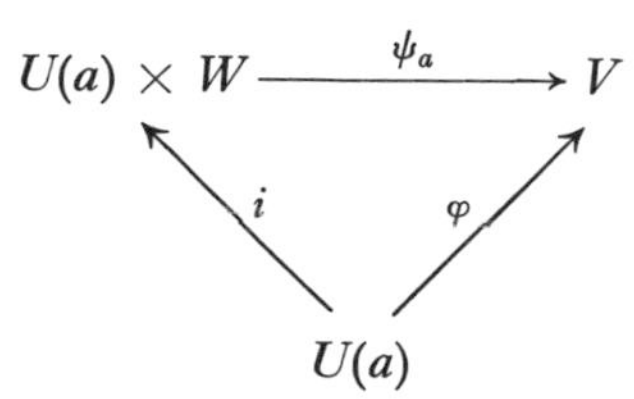

commutes (i denotes the inclusion map opposite 0).

Choose a countable open covering U_i ($i = 1, 2, \ldots$) of M such that each $\overline{U}_i$ is compact and contained in some $U(a_i)$. Since φ is surjective, it follows that $\bigcup_i \varphi(\overline{U}_i) \supset N$.

Now assume that $r > n$. Then the diagram implies that no $\varphi(\overline{U}_i)$ contains an open set. Thus, by the category theorem [4, Theorem 10.1, p. 249] N could not be Hausdorff. This contradiction shows that $n = r$.

Since $n = r$, φ is a local diffeomorphism. On the other hand, φ is bijective. Since it is a local diffeomorphism, Theorem I implies that its inverse is smooth. Thus φ is a diffeomorphism.

Q.E.D.

3.9. Quotient manifolds. *A quotient manifold* of a manifold M is a manifold N together with a smooth map $\pi: M \to N$ such that π and each linear map $(d\pi)_x: T_x(M) \to T_{\pi x}(N)$ is surjective (and thus $\dim M \geqslant \dim N$).

Lemma IV: Let $\pi: M \to N$ make N into a quotient manifold of M. Then the map π is open.

Proof: It is sufficient to show that, for each $a \in M$, there is a neighbourhood U of a such that the restriction of π to U is open. This follows at once from Theorem I, part (3), sec. 3.8.

Q.E.D.

Proposition V: Let $\pi: M \to N$ make N into a quotient manifold of M. Assume that $\varphi: M \to Q$, $\psi: N \to Q$ are maps into a third manifold Q such that the diagram

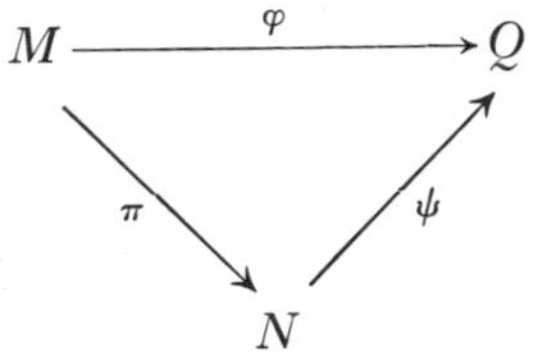

commutes. Then φ is smooth if and only if ψ is smooth.

Proof: Use the corollary part (3) to Theorem I, sec. 3.8.

Q.E.D.

Corollary: Let $\pi_1: M \to N_1$, $\pi_2: M \to N_2$ be quotient manifolds. Assume that $\pi_1 x = \pi_1 y$ $(x, y \in M)$ holds if and only if $\pi_2 x = \pi_2 y$ holds. Let $\varphi: N_1 \to N_2$ be the unique set bijection such that the diagram

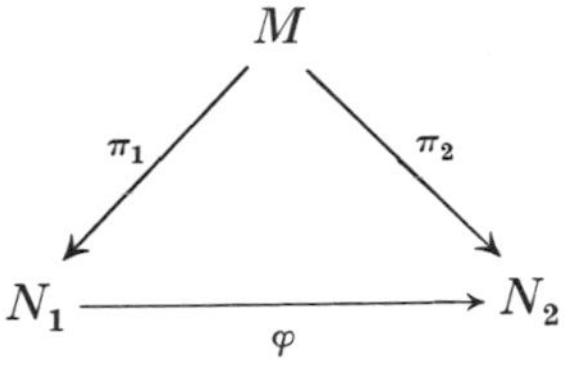

commutes. Then φ is a diffeomorphism.

3.10. Submanifolds. Let M be a manifold. An *embedded manifold* is a pair (N, φ), where N is a second manifold and $\varphi: N \to M$ is a smooth map such that the derivative

$$d\varphi: T_N \to T_M$$

is injective. In particular, since the maps $(d\varphi)_x: T_x(N) \to T_{\varphi(x)}(M)$ are injective, it follows that $\dim N \leq \dim M$.

Given an embedded manifold (N, φ), consider the subset $M_1 = \varphi(N)$. φ may be considered as a bijective map

$$\varphi_1: N \to M_1 .$$

This bijection defines a smooth structure on M_1, such that φ_1 becomes a diffeomorphism.

A *submanifold* of a manifold M is an embedded manifold (N, φ) such that $\varphi_1: N \to \varphi(N)$ is a homeomorphism, when $\varphi(N)$ is given the topology induced by the topology of M. If N is a subset of M and φ is the inclusion map, we say simply that N is a submanifold of M.

Not every embedded manifold is a submanifold, as the following example shows:

Let M be the 2-torus T^2 (cf. Example 3, sec. 1.4) and let $N = \mathbb{R}$. Define a map $\varphi: \mathbb{R} \to T^2$ by

$$\varphi(t) = \pi(t, \lambda t), \qquad t \in \mathbb{R},$$

where λ is an irrational number and $\pi: \mathbb{R}^2 \to T^2$ denotes the projection. Then $d\varphi: T_{\mathbb{R}} \to T_{T^2}$ is injective and so $(\mathbb{R}, \varphi)$ is an embedded manifold. Since λ is irrational, $\varphi(\mathbb{R}^+)$ is dense in T^2. In particular there are real numbers $a_i > 0$ such that $\varphi(a_i) \to \varphi(-1)$. Thus T^2 does not induce the standard topology in $\varphi(\mathbb{R})$.

Proposition VI: Let (N, i) be a submanifold of M. Assume that Q is a smooth manifold and

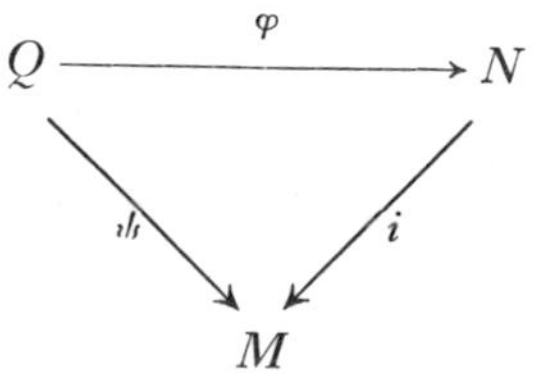

is a commutative diagram of maps. Then φ is smooth if and only if ψ is.

Proof: If φ is smooth then clearly so is ψ. Conversely, assume that ψ is smooth. Fix a point $a \in Q$ and set $b = \psi(a)$. Since di is injective, there are neighbourhoods U, V of b in N and M, respectively, and there is a smooth map $\chi: V \to U$ such that $\chi \circ i_U = \iota$ (cf. Corollary, Theorem I, sec. 3.8).

Since N is a submanifold of M, the map φ is continuous. Hence there is a neighbourhood W of a such that $\varphi(W) \subset U$. Then

$$i_U \circ \varphi_W = \psi_W,$$

where φ_W, ψ_W denote the restrictions of φ, ψ to W.

It follows that

$$\chi \circ \psi_W = \chi \circ i_U \circ \varphi_W = \varphi_W$$

and so φ is smooth in W; thus φ is a smooth map.

Q.E.D.

Corollary: Suppose submanifolds N_1 and N_2 of M coincide as sets. Then they coincide as submanifolds.

Examples: 1. An open subset of a manifold is a submanifold.

2. If (Q, ψ) is a submanifold of N, and (N, φ) is a submanifold of M then $(Q, \varphi \circ \psi)$ is a submanifold of M.

3. *Retracts*: A manifold N is called a *retract* of a manifold M if there are smooth maps $i: N \to M$, $\rho: M \to N$ so that

$$\rho \circ i = \iota_N .$$

ρ is called the retraction; i is called the inclusion. In this case (N, i) is a submanifold of M.

In particular (cf. sec. 3.4) the sphere S^n is a retract of the manifold $E - \{0\}$ (E an $(n + 1)$-dimensional Euclidean space). Hence it is a submanifold of E.

4. *Fibre bundles*: Let (E, π, B, F) be a smooth fibre bundle and fix $b \in B$. Let $\{(U_\alpha, \psi_\alpha)\}$ be a coordinate representation for the bundle and choose U_α to contain b. Then the bijection

$$\psi_{\alpha,b}: F \to F_b$$

defines a manifold structure on F_b (independent of the choice of coordinate representation).

Since the inclusion $F \to U_\alpha \times F$ opposite b and the projection $U_\alpha \times F \to F$ make F into a retract of $U_\alpha \times F$ it follows via the diffeomorphism ψ_α that F_b is a retract of the open set $\pi^{-1}U_\alpha$. In particular it is a submanifold of $\pi^{-1}U_\alpha$ and hence a submanifold of E.

5. Let $\xi = (E, \pi, B, F)$ be a vector bundle. Then the zero cross-section $o: B \to E$ makes B into a retract of E. Hence (B, o) is a closed submanifold of E.

On the other hand, set $\dot{F} = F - \{0\}$ and $\dot{F}_x = F_x - \{0\}$ $(x \in B)$. Then

$$\dot{E} = \bigcup_{x \in B} \dot{F}_x$$

is the complement of $o(B)$ in E; hence, it is an open submanifold. If $\{(U_\alpha, \psi_\alpha)\}$ is a coordinate representation for ξ, the ψ_α restrict to diffeomorphisms

$$U_\alpha \times \dot{F} \xrightarrow{\cong} \dot{\pi}^{-1}(U_\alpha),$$

where $\dot{\pi}: \dot{E} \to B$ is the restriction of π. Hence $\dot{\xi} = (\dot{E}, \dot{\pi}, B, \dot{F})$ is a fibre bundle, called the *deleted bundle for* ξ.

6. Assume the vector bundle ξ above has been given a Riemannian metric and let S_x (resp. S) denote the unit sphere of the Euclidean space F_x, $x \in B$ (resp. of F). Set

$$E_S = \bigcup_{x \in B} S_x$$

and let $\pi_S: E_S \to B$ be the restriction of π. Then $\xi_S = (E_S, \pi_S, B, S)$ is a fibre bundle.

In fact, let $\{(U_\alpha, \psi_\alpha)\}$ be a Riemannian coordinate representation for ξ. Then the maps ψ_α restrict to bijections

$$\varphi_\alpha: U_\alpha \times S \to \pi_S^{-1}(U_\alpha).$$

The bijections $\varphi_\alpha^{-1} \circ \varphi_\beta$ in $U_\alpha \cap U_\beta \times S$ are the restrictions of the diffeomorphisms $\psi_\alpha^{-1} \circ \psi_\beta$. Since S is a submanifold of F, $(U_\alpha \cap U_\beta) \times S$ is a submanifold of $U_\alpha \cap U_\beta \times F$; thus Proposition VI implies that $\varphi_\alpha^{-1} \circ \varphi_\beta$ is smooth. Now it follows from Proposition X, sec. 1.13, that ξ_S is a smooth bundle.

ξ_S is called the *sphere bundle* associated with ξ. The map $\rho: \dot{E} \to E_S$ given by

$$\rho(z) = \frac{z}{|z|}, \qquad z \in F_x, \quad x \in B,$$

is smooth, and makes E_S into a retract of $\dot{E}$. In particular, E_S is a submanifold of $\dot{E}$.

Finally, observe that an isometry $\varphi: \xi \to \xi'$ between Riemannian vector bundles restricts to a fibre preserving map

$$\varphi_S: E_S \to E'_S$$

which, by Proposition VI is smooth. In particular, since isomorphic bundles are isometric (Proposition VI, sec. 2.17), the bundle ξ_S is independent of the choice of metric in ξ.

§3. Vector fields

3.11. Vector fields. Definition: A *vector field* X on a manifold M is a cross-section in the tangent bundle τ_M. Thus a vector field X assigns to every point $x \in M$ a tangent vector $X(x)$ such that the map $M \to T_M$ so obtained is smooth. The vector fields on M form a module over the ring $\mathscr{S}(M)$, which will be denoted by $\mathscr{X}(M)$.

Proposition VII: There is a canonical isomorphism of $\mathscr{X}(M)$ onto the $\mathscr{S}(M)$-module, Der $\mathscr{S}(M)$, of derivations in the algebra $\mathscr{S}(M)$.

Proof: Let X be a vector field. For each $f \in \mathscr{S}(M)$, define a function $X(f)$ on M by

$$X(f)(x) = X(x)(f), \qquad x \in M.$$

$X(f)$ is smooth. To see this we may assume that $M = \mathbb{R}^n$. But then (cf. sec. 3.3)

$$X(f)(x) = f'(x; X(x))$$

is smooth.

Hence every vector field X on M determines a map θ_X: $\mathscr{S}(M) \to \mathscr{S}(M)$ given by

$$\theta_X(f) = X(f).$$

Obviously θ_X is a derivation in the algebra $\mathscr{S}(M)$. The assignment $X \mapsto \theta_X$ defines a homomorphism

$$\theta: \mathscr{X}(M) \to \text{Der}\, \mathscr{S}(M).$$

We show now that θ is an isomorphism.

Suppose $\theta_X = 0$, for some $X \in \mathscr{X}(M)$. Then

$$X(x)f = 0, \qquad x \in M, \quad f \in \mathscr{S}(M).$$

This implies that $X(x) = 0$; i.e. $X = 0$.

To prove that θ is surjective, let Φ be any derivation in $\mathscr{S}(M)$. Then Φ determines, for every point $x \in M$, the vector $\xi_x \in T_x(M)$, given by

$$\xi_x(f) = \Phi(f)(x), \qquad f \in \mathscr{S}(M).$$

Define $X: M \to T_M$ by $X(x) = \xi_x$. To show that this map is smooth, fix a point $a \in M$. Using a chart, it is easy to construct vector fields X_i $(i = 1, ..., n)$ and smooth functions f_i $(i = 1, ..., n; n = \dim M)$ on M such that $X_i(x) f_j = \delta_{ij}$, $x \in V$, (V some neighbourhood of a). Then the vectors $X_i(x)$ $(i = 1, ..., n)$ form a basis for $T_x(M)$ $(x \in V)$.

Hence, for each $x \in V$, there is a unique system of numbers λ_x^i $(i = 1, ..., n)$ such that

$$\xi_x = \sum_{i=1}^{n} \lambda_x^i X_i(x).$$

Applying ξ_x to f_i, we obtain

$$\lambda_x^i = \xi_x(f_i) = \Phi(f_i)(x), \qquad x \in V.$$

Hence

$$X(x) = \sum_{i=1}^{n} \Phi(f_i)(x)\, X_i(x), \qquad x \in V.$$

Since the $\Phi(f_i)$ are smooth functions on M, this equation shows that X is smooth in V; i.e. X is a vector field.

Finally, it follows from the definition that

$$\theta_X = \Phi.$$

Thus θ is surjective.

Q.E.D.

Corollary: The $\mathscr{S}(M)$-module Der $\mathscr{S}(M)$ is finitely generated and projective.

Proof: Apply the Proposition and the corollary of Theorem I, sec. 2.23.

Q.E.D.

3.12. Lie product. The $\mathscr{S}(M)$-module Der $\mathscr{S}(M)$ is a Lie algebra over $\mathbb{R}$, the Lie product being given by

$$[\theta_1, \theta_2] = \theta_1 \circ \theta_2 - \theta_2 \circ \theta_1 .$$

Note that the Lie product is not bilinear over $\mathscr{S}(M)$. In fact, we have

$$[\theta_1, f\theta_2] = f[\theta_1, \theta_2] + \theta_1(f)\, \theta_2, \qquad f \in \mathscr{S}(M), \quad \theta_1, \theta_2 \in \operatorname{Der} \mathscr{S}(M).$$

Identifying $\mathscr{X}(M)$ with Der $\mathscr{S}(M)$ under the isomorphism of Proposition VII we obtain a Lie algebra structure in $\mathscr{X}(M)$. The Lie product of two vector fields X and Y is given by

$$[X, Y](f) = X(Y(f)) - Y(X(f)), \qquad f \in \mathscr{S}(M).$$

It is bilinear over $\mathbb{R}$, skew symmetric, and satisfies the Jacobi identity. Moreover, for $f \in \mathscr{S}(M)$ we have

$$[fX, Y] = f[X, Y] - Y(f)X \qquad \text{and} \qquad [X, fY] = f[X, Y] + X(f)Y.$$

Examples: 1. Let M be an open subset of a vector space E. According to Example 1, sec. 3.5, we can write

$$T_M = M \times E.$$

A vector field X on M given by

$$X(x) = (x, h), \qquad x \in M,$$

for some fixed $h \in E$ will be called the *constant field corresponding to* $h \in E$.

If X is the constant field corresponding to h then

$$X(f)(x) = f'(x; h), \qquad x \in M, \quad f \in \mathscr{S}(M),$$

Thus if Y is a second constant vector field, corresponding to $k \in E$, then

$$YX(f)(x) = f''(x; h, k) = f''(x; k, h) = XY(f)(x),$$

whence $[X, Y] = 0$.

2. Let M be an open subset of a real vector space E and consider $X, Y \in \mathscr{X}(M)$ defined by $X(x) = h$ and $Y(x) = x$, where $h \in E$ is fixed. Then $[X, Y] = X$. In fact, let f be any *linear* function in E. Then

$$X(f)(x) = f'(x; h) = f(h), \qquad x \in E,$$

while

$$Y(f)(x) = f'(x; x) = f(x), \qquad x \in E;$$

i.e., $X(f)$ is the constant function $x \mapsto f(h)$, while $Y(f) = f$. It follows that

$$[X, Y](f) = X(f), \qquad f \in E^*.$$

This shows that $[X, Y] = X$.

3.13. φ-related vector fields. Let $\varphi\colon M \to N$ be a smooth map. We say that two vector fields $X \in \mathscr{X}(M)$ and $Y \in \mathscr{X}(N)$ are *φ-related*, $X \underset{\varphi}{\sim} Y$, if

$$Y(\varphi(x)) = d\varphi(X(x)), \qquad x \in M. \tag{3.2}$$

Lemma V: Two vector fields $X \in \mathscr{X}(M)$ and $Y \in \mathscr{X}(N)$ are φ-related if and only if

$$\varphi^*(Y(g)) = X(\varphi^*g), \qquad g \in \mathscr{S}(N). \tag{3.3}$$

Proof: In fact for $x \in M$, $g \in \mathscr{S}(N)$,

$$\varphi^*(Y(g))(x) = Y(g)(\varphi(x)) = Y(\varphi(x))g$$

and

$$X(\varphi^*g)(x) = X(x)\,\varphi^*g = [d\varphi(X(x))](g),$$

as follows from the definition of $d\varphi$.

Q.E.D.

Example: Let U be an open subset of M and let $X \in \mathscr{X}(M)$. Then X induces a vector field X_U on U, given by $X_U(x) = X(x)$, $x \in U$. X_U is called the *restriction* of X to U. If $i\colon U \to M$ is the inclusion map, then clearly $X_U \underset{i}{\sim} X$.

Proposition VIII: Suppose that $\varphi\colon M \to N$ is smooth and $X \underset{\varphi}{\sim} X_1$, $Y \underset{\varphi}{\sim} Y_1$. Then

(1) $\lambda X + \mu Y \underset{\varphi}{\sim} \lambda X_1 + \mu Y_1$, $\qquad \lambda, \mu \in \mathbb{R}$

(2) $\varphi^*f \cdot X \underset{\varphi}{\sim} f \cdot X_1$, $\qquad f \in \mathscr{S}(N)$,

(3) $[X, Y] \underset{\varphi}{\sim} [X_1, Y_1]$.

Proof: An easy consequence of Lemma V.

Q.E.D.

Remark: If $\varphi\colon M \to N$ is surjective, then for every $X \in \mathscr{X}(M)$, there is at most one $Y \in \mathscr{X}(N)$ such that $X \underset{\varphi}{\sim} Y$.

Now let $\varphi\colon M \xrightarrow{\cong} N$ be a diffeomorphism. Then the map

$$\varphi^*\colon \mathscr{S}(M) \leftarrow \mathscr{S}(N)$$

is an isomorphism. Hence every vector field X on M determines a vector field $\varphi_* X$ on N, given by

$$(\varphi_* X)(g) = (\varphi^*)^{-1}(X(\varphi^* g)), \qquad g \in \mathscr{S}(N).$$

Explicitly we have

$$(\varphi_* X)(\varphi(x)) = d\varphi(X(x)), \qquad x \in M.$$

This formula shows that the vector fields X and $\varphi_* X$ are φ-related.

It follows at once from the definition and Proposition VIII that the correspondence $X \mapsto \varphi_* X$ defines an isomorphism of Lie algebras,

$$\varphi_*: \mathscr{X}(M) \to \mathscr{X}(N);$$

in particular,

$$\varphi_*[X_1, X_2] = [\varphi_* X_1, \varphi_* X_2], \qquad X_1, X_2 \in \mathscr{X}(M).$$

If $\psi: N \to Q$ is a second diffeomorphism, then

$$(\psi \circ \varphi)_* = \psi_* \circ \varphi_* .$$

3.14. Product manifolds. Let M and N be manifolds and consider the product manifold $M \times N$. Recall (sec. 3.7) that $T_{M \times N} = T_M \times T_N$. Now let X be a vector field on M. Then a vector field $i_M X$ on $M \times N$ is given by

$$(i_M X)(x, y) = (X(x), 0), \qquad (x, y) \in M \times N.$$

Similarly, every vector field Y on N determines a vector field $i_N Y$ on $M \times N$ given by

$$(i_N Y)(x, y) = (0, Y(y)), \qquad (x, y) \in M \times N.$$

It follows from these definitions that

$$i_M X \underset{\pi_M}{\sim} X \qquad \text{and} \qquad i_N Y \underset{\pi_N}{\sim} Y.$$

Moreover, we have the relations

$$\begin{aligned} (i_M X)(\pi_M^* f) &= \pi_M^*(X(f)), \qquad (i_N Y)(\pi_M^* f) = 0, \qquad f \in \mathscr{S}(M) \\ (i_N Y)(\pi_N^* g) &= \pi_N^*(Y(g)), \qquad (i_M X)(\pi_N^* g) = 0, \qquad g \in \mathscr{S}(N). \end{aligned} \tag{3.4}$$

Lemma VI: Assume that Z is a vector field on $M \times N$ such that

$$Z(\pi_M^* f) = 0 \quad (f \in \mathscr{S}(M)) \qquad \text{and} \qquad Z(\pi_N^* g) = 0 \quad (g \in \mathscr{S}(N)).$$

Then $Z = 0$.

Proof: Let $a \in M$, $b \in N$ be arbitrary and set $Z(a, b) = \zeta$. Then we can write

$$\zeta = dj_b(\xi) + dj_a(\eta), \qquad \xi \in T_a(M), \quad \eta \in T_b(N)$$

(cf. sec. 3.7). It follows (via the relations of sec. 3.7) that

$$0 = \zeta(\pi_M^* f) = \xi(f), \qquad f \in \mathscr{S}(M),$$

whence $\xi = 0$. Similarly, we obtain $\eta = 0$ and so $\zeta = 0$.

Q.E.D.

Proposition IX: The maps

$$i_M: \mathscr{X}(M) \to \mathscr{X}(M \times N) \qquad \text{and} \qquad i_N: \mathscr{X}(N) \to \mathscr{X}(M \times N)$$

are homomorphisms of Lie algebras. Moreover,

$$[i_M X, i_N Y] = 0, \qquad X \in \mathscr{X}(M), \quad Y \in \mathscr{X}(N).$$

Proof: Let $X_1, X_2 \in \mathscr{X}(M)$. The relations (3.4) imply that the vector fields $[i_M X_1, i_M X_2]$ and $i_M[X_1, X_2]$ agree when applied to functions of the form $\pi_M^* f$ ($f \in \mathscr{S}(M)$) or $\pi_N^* g$ ($g \in \mathscr{S}(N)$). Thus Lemma VI shows that

$$[i_M X_1, i_M X_2] = i_M[X_1, X_2]$$

and so i_M is a homomorphism of Lie algebras. The rest of the proposition is proved in the same way.

Q.E.D.

Next, consider an arbitrary vector field Z on $M \times N$. Z determines vector fields Z^M and Z^N on $M \times N$ by the equations

$$Z(x, y) = Z^M(x, y) + Z^N(x, y), \qquad (x, y) \in M \times N,$$

where

$$Z^M(x, y) \in T_x(M) \qquad \text{and} \qquad Z^N(x, y) \in T_y(N).$$

Definition: The *M-product* of two vector fields Z_1 and Z_2 on $M \times N$, denoted by $(Z_1, Z_2)_M$, is the vector field on $M \times N$ given by

$$(Z_1, Z_2)_M = [Z_1^M, Z_2] + [Z_1, Z_2^M] - [Z_1, Z_2]^M.$$

The *N-product* of Z_1 and Z_2 is defined by

$$(Z_1, Z_2)_N = [Z_1^N, Z_2] + [Z_1, Z_2^N] - [Z_1, Z_2]^N.$$

The following lemma states obvious properties:

Lemma VII: Let Z_1, Z_2 be vector fields on $M \times N$. Then

(1) $(Z_1, Z_2)_M = -(Z_2, Z_1)_M$

(2) $(fZ_1, Z_2)_M = f(Z_1, Z_2)_M - Z_2^M(f) \cdot Z_1, \quad f \in \mathscr{S}(M \times N)$

(3) $(Z_1^M, Z_2^M)_M = [Z_1^M, Z_2^M]$

(4) $(Z_1^M, Z_2^M)_N = 0$

(5) $(Z_1, Z_2)_M + (Z_1, Z_2)_N = [Z_1, Z_2]$.

3.15. Differential equations. Let X be a vector field on a manifold M. An *orbit* for X is a smooth map $\alpha: I \to M$ ($I \subset \mathbb{R}$ some open interval) such that

$$\dot{\alpha}(t) = X(\alpha(t)), \qquad t \in I.$$

Proposition X: Let X be a vector field on M. Fix $a \in M$ and $t_0 \in \mathbb{R}$. Then there is an interval $I \ni t_0$ and an orbit $\alpha: I \to M$ of X such that $\alpha(t_0) = a$.

Moreover, if $\alpha, \beta: J \to M$ are orbits for X which agree at some $s_0 \in J$, then $\alpha = \beta$.

Proof: For the first statement of the proposition we may assume $M = \mathbb{R}^n$. In this case it is the standard Picard existence theorem (cf. [3, Theorem 7.1, p. 22]).

To prove the second part we show that the set of $s \in J$ for which $\alpha(s) = \beta(s)$ is both closed and open, and hence all of J. It is obviously closed. To show that it is open we may assume $M = \mathbb{R}^n$ and then apply the Picard uniqueness theorem (cf. [3, p. 22]).

Q.E.D.

Now consider the product manifold $\mathbb{R} \times M$. We call a subset $W \subset \mathbb{R} \times M$ *radial* if for each $a \in M$

$$W \cap (\mathbb{R} \times a) = I_a \times a \qquad \text{or} \qquad W \cap (\mathbb{R} \times a) = \varnothing$$

where I_a is an open interval on $\mathbb{R}$ containing the point 0. The union and finite intersection of radial sets is again radial.

Theorem II: Let X be a vector field on a manifold M. Then there is a radial neighbourhood W, of $0 \times M$ in $\mathbb{R} \times M$ and a smooth map $\varphi: W \to M$ such that

$$\dot{\varphi}_x(t) = X(\varphi_x(t)), \qquad (t, x) \in W, \qquad \text{and} \qquad \varphi_x(0) = x, \quad x \in M,$$

where φ_x is given by $\varphi_x(t) = \varphi(t, x)$. Moreover, φ is uniquely determined by X.

Remark: φ is called the *flow* generated by X. Each $\varphi_x: I_x \to M$ is an orbit of X satisfying $\varphi_x(0) = x$.

Proof: Let $\{(U_\alpha, u_\alpha)\}$ be an atlas for M. The Picard existence theorem [3, p. 22] implies our theorem for each U_α. Hence there are radial neighbourhoods W_α of $0 \times U_\alpha$ in $\mathbb{R} \times U_\alpha$ and there are smooth maps $\varphi_\alpha: W_\alpha \to U_\alpha$ such that

$$\dot{\varphi}_\alpha(t, x) = X(\varphi_\alpha(t, x)), \qquad (t, x) \in W_\alpha$$

and

$$\varphi_\alpha(0, x) = x, \qquad x \in U_\alpha .$$

Now set $W = \bigcup_\alpha W_\alpha$. Then W is a radial neighbourhood of $0 \times M$ in $\mathbb{R} \times M$. Moreover, $W_\alpha \cap W_\beta$ is a radial neighbourhood of $0 \times (U_\alpha \cap U_\beta)$; if $x \in U_\alpha \cap U_\beta$, then $W_\alpha \cap W_\beta \cap (\mathbb{R} \times x)$ is an interval I containing 0. Clearly $\varphi_\alpha, \varphi_\beta: I \times x \to M$ are orbits of X agreeing at 0, and so by Proposition X they agree in I. It follows that they agree in $W_\alpha \cap W_\beta$. Thus the φ_α define a smooth map $\varphi: W \to M$ which has the desired properties.

The uniqueness of φ is immediate from Proposition X.

Q.E.D.

Corollary: If (t, x), $(s, \varphi(t, x))$, and $(t + s, x)$ are all in W, then

$$\varphi(s, \varphi(t, x)) = \varphi(t + s, x).$$

Proof: Since W is radial there is an open interval $I \ni 0, s$ and such that

$$(t + I) \times x \subset W \qquad \text{and} \qquad I \times \varphi(t, x) \subset W.$$

Thus orbits $\alpha, \beta: I \to M$ of X are given by

$$\alpha(u) = \varphi(u, \varphi(t, x)) \qquad \text{and} \qquad \beta(u) = \varphi(t + u, x), \qquad u \in I.$$

Since $\alpha(0) = \varphi(t, x) = \beta(0)$, Proposition X implies that $\alpha = \beta$; in particular, $\alpha(s) = \beta(s)$.

Q.E.D.

Remark: If we write $\varphi(t, x) = \varphi_t(x)$, then the corollary reads

$$(\varphi_s \circ \varphi_t)(x) = \varphi_{s+t}(x).$$

For this reason φ is often called the *local one-parameter group generated by X*.

§4. Differential forms

3.16. One-forms. A *one-form* on a manifold M is a cross-section in the cotangent bundle; i.e., a one-form assigns (smoothly) to every point $x \in M$ a linear function $\omega(x) \in T_x(M)^*$.

The one-forms on M are an $\mathscr{S}(M)$-module, which will be denoted by $A^1(M)$. The duality between $T_x(M)$ and $T_x(M)^*$ induces an $\mathscr{S}(M)$-bilinear map $A^1(M) \times \mathscr{X}(M) \to \mathscr{S}(M)$ given by

$$\langle \omega, X \rangle(x) = \langle \omega(x), X(x) \rangle, \qquad \omega \in A^1(M), \quad X \in \mathscr{X}(M).$$

It follows from the remark following Corollary II to Proposition XIII, sec. 2.24, that the associated map

$$A^1(M) \to \operatorname{Hom}_M(\mathscr{X}(M); \mathscr{S}(M))$$

is an isomorphism.

Now let $\varphi: M \to N$ be a smooth map. Then the bundle map $d\varphi: \tau_M \to \tau_N$ yields a dual map

$$(d\varphi)^*: \operatorname{Sec} \tau_M^* \leftarrow \operatorname{Sec} \tau_N^*$$

(cf. sec. 2.15). We shall denote this map by φ^*,

$$\varphi^*: A^1(M) \leftarrow A^1(N).$$

Explicitly, we have

$$(\varphi^*\omega)(x; \xi) = \omega(\varphi(x); d\varphi(\xi)), \qquad x \in M, \quad \xi \in T_x(M).$$

According to sec. 2.15

$$\varphi^*(f_1\omega_1 + f_2\omega_2) = \varphi^*f_1 \cdot \varphi^*\omega_1 + \varphi^*f_2 \cdot \varphi^*\omega_2, \qquad f_1, f_2 \in \mathscr{S}(N), \quad \omega_1, \omega_2 \in A^1(N).$$

3.17. The gradient. Every smooth function f on a manifold M determines an $\mathscr{S}(M)$-module homomorphism

$$\varphi_f: \mathscr{X}(M) \to \mathscr{S}(M)$$

given by $\varphi_f: X \mapsto X(f)$ $(X \in \mathscr{X}(M))$. In view of the isomorphism of the last section, there exists a unique $\delta f \in A^1(M)$ such that

$$X(f) = \langle \delta f, X \rangle, \qquad X \in \mathscr{X}(M).$$

We call δf the *gradient* of f.

From the fact that vector fields act as (real linear) derivations in $\mathscr{S}(M)$ we obtain that

$$\delta(\lambda f + \mu g) = \lambda \delta f + \mu \delta g, \qquad \lambda, \mu \in \mathbb{R}, \quad f, g \in \mathscr{S}(M),$$

and

$$\delta(f \cdot g) = \delta f \cdot g + f \cdot \delta g.$$

Next, consider a smooth map $\varphi: M \to N$. For the induced map $\varphi^*: A^1(M) \leftarrow A^1(N)$ we have the relation

$$\varphi^* \delta f = \delta(\varphi^* f), \qquad f \in \mathscr{S}(N) \tag{3.5}$$

as follows from a simple computation.

In particular, if $i: U \to M$ is the inclusion map of an open subset $U \subset M$ we have

$$i^* \delta f = \delta i^* f.$$

But $i^* \delta f$ is the restriction of the one-form δf to U. Thus it follows that

$$\operatorname{carr} \delta f \subset \operatorname{carr} f, \qquad f \in \mathscr{S}(M).$$

Examples: **1** A vector field d/dt on $\mathbb{R}$ is given by $f \mapsto f'$, $f \in \mathscr{S}(\mathbb{R})$. If we write $T_{\mathbb{R}} = \mathbb{R} \times \mathbb{R}$, then d/dt is the vector field $s \mapsto (s, 1)$. On the other hand, the identity map $\iota: \mathbb{R} \to \mathbb{R}$ is a smooth function. Hence its gradient is a one-form; we denote it by δt. A simple computation shows that

$$\langle \delta t, d/dt \rangle(s) = 1, \qquad s \in \mathbb{R}.$$

2 Let $f \in \mathscr{S}(M)$. Then $f = \iota \circ f = f^* \iota$ and hence

$$\delta f = f^* \delta t.$$

On the other hand, the derivative of f is the bundle map

$$df: T_M \to T_{\mathbb{R}} = \mathbb{R} \times \mathbb{R}.$$

Using Example 1 we obtain, for $\xi \in T_x(M)$:

$$df(\xi) = (f(x), \delta f(x; \xi)).$$

In particular it follows that $\delta f = 0$ if and only if $df = 0$. Thus Proposition II, sec. 3.3, yields the following

Proposition XI: The gradient of a constant function is zero. Conversely if $f \in \mathscr{S}(M)$ satisfies $\delta f = 0$ and if M is connected, then f is constant.

Corollary: If $f \in \mathscr{S}(M)$ is constant in some open subset $U \subset M$, then

$$\delta f(x) = 0, \qquad x \in U.$$

Example 3: Let $f \in \mathscr{S}(U)$, $g \in \mathscr{S}(M)$ (U an open subset of M). Assume $\operatorname{carr} g \subset U$. Then considering $\delta(fg)$, $f\delta g$, and $g\delta f$ as one-forms on U we see they all have carrier contained in $\operatorname{carr} g$. Hence they extend to M (put them equal to 0 outside U). Denote their extensions again by $\delta(fg)$, $f\delta g$, and $g\delta f$, and observe that

$$\delta(fg) = f\delta g + g\delta f.$$

Proposition XII: The $\mathscr{S}(M)$-module $A^1(M)$ is generated by gradients.

Proof: Consider first the case that M is an open subset of $\mathbb{R}^n$. Fix a basis e_i ($i = 1, \ldots, n$) of $\mathbb{R}^n$ and define vector fields $X_i \in \mathscr{X}(M)$, by

$$X_i(x) = (x, e_i), \qquad x \in M, \quad e_i \in \mathbb{R}^n.$$

It is an immediate consequence of Example 1, sec. 3.5, that $\mathscr{X}(M)$ is a module with basis X_i ($i = 1, \ldots, n$).

Now let e^{*i} ($i = 1, \ldots, n$) be the dual basis in $(\mathbb{R}^n)^*$. Considering the e^{*i} as smooth functions on M we see that

$$\langle \delta e^{*i}, X_j \rangle = e^{*i}(e_j) = \delta^i_j .$$

It follows that the δe^{*i} form a basis for the module $A^1(M)$. In particular, $A^1(M)$ is generated by gradients.

Now let M be any manifold. According to sec. 1.3 and the corollary to Theorem I, sec. 1.2, M admits a finite atlas $\{U_\alpha\}$ ($\alpha = 1, \ldots, k$). Let p_α ($\alpha = 1, \ldots, k$) be a partition of unity subordinate to this covering. Since, for $\omega \in A^1(M)$,

$$\omega = \sum_{\alpha=1}^{k} p_\alpha \omega,$$

and $\operatorname{carr}(p_\alpha \omega) \subset U_\alpha$, it is sufficient to consider the case $\operatorname{carr} \omega \subset U$, where (U, u) is a chart on M.

By the first part of the proof we can define smooth functions $f_i \in \mathscr{S}(U)$ such that the δf_i form a basis of $A^1(U)$ over $\mathscr{S}(U)$. Hence, if ω_U denotes the restriction of ω to U, we have

$$\omega_U = \sum_{i=1}^{n} h_i \delta f_i, \qquad h_i \in \mathscr{S}(U).$$

Since the δf_i form a basis for $A^1(U)$, it follows that

$$\operatorname{carr} h_i \subset \operatorname{carr} \omega \subset U, \qquad i = 1, \ldots, n,$$

so that we can extend the h_i to M by setting $h_i(x) = 0$, $x \notin \operatorname{carr} \omega$. With this convention, we have

$$\omega = \sum_i h_i \, \delta f_i = \sum_i \delta(h_i f_i) - \sum_i f_i \, \delta h_i . \tag{3.6}$$

Finally, since $\operatorname{carr} h_i \subset U$, it follows that $h_i f_i \in \mathscr{S}(M)$. On the other hand, choose $g \in \mathscr{S}(M)$ so that $g(x) = 1$ ($x \in \operatorname{carr} \omega$) and $\operatorname{carr} g \subset U$. Then $gf_i \in \mathscr{S}(M)$ and

$$gf_i \cdot \delta h_i = f_i \cdot \delta h_i .$$

Thus (3.6) can be rewritten as

$$\omega = \sum_i \delta(h_i f_i) - \sum_i (gf_i) \, \delta h_i .$$

This completes the proof.

Q.E.D.

3.18. Tensor fields. Let M be a manifold and consider the cotangent bundle τ_M^*. A *covariant tensor field of degree p* is a cross-section in the vector bundle

$$\tau^p(M) = \underbrace{\tau_M^* \otimes \cdots \otimes \tau_M^*}_{p \text{ factors}} .$$

Thus a covariant tensor field of degree p assigns to each point $x \in M$ an element of the space $T_x(M)^* \otimes \cdots \otimes T_x(M)^*$. The covariant tensor fields of degree p on M form an $\mathscr{S}(M)$-module, which we denote by $\mathscr{X}^p(M)$. (In particular, $\mathscr{X}^1(M) = A^1(M)$.) We extend the definition by putting $\mathscr{X}^0(M) = \mathscr{S}(M)$. By the corollary to Theorem I, sec. 2.23, each $\mathscr{X}^p(M)$ is finitely generated and projective.

The *product* of two covariant tensor fields $\Phi \in \mathscr{X}^p(M)$, $\Psi \in \mathscr{X}^q(M)$ is defined to be the covariant tensor field of degree $p + q$ given by

$$(\Phi \cdot \Psi)(x) = \Phi(x) \otimes \Psi(x), \qquad x \in M.$$

The assignment $(\Phi, \Psi) \mapsto \Phi \cdot \Psi$ defines an $\mathscr{S}(M)$-bilinear map

$$\mathscr{X}^p(M) \times \mathscr{X}^q(M) \to \mathscr{X}^{p+q}(M).$$

In this way the direct sum $\sum_{p \geqslant 0} \mathscr{X}^p(M)$ becomes an algebra over $\mathscr{S}(M)$. Proposition XIV, sec. 2.24, implies that this is the $\mathscr{S}(M)$-tensor algebra over $A^1(M)$ $(= \mathscr{X}^1(M))$.

Analogously, we define a *contravariant* tensor field of degree p on M to be an element of Sec $\tau_p(M)$, where $\tau_p(M) = \tau_M \otimes \cdots \otimes \tau_M$ (p factors).

The scalar product between $T_x(M)^*$ and $T_x(M)$ induces a bilinear map $\mathscr{X}^p(M) \times \mathscr{X}_p(M) \to \mathscr{S}(M)$, given by

$$\langle \Phi, \Psi \rangle(x) = \langle \Phi(x), \Psi(x) \rangle, \qquad x \in M.$$

The associated maps

$$\mathscr{X}^p(M) \to \mathrm{Hom}_M(\mathscr{X}_p(M); \mathscr{S}(M)), \qquad \mathscr{X}_p(M) \to \mathrm{Hom}_M(\mathscr{X}^p(M); \mathscr{S}(M))$$

are isomorphisms (cf. sec. 2.24).

Finally, a *mixed tensor field* of type (p, q) is an element of $\mathscr{X}^p_q(M) = \mathrm{Sec}\, \tau^p_q(M)$, where $\tau^p_q(M) = \tau^p(M) \otimes \tau_q(M)$. We may identify $\mathscr{X}^p_q(M)$ with $\mathrm{Hom}_M(\mathscr{X}^q_p(M); \mathscr{S}(M))$.

Example: The *unit tensor field* t is the tensor field of type (1,1) given by

$$t(x; \eta^*, \xi) = \langle \eta^*, \xi \rangle, \qquad \xi \in T_x(M), \quad \eta^* \in T_x(M)^*.$$

Since (cf. sec. 2.24)

$$\mathscr{X}^1_1(M) = \mathrm{Sec}(\tau_M^* \otimes \tau_M) = \mathrm{Sec}\, \tau_M^* \otimes_M \mathrm{Sec}\, \tau_M ,$$

we may write

$$t = \sum_{i=1}^m \omega^i \otimes X_i , \qquad \omega^i \in A^1(M), \quad X_i \in \mathscr{X}(M).$$

3.19. Differential forms. Recall that the exterior algebra bundle $\wedge \tau_M^*$ over the cotangent bundle τ_M^* is the Whitney sum of bundles $\wedge^p \tau_M^*$, $p = 0, \ldots, n$, with $\wedge^0 \tau_M^* = M \times \mathbb{R}$ (cf. sec. 2.11). The fibres of $\wedge^p \tau_M^*$ are the vector spaces $\wedge^p T_x(M)^*$, which may be identified with the spaces of skew-symmetric p-linear maps

$$T_x(M) \times \cdots \times T_x(M) \to \mathbb{R}.$$

Definition: A *differential form of degree p* on M (a p-form) is an element of $\operatorname{Sec}(\wedge^p \tau_M^*)$. We denote the $\mathscr{S}(M)$-module of p-forms by $A^p(M)$ (cf. sec. 2.6). Thus if $\Phi \in A^p(M)$, then $\Phi(x)$ may be regarded as a skew-symmetric p-linear function in $T_x(M)$, ($x \in M$).

The *exterior product* of two differential forms $\Phi \in A^p(M)$ and $\Psi \in A^q(M)$ is defined to be the differential form $\Phi \wedge \Psi \in A^{p+q}(M)$ given by

$$(\Phi \wedge \Psi)(x) = \Phi(x) \wedge \Psi(x), \qquad x \in M.$$

Explicitly, if $\xi_i \in T_x(M)$, then

$$(\Phi \wedge \Psi)(x; \xi_1, \ldots, \xi_{p+q})$$
$$= \frac{1}{p!q!} \sum_\sigma \epsilon_\sigma \Phi(x; \xi_{\sigma(1)}, \ldots, \xi_{\sigma(p)}) \Psi(x; \xi_{\sigma(p+1)}, \ldots, \xi_{\sigma(p+q)}), \quad p, q \geqslant 1,$$

where σ runs over the symmetric group S^{p+q} and $\epsilon_\sigma = 1$ (-1) if σ is an even (odd) permutation. The product map $(\Phi, \Psi) \mapsto \Phi \wedge \Psi$ makes the direct sum

$$A(M) = \sum_{p=0}^{n} A^p(M), \qquad n = \dim M$$

into an anticommutative graded algebra over the ring $\mathscr{S}(M)$. It follows from Proposition XV of sec. 2.24 that the algebra $A(M)$ is an exterior algebra over the $\mathscr{S}(M)$-module $A^1(M)$.

$A(M)$, considered as an algebra over $\mathbb{R}$, is generated by gradients and functions since, by Proposition XII, sec. 3.17, the real vector space $A^1(M)$ is generated by gradients and functions.

A p-form Φ on M determines a p-linear (over $\mathscr{S}(M)$) skew-symmetric map $\mathscr{X}(M) \times \cdots \times \mathscr{X}(M) \to \mathscr{S}(M)$ given by

$$(X_1, \ldots, X_p) \mapsto \Phi(x; X_1(x), \ldots, X_p(x)), \qquad x \in M.$$

In this way we obtain an $\mathscr{S}(M)$-homomorphism

$$A^p(M) \to A^p(\mathscr{X}(M); \mathscr{S}(M)).$$

Applying the isomorphisms of sec. 2.24 following Proposition XV we see that this map is an isomorphism,

$$A^p(M) \cong A^p(\mathscr{X}(M)); \mathscr{S}(M)).$$

Henceforth we shall identify these modules without further reference.

Example: Suppose that O is an open subset of a vector space E. Then (cf. sec. 3.5) we have $\tau_O = O \times E$, whence

$$\wedge\tau_O^* = O \times \wedge E^*.$$

Thus cross-sections in $\wedge\tau_O^*$ may be interpreted as smooth maps $O \to \wedge E^*$; i.e.

$$A(O) = \mathscr{S}(O; \wedge E^*) = \mathscr{S}(O) \otimes_{\mathbb{R}} \wedge E^*.$$

Next, let $\varphi: M \to N$ be a smooth map. Then every p-form Ψ on N determines a p-form on M, denoted by $\varphi^*\Psi$ and given by

$$(\varphi^*\Psi)(x; \xi_1, \ldots, \xi_p) = \Psi(\varphi(x); (d\varphi)\,\xi_1, \ldots, (d\varphi)\,\xi_p).$$

To see that $\varphi^*\Psi$ is indeed a p-form observe that

$$\varphi^*\Psi = (\wedge^p d\varphi)^*\,\Psi,$$

where $d\varphi: \tau_M \to \tau_N$ is the bundle map induced by φ.

Proposition XIII: (1) φ^* is a homomorphism of graded $\mathbb{R}$-algebras. In particular,

$$\varphi^*(f \cdot \Phi) = \varphi^*f \cdot \varphi^*\Phi, \qquad f \in \mathscr{S}(N), \quad \Phi \in A(N).$$

(2) If $\psi: N \to Q$ is a smooth map into a third manifold, then

$$(\psi \circ \varphi)^* = \varphi^* \circ \psi^*.$$

(3) For the identity map $\iota_M: M \to M$, we have

$$\iota_M^* = \iota_{A(M)}\,.$$

3.20. Examples: 1. Let U be an open subset of a manifold M. Writing $\tau_U = \tau_M|_U$ (cf. sec. 3.5) we obtain

$$\wedge\tau_U^* = \wedge\tau_M^*|_U\,.$$

Thus if $\Phi \in A(M)$, its restriction to U is a differential form on U. If $i: U \to M$ denotes the inclusion map we have, clearly,

$$i^*\Phi = \Phi|_U\,.$$

2. *Products*: Let M and N be manifolds. Since $T_{(x,y)}(M \times N) =$

$T_x(M) \oplus T_y(N)$, the fibre of $\wedge \tau_{M\times N}^*$ at (x, y) is the vector space $\wedge T_x(M)^* \otimes \wedge T_y(N)^*$. The direct decompositions

$$\wedge T_x(M)^* \otimes \wedge T_y(N)^* = \sum_{p,q} \wedge^p T_x(M)^* \otimes \wedge^q T_y(N)^*$$

define a decomposition of $\wedge \tau_{M\times N}^*$ as a Whitney sum of subbundles. The corresponding decomposition of $A(M \times N)$ as a direct sum of submodules is written

$$A(M \times N) = \sum_{p,q} A^{p,q}(M \times N).$$

Evidently this makes $A(M \times N)$ into a bigraded algebra.

Next observe that the projections $\pi_M: M \times N \to M$, $\pi_N: M \times N \to N$ define homomorphisms

$$\pi_M^*: A(M) \to A(M \times N), \qquad \pi_N^*: A(N) \to A(M \times N).$$

If $\Phi \in A(M)$, $\Psi \in A(N)$, we establish the following *notation convention*:

$$\Phi \times \Psi = \pi_M^*\Phi \wedge \pi_N^*\Psi \in A(M \times N).$$

If $\Phi \in A^p(M)$, $\Psi \in A^q(N)$, then $\Phi \times \Psi \in A^{p,q}(M \times N)$.

3. *Quotient manifolds*: Let $\pi: M \to N$ make N into a quotient manifold of M. Then each $(d\pi)_x$ is surjective; hence so is each $\wedge (d\pi)_x$. It follows that the dual maps

$$\wedge(d\pi)_x^* : \wedge T_x(M)^* \leftarrow \wedge T_{\pi(x)}(N)^*$$

are injective. Hence $\pi^*: A(M) \leftarrow A(N)$ is injective.

4. *Involutions*: Let ω be an involution of N $(\omega^2 = \iota_N)$. Then ω^* is an involution of $A(N)$. Hence $A(N)$ is the direct sum of the graded subspaces $A_+(N)$ and $A_-(N)$, where

$$A_+(N) = \{\Phi \mid \omega^*\Phi = \Phi\}, \qquad A_-(N) = \{\Phi \mid \omega^*\Phi = -\Phi\}.$$

Since ω^* is a homomorphism, $A_+(N)$ is a graded subalgebra of $A(N)$.

Now assume that $\pi: N \to M$ is a surjective local diffeomorphism such that, for $x \in M$, $\pi^{-1}(x)$ is of the form

$$\pi^{-1}(x) = \{z, \omega(z)\}.$$

We shall show that π^* can be considered as an isomorphism

$$A(M) \xrightarrow{\cong} A_+(N).$$

In fact by Example 3 above, π^* is injective. Since $\pi \circ \omega = \pi$, we have $\omega^* \circ \pi^* = \pi^*$; hence $\operatorname{Im} \pi^* \subset A_+(N)$. Finally, let $\Phi \in A_+(N)$. Fix $x \in M$ and suppose $\pi(z) = x$. Since $(d\pi)_z$ is an isomorphism there is a unique $\Psi_x \in \wedge T_x(M)^*$ such that

$$\wedge (d\pi)_z^* (\Psi_x) = \Phi(z). \tag{3.7}$$

Since $\omega^*\Phi = \Phi$ and $\pi^{-1}(x) = \{z, \omega(z)\}$, it follows that this relation holds with z replaced by $\omega(z)$.

In particular let U, V be open sets in M, N so that the restriction π_V of π to V is a diffeomorphism onto U (Theorem I, sec. 3.8). Then $\Psi_x = (\pi_V^{-1})^* \Phi(x)$, $x \in U$. Hence $x \mapsto \Psi_x$ is smooth in U; hence it is smooth in M. Let $\Psi \in A(M)$ be the differential form defined in this way. Then relation (3.7) implies that $\pi^*\Psi = \Phi$; i.e. $A_+(N) \subset \operatorname{Im} \pi^*$.

Remark: The situation discussed in Example 4 arises in the case of the double cover induced by a vector bundle (cf. sec. 2.20).

§5. Orientation

3.21. Orientable manifolds. An n-manifold M is called *orientable* if the tangent bundle τ_M is orientable (cf. sec. 2.16); in other words, M is orientable if there exists an n-form Δ on M such that

$$\Delta(x) \neq 0, \qquad x \in M.$$

An *orientation* of M is an equivalence class of such n-forms under the equivalence relation

$$\Delta_1 \sim \Delta_2 \quad \text{if and only if} \quad \Delta_1 = f \cdot \Delta_2 ,$$

where f is a smooth function on M such that $f(x) > 0$, $x \in M$.

If M is an orientable manifold, the choice of an orientation, or representing n-form, is said to *orient* M. An element of the representing class is called a *positive n-form* on the oriented manifold M. A basis $\xi_1, \ldots, \xi_n$ of a tangent space $T_x(M)$ of an oriented manifold is called *positive*, if

$$\Delta(x; \xi_1, \ldots, \xi_n) > 0,$$

where Δ represents the orientation. Because $A^n(M) = \text{Sec} \wedge^n \tau_M^*$ and $\dim \wedge^n T_x^*(M) = 1$, if Δ orients M, then every $\Phi \in A^n(M)$ is uniquely of the form $\Phi = f \cdot \Delta$, $f \in \mathscr{S}(M)$.

Examples: 1. Let Δ denote a determinant function in $\mathbb{R}^n$. Then Δ may be considered as an element of $A^n(\mathbb{R}^n)$ which orients $\mathbb{R}^n$. Thus the definition of orientation for $\mathbb{R}^n$ coincides with that given in [5, p. 127].

2. Let S^n denote the unit sphere of an $(n + 1)$-dimensional Euclidean space E and recall that we may identify $T_x(S^n)$ with the subspace $x^\perp \subset E$ $(x \in S^n)$ (cf. sec. 3.4). Let Δ denote a determinant function in E. Then the n-form $\Omega \in A^n(S^n)$ given by

$$\Omega(x; \xi_1, \ldots, \xi_n) = \Delta(x, \xi_1, \ldots, \xi_n), \qquad x \in S^n, \quad \xi_i \in T_x(S^n),$$

orients S^n. Evidently, for $\lambda > 0$, $\lambda\Delta$ induces the same orientation in S^n. Thus the orientation of S^n depends only on the orientation of E. It is called the *orientation of S^n induced by the orientation of E*.

3. Let $\mathbb{R}P^n$ be the real n-dimensional projective space (i.e. the n-sphere with antipodes identified). We shall show that

$$\mathbb{R}P^n \text{ is } \begin{cases} \text{orientable if } n \text{ is odd} \\ \text{nonorientable if } n \text{ is even.} \end{cases}$$

Consider S^n as the unit sphere of $\mathbb{R}^{n+1}$ (with respect to some inner product) and let Δ be a determinant function in $\mathbb{R}^{n+1}$. Then

$$\Omega(x; \xi_1, ..., \xi_n) = \Delta(x, \xi_1, ..., \xi_n), \qquad x \in S^n, \quad \xi_i \in T_x(S^n),$$

orients S^n. Let $\sigma: S^n \to S^n$ be the diffeomorphism given by $\sigma(x) = -x$. Then we have

$$\sigma^*\Omega = (-1)^{n+1}\Omega. \tag{3.8}$$

Thus, if n is odd, $\sigma^*\Omega = \Omega$. It follows from Example 4, sec. 3.20, that $\Omega = \pi^*\hat{\Omega}$, for some $\hat{\Omega} \in A^n(\mathbb{R}P^n)$, where $\pi: S^n \to \mathbb{R}P^n$ is the projection. Since π is a local diffeomorphism, it follows that

$$\hat{\Omega}(x) \neq 0, \qquad x \in \mathbb{R}P^n.$$

Hence $\hat{\Omega}$ orients $\mathbb{R}P^n$.

On the other hand, consider the case that n is even. Let $\hat{\Phi} \in A^n(\mathbb{R}P^n)$. Then

$$\sigma^*\pi^*\hat{\Phi} = \pi^*\hat{\Phi}. \tag{3.9}$$

Since Ω orients S^n, we can write

$$\pi^*\hat{\Phi} = f \cdot \Omega, \qquad f \in \mathscr{S}(S^n).$$

Since n is even, we obtain from (3.8) and (3.9) that $\sigma^*f = -f$; i.e. $f(-x) = -f(x)$, $x \in S^n$. Now the connectivity of S^n implies that $f(x) = 0$ for some $x \in S^n$. It follows that

$$\hat{\Phi}(\pi x) = 0.$$

Thus $\mathbb{R}P^n$ is not orientable.

4. Consider the equivalence relation in $\mathbb{R}^2$ given by

$$(x, y) \sim (x + k, (-1)^k y + h), \qquad h, k \in \mathbb{Z}, \qquad x, y \in \mathbb{R}.$$

It is easy to see that the quotient space under this relation is a connected two-manifold K (Klein bottle). It will be shown that K is nonorientable.

In fact, let $\pi: \mathbb{R}^2 \to K$ be the map which assigns to every pair $(x, y) \in \mathbb{R}^2$ its equivalence class. Then the diagram

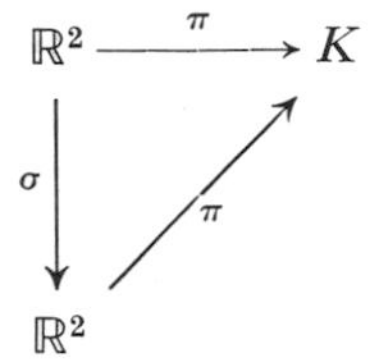

commutes, where σ is the map given by

$$\sigma(x, y) = (x + 1, -y).$$

Now assume that Φ is a two-form on K. Then $\pi^*\Phi$ is a two-form on $\mathbb{R}^2$ which satisfies

$$\sigma^*\pi^*\Phi = \pi^*\Phi.$$

Choosing a fixed determinant function Δ in $\mathbb{R}^2$ we can write

$$\pi^*\Phi = f \cdot \Delta, \qquad f \in \mathscr{S}(\mathbb{R}^2).$$

Since $\sigma^*\Delta = -\Delta$, it follows that $\sigma^*f = -f$; i.e. $f(x + 1, -y) = -f(x, y)$. Hence f must have a zero, (x_0, y_0). It follows that

$$\Phi(\pi(x_0, y_0)) = 0.$$

5. Let U be an open subset of a manifold M which is oriented by $\Delta \in A^n(M)$. Then the restriction of Δ to U orients U. The orientation of U so obtained depends only on the original orientation of M, and is called the *induced orientation.*

6. Let $\{U_\alpha\}$ be a locally finite open cover of a manifold M. Suppose each U_α is oriented, and that the orientations in U_α and U_β induce the same orientation in $U_\alpha \cap U_\beta$ for each pair U_α, U_β. Then there is a unique orientation of M which induces the given orientations in the U_α. (A simple partition of unity argument.)

7. The real line $\mathbb{R}$ has a canonical orientation; namely, it is oriented by the one-form δt (cf. Example 1, sec. 3.17).

8. If M and N are orientable manifolds, then so is $M \times N$. In fact, let $\Delta_M \in A^n(M)$ and $\Delta_N \in A^r(N)$ orient these manifolds. Then

$$\Delta_M \times \Delta_N \in A^{n+r}(M \times N)$$

(cf. Example 2, sec. 3.20) orients $M \times N$. The orientation represented by $\Delta_M \times \Delta_N$ depends only on the orientations represented by Δ_M and Δ_N; it is called the *product orientation.*

If ξ_i $(i = 1, ..., n)$ and η_j $(j = 1, ..., r)$ are positive bases of $T_x(M)$, $T_y(N)$, then the vectors ζ_k, given by

$$\zeta_i = (\xi_i, 0), \quad i = 1, ..., n; \qquad \zeta_{j+n} = (0, \eta_j), \quad j = 1, ..., r,$$

form a positive basis of $T_{(x,y)}(M \times N)$.

9. Let M be a connected nonorientable manifold and denote by $\tilde{M}$ the double cover of M obtained via the tangent bundle τ_M (cf. sec. 2.20). According to Proposition IX, sec. 2.20, $\tilde{M}$ is connected. We show that $\tilde{M}$ is an orientable manifold.

In fact, the projection $\pi: \tilde{M} \to M$ is a local diffeomorphism. Hence the derivative

$$d\pi: \tau_{\tilde{M}} \to \tau_M$$

restricts to a linear isomorphism in each fibre. Thus (cf. the example of sec. 2.5) $\tau_{\tilde{M}}$ is strongly isomorphic to the pull-back of τ_M; and so by Proposition X, sec. 2.20, it is orientable. Thus $\tilde{M}$ is orientable.

3.22. Orientation preserving maps. Let $\varphi: M \to N$ be a local diffeomorphism between oriented manifolds. Let Δ_M and Δ_N be n-forms on M and N, respectively, which represent the orientations. Then we have

$$\varphi^* \Delta_N = f \cdot \Delta_M, \qquad f \in \mathscr{S}(M),$$

where $f(x) \neq 0$, $x \in M$. The mapping φ is called *orientation preserving* if $f(x) > 0$ $(x \in M)$, and *orientation reversing* if $f(x) < 0$ $(x \in M)$. If $\psi: N \to Q$ is a local diffeomorphism into a third oriented manifold, then $\psi \circ \varphi$ preserves the orientation if the maps φ and ψ both preserve (or both reverse) the orientation. If M is connected, φ either preserves or reverses orientations.

Now let φ be a diffeomorphism of a connected orientable manifold M onto itself. Then whether or not φ is orientation preserving with respect to a single orientation of M is independent of the choice of orientation. If every diffeomorphism of M is orientation preserving, M is called *irreversible* (see sec. 5.16).

Examples: 1. The map $\sigma: S^n \to S^n$ given by $\sigma(x) = -x$ is orientation preserving if n is odd, and orientation reversing if n is even, as follows from Example 3, sec. 3.21.

2. Let $\dot{E} = E - \{0\}$ where E is an n-dimensional Euclidean space. Consider the diffeomorphism $\varphi\colon \dot{E} \to \dot{E}$ given by

$$\varphi(x) = \frac{x}{|x|^2}, \qquad x \in E.$$

Then φ reverses the orientation.

In fact, it is easy to verify that the derivative of φ is given by

$$d\varphi(\xi) = \frac{|x|^2 \xi - 2\langle x, \xi\rangle x}{|x|^4}, \qquad x \in \dot{E}, \quad \xi \in T_x(\dot{E}).$$

In particular, if $|x| = 1$, we have

$$d\varphi(\xi) = \xi - 2\langle x, \xi\rangle x.$$

This linear map is the reflection of E in the space which is orthogonal to x. This shows that

$$\det((d\varphi)_x) < 0$$

and hence φ reverses the orientation.

3. Let E be a Euclidean space of dimension $n + 1$ ($n \geqslant 1$) and let $\dot{E} = E - \{0\}$. Let $\mathbb{R}^+ = \{t \in \mathbb{R} \mid t > 0\}$ and consider the diffeomorphism

$$\alpha\colon \mathbb{R}^+ \times S^n \to \dot{E}$$

given by

$$\alpha(t, x) = tx,$$

where S^n is the unit sphere of E.

Let S^n, $\dot{E}$ have the orientations induced from a given orientation of E (cf. Examples 2, and 5, sec. 3.21) and let $\mathbb{R}^+$ have the orientation defined by the one-form $\delta t \in A^1(\mathbb{R}^+)$ (Example 7, sec. 3.21). Finally, let $\mathbb{R}^+ \times S^n$ have the product orientation. Then α is orientation preserving.

In fact, since $\mathbb{R}^+ \times S^n$ is connected, it is sufficient to prove that α is orientation preserving at some point; i.e., it is sufficient to prove that if $\xi_0, \dots, \xi_n$ is a positive basis for $\mathbb{R}^+ \times S^n$ at $(1, x)$, then

$$d\alpha(\xi_0), \dots, d\alpha(\xi_n)$$

is a positive basis for E.

But such a positive basis is given by the vectors

$$(d/dt, 0), (0, \xi_1), \dots, (0, \xi_n),$$

where $\xi_1, ..., \xi_n$ is a positive basis for $T_x(S^n)$. A short computation shows that

$$(d\alpha)_{(1,x)}(d/dt, 0) = x$$

and

$$(d\alpha)_{(1,x)}(0, \xi_i) = \xi_i, \qquad i = 1, ..., n.$$

Thus if Δ is a determinant function representing the orientation of E, then

$$\Delta(d\alpha(d/dt, 0), d\alpha(0, \xi_1), ..., d\alpha(0, \xi_n)) = \Delta(x, \xi_1, ..., \xi_n) = \Omega(x; \xi_1, ..., \xi_n) > 0$$

and so

$$d\alpha(d/dt, 0), d\alpha(0, \xi_1), ..., d\alpha(0, \xi_n)$$

is a positive basis for E.

4. Let M be an n-manifold and consider the $2n$-manifold T_M. It is orientable. To see this, let $\pi: T_M \to M$ be the projection, and let $\{(U_\alpha, u_\alpha)\}$ be a locally finite atlas for M. Then each

$$du_\alpha: \pi^{-1}U_\alpha \to u_\alpha(U_\alpha) \times \mathbb{R}^n$$

is a diffeomorphism. Hence $\{(\pi^{-1}(U_\alpha), du_\alpha)\}$ is a locally finite atlas for T_M.

Now assign a fixed orientation to $\mathbb{R}^n$, and let $\mathbb{R}^n \times \mathbb{R}^n$ have the product orientation. Then there is a unique orientation in $\pi^{-1}U_\alpha$ such that du_α is orientation preserving. In view of Example 6, sec. 3.21, we have only to show that the orientations in $\pi^{-1}(U_\alpha \cap U_\beta)$ induced from those given in $\pi^{-1}U_\alpha$ and $\pi^{-1}U_\beta$ coincide. In other words, we must prove that the identification maps

$$\psi_{\beta\alpha} = du_\beta \circ du_\alpha^{-1}: u_\alpha(U_\alpha \cap U_\beta) \times \mathbb{R}^n \to u_\beta(U_\alpha \cap U_\beta) \times \mathbb{R}^n$$

are orientation preserving. Hence it is sufficient to prove the following

Lemma VIII: Let $\varphi: U \to V$ be a diffeomorphism between two open subsets U, V of a vector space E, and let

$$\varphi'(a): E \to E$$

be the derivative of φ at a point $a \in U$. Consider the map

$$\Phi: U \times E \to V \times E$$

given by

$$\Phi(x, \xi) = (\varphi(x), \varphi'(x; \xi)), \qquad x \in U, \quad \xi \in E.$$

Then the derivative of Φ,

$$\Phi'(x, y): E \oplus E \to E \oplus E$$

has positive determinant for every $x \in U$ and $y \in E$.

Proof: It follows from the definition of Φ that

$$\Phi'(x, y)(\xi, \eta) = (\varphi'(x; \xi), \varphi''(x; y, \xi) + \varphi'(x; \eta)), \qquad \xi, \eta \in E.$$

This relation implies that

$$\det \Phi'(x, y) = [\det \varphi'(x)]^2 > 0$$

and so the lemma is proved.

Q.E.D.

Example 5: Let M and N be connected oriented n-manifolds. Assume that $\varphi: M \to N$ and $\psi: M \to N$ are diffeomorphisms which are connected by a homotopy $H: \mathbb{R} \times M \to N$ such that every map $H_t: M \to N$ is a diffeomorphism. Then φ and ψ both preserve or both reverse the orientations.

In fact, let Δ_N represent the orientation of N. Let $x \in M$ be fixed and define $f \in \mathscr{S}(\mathbb{R} \times M)$ by

$$f(t, x) \cdot (\varphi^*\Delta_N)(x) = (H_t^*\Delta_N)(x), \qquad x \in M.$$

Then f is never zero and $f(0, x) = 1$; because M is connected, it follows that $f(t, x) > 0$ for all $t \in \mathbb{R}$, $x \in M$. Hence $\psi^*\Delta_N \sim \varphi^*\Delta_N$ (set $t = 1$).

Problems

1. **Local coordinates.** A *local coordinate system* on a smooth manifold M is a chart (U, u) together with a basis $e_1, ..., e_n$ of $\mathbb{R}^n$. For such a system the functions x^i in U given by $x^i(x) = \langle e^{*i}, u(x)\rangle$ are called *coordinate functions.*

(i) If $(U_\alpha, u_\alpha, e_i)$ and (U_β, u_β, f_i) are two overlapping coordinate systems and $u_{\alpha\beta}$ is the identification map, relate the corresponding coordinate functions in terms of $u_{\alpha\beta}$.

(ii) Let (U, u, e_i) be a local coordinate system on M. Consider the vector fields $\partial/\partial x^i$ $(i = 1, ..., n)$ in U corresponding under u to the constant vector fields e_i in $u(U)$. Show that, for $a \in U$, $(\partial/\partial x^i)(a)$ and $(\delta x^i)(a)$ is a pair of dual bases of $T_a(M)$ and $T_a(M)^*$.

(iii) Given overlapping coordinate systems (U, u, e_i) and (V, v, f_i) with corresponding bases $\partial/\partial x^i$ and $\partial/\partial y^j$ of $T_a(U \cap V)$, use $(u \circ v^{-1})'$ to find the matrix which expresses one basis in terms of the other.

(iv) Let $\Phi \in A^p(M)$ and let (U, u, e_i) be a local coordinate system. The functions $\Phi_{i_1 \cdots i_p}(x) = \Phi(x; \partial/\partial x^{i_1}, ..., \partial/\partial x^{i_p})$ $(i_1 < \cdots < i_p)$ are called the *components* of Φ with respect to the given local coordinate system. If (V, v, f_i) is a second system and $x \in U \cap V$, express one set of components at x in terms of the other.

2. **Vector fields.** (i) E is a Euclidean space, $\dot{E} = E - \{0\}$. Define $X, Y \in \mathscr{X}(\dot{E})$ by $X(x) = a$ and $Y(x) = x/|x|$ (a fixed). Compute $[X, Y]$.

(ii) Show that the Lie product of two vector fields in $\mathbb{R}^n$ is given by

$$[X, Y](x) = X'(x; Y(x)) - Y'(x; X(x)).$$

Conclude that the Lie product of two constant vector fields is zero.

(iii) Let $X \in \mathscr{X}(M)$ satisfy $X(a) \neq 0$. Find a local coordinate system (U, u, e_i) about a such that, in U, $X = \partial/\partial x^1$. Hence show that if $f \in \mathscr{S}(M)$, there is a $g \in \mathscr{S}(U)$ such that $X(g) = f$ in U.

(iv) Suppose $\dim M = 2$, $X, Y \in \mathscr{X}(M)$ and for each $a \in M$, $X(a)$ and $Y(a)$ are linearly independent. For each $a \in M$ find a neighbourhood $U(a)$ and functions $f, g \in \mathscr{S}(U)$ such that $f(x)$ and $g(x)$ are never zero and $[fX, gY] = 0$.

(v) Assume M compact. Show that the flow of any $X \in \mathscr{X}(M)$ can be defined in all of $\mathbb{R} \times M$.

(vi) Let $X \in \mathscr{X}(M)$, M arbitrary. Find $f \in \mathscr{S}(M)$ with $f(x) > 0$ for all x, such that the flow generated by fX can be defined in all of $\mathbb{R} \times M$.

3. Consider the map $\varphi: \mathbb{R}^2 \to \mathbb{R}^3$ given by

$$\varphi(t, \tau) = \{(b + a \cos t) \cos \tau, (b + a \cos t) \sin \tau, a \sin t\}, \qquad t, \tau \in \mathbb{R},$$

where $b > a > 0$ are real numbers.

(i) Show that each map $(d\varphi)_{(t,\tau)}$ is injective.

(ii) Show that $\operatorname{Im} \varphi$ is a smooth manifold and is diffeomorphic to the 2-torus.

4. Construct a quotient manifold $\pi: M \to N$, where π is *not* the projection of a fibre bundle.

5. Cayley numbers. Consider $\mathbb{R}^8 = \mathbb{H} \oplus \mathbb{H}$ with the induced inner product. Define a multiplication in $\mathbb{R}^8$ by

$$(p, q)(p', q') = (pp' - \bar{q}'q, q'p + q\bar{p}'), \qquad p, q \in \mathbb{H},$$

where $\bar{p}$ denotes the conjugate of p.

(i) Show that, for $x_1, x_2, y_1, y_2 \in \mathbb{R}^8$,

$$\langle x_1 y_1, x_2 y_2 \rangle + \langle x_2 y_1, x_1 y_2 \rangle = 2\langle x_1, x_2 \rangle \langle y_1, y_2 \rangle.$$

(ii) Use (i) to prove that the product defined above makes $\mathbb{R}^8$ into a (nonassociative) division algebra over $\mathbb{R}$. It is called the algebra of *Cayley numbers.*

(iii) Show that if 1 is the identity of $\mathbb{H}$ then $e = (1, 0)$ is the identity of the Cayley numbers. Regard S^6 as the unit sphere of $e^\perp$ and make τ_{S^6} into a complex vector bundle.

(iv) Use the Cayley numbers to construct a smooth bundle (S^{15}, π, S^8, S^7).

6. Jet bundles for functions. M is an n-manifold. Let $a \in M$ and let $\mathscr{I}_a(M) \subset \mathscr{S}_a(M)$ be the ideal of germs which vanish at a.

(i) Show that $\mathscr{I}_a(M)$ is the unique maximal ideal in $\mathscr{S}_a(M)$.

(ii) Denote $\mathscr{I}_a(M) \cdot \cdots \cdot \mathscr{I}_a(M)$ (p factors) by $\mathscr{I}_a^p(M)$. Let $f \in \mathscr{S}(M)$. Show that the following are equivalent: (a) The germ of f at a is in $\mathscr{I}_a^p(M)$. (b) For any $X_1, \ldots, X_q \in \mathscr{X}(M)$, $(q \leqslant p)$,

$$X_1(X_2(\cdots(X_q f))(a) = 0.$$

(c) If (U, u) is a chart containing a, then the first p derivatives of $f \circ u^{-1}$ are zero at $u(a)$.

(iii) Show that the spaces $\mathscr{S}_a(M)/\mathscr{I}_a^p(M)$ $(a \in M)$ are the fibres of a vector bundle $\mathscr{J}^p(M)$ over M whose typical fibre is the space $\sum_{j=0}^p \vee^j(\mathbb{R}^n)^*$. $\mathscr{J}^p(M)$ is called the *pth jet bundle over M*. Show that $\operatorname{rank} \mathscr{J}^p(M) = \binom{r+p}{p}$.

(iv) Show that each $f \in \mathscr{S}(M)$ determines a cross-section $j^p(f)$ in $\mathscr{J}^p(M)$, its *pth jet*. Construct an isomorphism $\mathscr{J}^1(M) \cong (M \times \mathbb{R}) \oplus \tau_M^*$ and show that $j^1(f)(x) = (f(x), (\delta f)(x))$.

(v) Let $\mathscr{J}_p(M)$ be the bundle dual to $\mathscr{J}^p(M)$. If $\sigma \in \operatorname{Sec} \mathscr{J}_p(M)$, define $D: \mathscr{S}(M) \to \mathscr{S}(M)$ by $Df = \langle \sigma, j^p(f) \rangle$. D is called a *pth order differential operator*. Suppose $X_1, \ldots, X_p \in \mathscr{X}(M)$ and show that there is a unique $\sigma \in \operatorname{Sec} \mathscr{J}_p(M)$ such that $Df = X_1(\cdots(X_p f))$. Show that with respect to a local coordinate system (U, u, e_i) a *pth order differential operator* has the form

$$\sum_{q=0}^{p} \sum_{j_1 < \cdots < j_q} f_{j_1 \cdots j_q} \frac{\partial}{\partial x^{j_1}} \cdots \frac{\partial}{\partial x^{j_q}},$$

with $f_{j_1 \cdots j_q} \in \mathscr{S}(U)$.

(vi) Construct exact sequences

$$0 \to \vee^p \tau_M^* \to \mathscr{J}^p(M) \to \mathscr{J}^{p-1}(M) \to 0$$

and

$$0 \to \mathscr{J}_{p-1}(M) \to \mathscr{J}_p(M) \to \vee^p \tau_M \to 0.$$

If $\rho: \mathscr{J}_p(M) \to \vee^p \tau_M$ is the projection and $\sigma \in \operatorname{Sec} \mathscr{J}_p(M)$, then $\rho \circ \sigma$ is a cross-section in the bundle $\vee^p \tau_M$. It is called the *symbol* of the differential operator D.

(vii) Show that a smooth map $\varphi: M \to N$ induces bundle maps $\mathscr{J}_p(M) \to \mathscr{J}_p(N)$ which dualize to maps $\operatorname{Sec} \mathscr{J}^p(M) \leftarrow \operatorname{Sec} \mathscr{J}^p(N)$. Interpret these via $d\varphi$ and φ^* when $p = 1$. Express (for general p) these maps in terms of a coordinate system.

7. Jet bundles of vector bundles. Fix a vector bundle ξ over M.

(i) Carry out the constructions of problem 6, replacing $\mathscr{S}(M)$ by Sec ξ, to obtain a vector bundle $\mathscr{J}^p(\xi)$ whose typical fibre is $\sum_{j=0}^p \vee^j(\mathbb{R}^n)^* \otimes F$ (F, typical fibre of ξ). Show that each $\sigma \in \operatorname{Sec} \xi$ determines a cross-section $j^p(\sigma)$ in $\mathscr{J}^p(\xi)$, its *pth jet*.

(ii) If η is a second vector bundle over M with typical fibre H, a *pth order differential operator from* ξ *to* η is a map D: Sec $\xi \to$ Sec η of the form $D(\sigma) = \varphi(j^p\sigma)$ where $\varphi \in \operatorname{Sec} L(\mathscr{J}^p(\xi); \eta)$. Express D in local coordinates.

(iii) Construct canonical exact sequences of bundles

$$0 \to \vee^p \tau_M^* \otimes \xi \to \mathscr{J}^p(\xi) \to \mathscr{J}^{p-1}(\xi) \to 0$$

and

$$0 \to L(\mathscr{J}^{p-1}(\xi); \eta) \to L(\mathscr{J}^p(\xi); \eta) \xrightarrow{\rho} \vee^p \tau_M \otimes L(\xi; \eta) \to 0.$$

If $\varphi \in \operatorname{Sec} L(\mathscr{J}^p(\xi); \eta)$, its *symbol* is $\rho \circ \varphi$; $\rho\varphi(x)$ is a p-linear map from $T_x^*(M)$ to $L(F_x; H_x)$. If for each $x \in M$ and nonzero $h^* \in T_x^*(M)$ $\rho\varphi(x; h^*, \ldots, h^*)$ is an isomorphism, D is called elliptic.

(iv) Let D: Sec $\xi \to$ Sec η be a *pth* order elliptic differential operator. Let $\tilde{\xi}$, $\tilde{\eta}$ denote the pullback of ξ, η to the deleted bundle $\dot{\tau}_M^*$. Show that $\tilde{\xi}$ and $\tilde{\eta}$ are strongly isomorphic.

8. Distributions. A *distribution* on M is a subbundle ξ of τ_M. It is called *involutive*, if, whenever $X, Y \in \operatorname{Sec} \xi$, then $[X, Y] \in \operatorname{Sec} \xi$.

(i) Let ξ be a distribution on M with fibre F_x at x and let $X \in \mathscr{X}(M)$ have orbits $\varphi_t(x)$. Show that the conditions (a) $[X, Y] \in \operatorname{Sec} \xi$, if $Y \in \operatorname{Sec} \xi$ and (b) $d\varphi_t: F_x \to F_{\varphi(x)}$ for all x and for sufficiently small t, are equivalent.

(ii) Show that ξ is involutive if and only if for each point $a \in M$ there is a submanifold N_a of M containing a such that $T_x(N_a) = F_x$ (local Frobenius theorem).

(iii) Show that if ξ is involutive, then M is the disjoint union of maximal connected embedded manifolds N_α with $T_x(N_\alpha) = F_x$ (global Frobenius theorem). In particular, show that the N_α are second countable.

9. Second tangent bundle. Consider the tangent bundles

$$\tau_M = (T_M, \pi, M, \mathbb{R}^n) \qquad \text{and} \qquad \tau_M^2 = (T_M^2, \pi_T, T_M, \mathbb{R}^{2n}).$$

(i) Show that $\pi \circ d\pi = \pi \circ \pi_T$. If $\varphi: M \to N$ is smooth, show that $d(d\varphi)$ commutes with $d\pi$ and π_T.

(ii) Let U be open in $\mathbb{R}^n$. Write

$$T_U^2 = T(U \times \mathbb{R}^n) = T_U \times T_{\mathbb{R}^n} = (U \times \mathbb{R}^n) \times (\mathbb{R}^n \times \mathbb{R}^n).$$

Define an involution ω_U of U by

$$\omega_U(x; \xi, \eta, \zeta) = (x, \eta, \xi, \zeta).$$

Show that for $\varphi: U \to V$ (V open in $\mathbb{R}^n$)

$$d(d\varphi) \circ \omega_U = \omega_V \circ d(d\varphi).$$

Thus obtain a canonical involution ω_M of T_M^2 such that for $\psi: M \to N$,

$$d(d\psi) \circ \omega_M = \omega_N \circ d(d\varphi).$$

Show that $\pi_T \circ \omega_M = d\pi$. Is there an intrinsic definition of ω_M?

(iii) Let $X \in \mathscr{X}(M)$. Show that $\omega_M \circ dX: T_M \to T_M^2$ is a vector field on T_M. Relate its orbits to the orbits of X.

(iv) Let $j_x: T_x(M) \to T_M$ be the inclusion. Regard $(dj_x)_z$ as a linear injection $T_x(M) \to T_z(T_M)$ ($x = \pi z$). Show that $\operatorname{Im}(dj_x)_z = \ker(d\pi)_z$. Prove that these spaces are the fibres of a subbundle of τ_M^2. Denote its total space by V_{T_M}. If $X \in \mathscr{X}(M)$, show that $\hat{X}(z) = (dj_x)_z X(x)$ ($x = \pi z$) defines a vector field on T_M. Show that for $X, Y \in \mathscr{X}(M)$

$$\widehat{[X, Y]} \circ Y = dY \circ X - \omega_M \circ dX \circ Y.$$

Generalize as far as possible to the tangent bundle of the total manifold of any vector bundle.

10. Sprays. A *spray* on M is a vector field Z on τ_M such that $d\pi \circ Z = \iota$. A spray is called *affine*, if $(\mu_t)_* Z = (1/t)Z$ ($t \neq 0$), where μ_t is the diffeomorphism $\xi \mapsto t\xi$ of T_M.

(i) Show that M admits affine sprays.

(ii) Let Z be an affine spray with flow ψ. Show that for sufficiently small t, τ and for $\xi \in T_M$, $\psi(t, \tau\xi) = \tau\psi(t\tau, \xi)$.

(iii) For ξ sufficiently close to zero show that $\psi(1, \xi)$ is defined and set $\exp \xi = \pi\psi(1, \xi)$. Show that exp is a smooth map from a neighbourhood of $o(M)$ (o, the zero vector field) in T_M to M.

(iv) Let $(\exp)_x$ denote the restriction of exp to $T_x(M)$. Show that its derivative at zero is the identity map.

(v) If M is compact, show that exp can be defined in all of T_M. *Hint*: See Appendix A.

11. Measure zero. A subset $A \subset \mathbb{R}^n$ has *Lebesgue measure zero* if for every $\epsilon > 0$ there is a countable covering of A by Euclidean n-balls B_i such that $\sum_i \text{volume}(B_i) < \epsilon$.

(i) Show that a smooth map between open subsets of $\mathbb{R}^n$ preserves sets of measure zero.

(ii) Show that a countable union of sets of measure zero has again measure zero.

A subset A of a manifold M is said to have *measure zero*, if there is an atlas $\{(U_\alpha, u_\alpha)\}$ for M such that each set $u_\alpha(U_\alpha \cap A)$ has zero measure.

(iii) Show that this definition is independent of the choice of the atlas.

(iv) Show that the countable union of sets $A_i \subset M$ of measure zero has again measure zero.

(v) Show that a smooth map between n-manifolds preserves sets of measure zero.

12. Critical points. Let $\varphi: M \to N$ be a smooth map with $\dim M = m$, $\dim N = n$. We call $a \in M$ a *regular point*, if $(d\varphi)_a$ is surjective; otherwise a is called *critical*. The set of critical points is written Crit φ. A point $b \in N$ is called a *regular value*, if all points of $\varphi^{-1}(b)$ are regular or if $\varphi^{-1}(b)$ is empty; otherwise b is called a *critical value*. The set of critical values is written $CV(\varphi)$.

(i) If b is a regular value for φ show that $\varphi^{-1}(b)$ is a closed submanifold of M.

(ii) Let $Q \subset M$ be a submanifold of M and let ψ denote the restriction of φ to Q. Show that if $a \in Q$ is a critical point for φ, it is a critical point for ψ.

13. Sard's theorem. Sard's theorem asserts that, for a smooth map $\varphi: M \to N$, the set of critical values has measure zero.

First, let $\varphi: \mathbb{R}^m \to \mathbb{R}^n$ be a smooth map. Write $x = (x^1, \ldots, x^m)$ and $\varphi(x) = (\Phi^1(x), \ldots, \Phi^n(x))$. Assume that, for some p $(0 \leqslant p < n)$, $\Phi^i(x) = x^i$ $(i = 1, \ldots, p)$.

(i) Show that the conditions rank $\varphi'(x) = p$ and $(\partial\Phi^i/\partial x^j)(x) = 0$ $(i, j \geqslant p + 1)$, on $x \in \mathbb{R}^n$, are equivalent.

(ii) Set $W = \{x \in \mathbb{R}^m \mid \operatorname{rank} \varphi'(x) = p\}$. Write $W = U \cup V$, where U consists of those points $x \in W$ such that

$$\frac{\partial\Phi^j}{\partial x^{i_1} \cdots \partial x^{i_q}}(x) = 0 \qquad (i_1, ..., i_q, j \geqslant p + 1)$$

and V consists of the other points. Let $K \subset \mathbb{R}^m$ be compact and choose an integer r satisfying $r(n - p) > m - p$. Show that, for $x \in K \cap U$, $y \in K$

$$|\varphi x - \varphi y| \leqslant \alpha_K\{\sup_{i \leqslant p} |x^i - y^i| + \sup_{i \geqslant p+1} |x^i - y^i|^r\},$$

where α_K is a constant depending only on K.

(iii) Show that V is contained in the union of countable many $(m - 1)$-dimensional submanifolds of $\mathbb{R}^m$.

(iv) Given $\epsilon = 1/N$, divide each unit box of $\mathbb{R}^m$ into boxes whose first p diameters are ϵ^r and whose last $(m - p)$ diameters are ϵ. Conclude that $\varphi(U)$ has measure zero.

(v) Prove Sard's theorem by induction on m.

14. Let $\varphi: E \to F$ be a smooth map between vector spaces, where $\dim E = m$, $\dim F = n$. If $n \geqslant 2m$, show that for some $\psi \in L(E; F)$ arbitrarily close to zero, $\varphi + \psi$ is an immersion. *Hint*: Apply Sard's theorem to the maps

$$L(E; F; m) \times E \to L(E; F),$$

given by $(\chi, x) \mapsto \chi - \varphi'(x)$, where $L(E; F; m)$ is the manifold of linear maps $E \to F$ of rank m (cf. problem 14, Chap. I).

15. Let $\varphi: M \to N$ be a smooth map, where $\dim M = m$, $\dim N = n$.

(i) If $n \geqslant 2m$, show that φ is homotopic to an immersion.

(ii) If $n > 2m$ show that φ is homotopic to an embedding.

16. Prove *Whitney's embedding theorem*: Every n-manifold can be embedded into $\mathbb{R}^{2n+1}$ as a closed submanifold.

17. (i) Show that the map $\varphi: S^2 \to \mathbb{R}^4$ given by

$$\varphi(x_1, x_2, x_3) = (x_1^2 - x_2^2)\, e_1 + x_1x_2e_2 + x_1x_3e_3 + x_2x_3e_4$$
$$x = (x_1, x_2, x_3) \in S^2$$

induces an embedding of $\mathbb{R}P^2$ in $\mathbb{R}^4$ (e_1, e_2, e_3, e_4, a basis of $\mathbb{R}^4$).

(ii) Use the embedding in (i) to construct an immersion of $\mathbb{R}P^2$ in $\mathbb{R}^3$.

18. Morse functions. Let $f \in \mathscr{S}(M)$. $a \in M$ is called a *critical point* for f, if $\delta f(a) = 0$.

(i) Suppose a is a critical point for f. If $X, Y \in \mathscr{X}(M)$, show that $X(Yf)(a)$ depends only on $X(a)$ and $Y(a)$ and defines a symmetric bilinear function in $T_a(M)$, the *Hessian* of f at a. Phrase this in jet bundle terminology.

(ii) A critical point is called *nondegenerate*, if the Hessian of f at a is nondegenerate. Show that the nondegenerate critical points are isolated.

(iii) Given a nondegenerate critical point of f, construct a local coordinate system such that near a

$$f(x) = \sum_{i=1}^{p} x_i^2 - \sum_{j=p+1}^{n} x_j^2$$

(Morse lemma).

(iv) A function all of whose critical points are nondegenerate is called a *Morse function*. Given $g \in \mathscr{S}(M)$ and $\epsilon > 0$ construct a Morse function f such that

$$|f(x) - g(x)| < \epsilon, \qquad x \in M.$$

19. Normal bundle. Let $\varphi: N \to M$ be an immersion.

(i) Show that $(d\varphi)\, T_N$ is a subbundle of $T_M|_N$. The corresponding quotient bundle is called the *normal bundle of* N (with respect to φ).

(ii) Find the normal bundle of S^n in $\mathbb{R}^{n+1}$.

20. Tubular neighbourhoods. Let N be a closed submanifold of M. Construct a diffeomorphism φ from its normal bundle onto a neighbourhood U of N such that $\varphi(o_x) = x$, $x \in N$. U is called a tubular neighbourhood of N. *Hint*: Use the exponential map of problem 10 (iii).

21. Orientations. (i) Show that the product of two manifolds is orientable if and only if both manifolds are.

(ii) Let M be orientable. When is the diffeomorphism of $M \times M$ given by $(x, y) \mapsto (y, x)$ orientation preserving?

(iii) When is the diffeomorphism ω_M of T_M^2 defined in problem 9 (ii) orientation preserving?

(iv) Show that the equations

$$\varphi(u, v) = \left(\cos v + u \sin\left(\frac{v}{2}\right)\right) e_1 + \left(\sin v - u \cos\left(\frac{v}{2}\right)\right) e_2 + u e_3, \qquad -\tfrac{1}{2} < u < \tfrac{1}{2}, \quad v \in \mathbb{R},$$

define a nonorientable submanifold of $\mathbb{R}^3$.

22. Manifolds-with-boundary. A Euclidean *half space* H is the closed subset of a Euclidean space $\mathbb{R}^n$ given by $\langle x, a \rangle \geqslant 0$, where $a \in \mathbb{R}^n$ is a fixed nonzero vector. The $(n - 1)$-dimensional subspace of $\mathbb{R}^n$ given by $\langle x, a \rangle = 0$ is called the boundary of H. If O is an open subset of H and $\varphi: O \to H_1$ is a map of O into another half space, then φ is called smooth, if it extends to a smooth map $U \to H_1$, where U is an open subset of $\mathbb{R}^n$ containing O.

A *manifold-with-boundary* is a second countable Hausdorff space M which admits an open covering U_α with the following properties: (a) For each α there is a homeomorphism $u_\alpha : U_\alpha \to \hat{U}_\alpha$, where $\hat{U}_\alpha$ is an open subset of a half space. (b) The identification maps

$$u_\beta \circ u_\alpha^{-1} : u_\alpha(U_{\alpha\beta}) \to u_\beta(U_{\alpha\beta})$$

are diffeomorphisms. A map between manifolds-with-boundary is called *smooth* if it is locally smooth.

(i) With the same definition of tangent space as given in sec. 3.1 construct the tangent bundle of a manifold-with-boundary. Generalize the results of sec. 2, Chap. III, to this case.

(ii) Let $a \in M$ and let (U_α, u_α) be a chart such that $a \in U_\alpha$. Show that the property $u_\alpha(a) \in F$ (F, boundary of H) is independent of the choice of (U_α, u_α). The points a for which $u_\alpha(a) \in F$ are called the boundary points of M. Show that the set of boundary points of M (with the induced topology) is an $(n - 1)$-dimensional submanifold of M. It is called the boundary of M and is denoted by ∂M. The open subset $M - \partial M$ is called the interior of M.

(iii) Construct a cross-section σ of $\tau_M|_{\partial M}$ which "points out of M." Show that $\sigma \oplus \tau_{\partial M} = \tau_M|_{\partial M}$. If M is orientable, use σ to orient ∂M.

(iv) Show that ∂M has a neighbourhood in M diffeomorphic to $I \times \partial M$, where $I = \{t \in \mathbb{R} \mid 0 < t \leqslant 1\}$ (see part i).

23. Let M and N be compact manifolds-with-boundary.

(i) Let $A \subset \partial M$, $B \subset \partial N$ be unions of boundary components chosen so that there is a diffeomorphism $\varphi: A \xrightarrow{\cong} B$. Consider the disjoint union $M \cup N$ and divide out by the equivalence relation $x \sim \varphi(x)$, $x \in A$. Show that the quotient space is a quotient manifold of $M \cup N$.

(ii) Let M be a compact manifold-with-boundary. Set $N = M$, $A = B = \partial M$ and carry out the procedure of (i) to obtain a compact manifold (without boundary). It is called the *double* of M. Show that the diffeomorphism of $M \cup M$ interchange induces an involution ω of the double of M. If the double is orientable, does ω preserve or reverse orientations?

24. Suppose M and N are manifolds (without boundary). Delete open balls from M and N to construct manifolds-with-boundary. Identify the boundary spheres via a diffeomorphism to obtain a manifold (without boundary). This manifold is called the *connected* sum of M and N and is denoted by $M \# N$. Modify the construction if M and N have boundaries.

25. Let (E, π, B, F) be a Riemannian vector bundle.

(i) Show that the vectors of length $\leqslant 1$ form a manifold-with-boundary.

(ii) Show that the Hopf fibering (problem 10, Chap. I) is the unit sphere bundle of the canonical vector bundle over $\mathbb{C}P^n$ (problem 8, Chap. II). Use this to find a manifold which is not diffeomorphic to the ball and whose boundary is diffeomorphic to a sphere.

(iii) Let $\hat{M}$ be obtained from a $2n$-manifold M by deleting a ball $B(a)$ and replacing it by the manifold constructed in (ii). Construct a smooth map $\varphi: \hat{M} \to M$ such that

$$\varphi^{-1}(a) = \mathbb{C}P^{n-1} \quad \text{and} \quad \varphi: \hat{M} - \mathbb{C}P^{n-1} \to M - \{a\}$$

is a diffeomorphism.

Chapter IV

Calculus of Differential Forms

§1. The operators i, θ, δ

4.1. The substitution operator. Given a p-form Φ ($p \geqslant 1$) and a vector field X on a manifold M, we define a $(p-1)$-form $i(X)\Phi$ by

$$(i(X)\Phi)(X_1, \ldots, X_{p-1}) = \Phi(X, X_1, \ldots, X_{p-1}), \qquad X_i \in \mathscr{X}(M),$$

or, equivalently,

$$(i(X)\Phi)(x; \xi_1, \ldots, \xi_{p-1}) = \Phi(x; X(x), \xi_1, \ldots, \xi_{p-1}), \qquad x \in M, \quad \xi_i \in T_x(M).$$

(Observe that we are regarding $A^p(M)$ as the module of skew p-linear maps from $\mathscr{X}(M)$ to $\mathscr{S}(M)$.) The definition is extended to $A^0(M)$ by putting

$$i(X)f = 0, \qquad f \in \mathscr{S}(M).$$

For a one-form ω, we have

$$i(X)\omega = \langle \omega, X \rangle.$$

Thus in particular, for a gradient δf,

$$i(X)\,\delta f = X(f)$$

(cf. sec. 3.17).

The map $i(X)$: $A(M) \to A(M)$ defined in this way is called the *substitution operator* induced by X. It is homogeneous of degree -1, and satisfies

$$i(X)(f \cdot \Phi + g \cdot \Psi) = f \cdot i(X)\Phi + g \cdot i(X)\Psi$$

and

$$i(X)(\Phi \wedge \Psi) = i(X)\,\Phi \wedge \Psi + (-1)^p\,\Phi \wedge i(X)\Psi,$$

$$f, g \in \mathscr{S}(M), \quad \Phi \in A^p(M), \quad \Psi \in A(M).$$

Thus, for each $X \in \mathscr{X}(M)$, $i(X)$ is an antiderivation in the algebra $A(M)$.

If Y is a second vector field on M, we have

$$i(f \cdot X + g \cdot Y) = f \cdot i(X) + g \cdot i(Y)$$

and

$$i(X)\, i(Y) = -i(Y)\, i(X) \qquad (f, g \in \mathscr{S}(M)).$$

Lemma I: If $\Phi \in A^p(M)$ $(p \geqslant 1)$ satisfies $i(X)\Phi = 0$ for every $X \in \mathscr{X}(M)$, then $\Phi = 0$.

4.2. The Lie derivative. Fix a vector field $X \in \mathscr{X}(M)$. Given a p-form $\Phi \in A^p(M)$ $(p \geqslant 1)$ define a map

$$\underbrace{\mathscr{X}(M) \times \cdots \times \mathscr{X}(M)}_{p \text{ factors}} \to \mathscr{S}(M)$$

by

$$(X_1, ..., X_p) \mapsto X(\Phi(X_1, ..., X_p)) - \sum_{j=1}^{p} \Phi(X_1, ..., [X, X_j], ..., X_p).$$

This map is obviously skew-symmetric and p-linear over $\mathbb{R}$. Moreover, the relations

$$X(f \cdot g) = X(f) \cdot g + f \cdot X(g)$$

and

$$[X, f \cdot Y] = f \cdot [X, Y] + X(f) \cdot Y, \qquad f, g \in \mathscr{S}(M),$$

(cf. sec. 3.12) imply that it is p-linear over $\mathscr{S}(M)$. Thus it defines a p-form on M.

Definition: Let $X \in \mathscr{X}(M)$. Then the *Lie derivative* with respect to X is the real linear map $\theta(X)$: $A(M) \to A(M)$, homogeneous of degree zero, given by

$$(\theta(X)\Phi)(X_1, ..., X_p) = X(\Phi(X_1, ..., X_p)) - \sum_{j=1}^{p} \Phi(X_1, ..., [X, X_j], ..., X_p),$$

$$\Phi \in A^p(M), \quad p \geqslant 1, \quad X_j \in \mathscr{X}(M),$$

and

$$\theta(X)f = X(f), \qquad f \in \mathscr{S}(M).$$

Remark: If $\omega \in A^1(M)$ we have

$$\langle \theta(X)\,\omega, Y \rangle + \langle \omega, [X, Y] \rangle = X(\langle \omega, Y \rangle), \qquad Y \in \mathscr{X}(M).$$

Example: Let O be an open subset of a vector space E. As in the example of sec. 3.19, write

$$A(O) = \mathscr{S}(O; \wedge E^*).$$

Let $\Phi \in A^p(O)$ be given by $\Phi(x) = f(x) \cdot a$ where $a \in \wedge^p E^*$ and $f \in \mathscr{S}(O)$. Let X be a constant vector field on O. Then from Example 1 of sec. 3.12 it is easy to see that

$$(\theta(X)\Phi)(x) = (X(f))(x) \cdot a.$$

In particular, if X is given by $X(x) = (x, h)$ for some fixed $h \in E$, then

$$(\theta(X)\Phi)(x) = f'(x; h) \cdot a.$$

Proposition I: The Lie derivative has the following properties:

(1) $\theta(X)\,\delta f = \delta\theta(X) f = \delta X(f)$

(2) $i([X, Y]) = \theta(X)\, i(Y) - i(Y)\, \theta(X)$ $\qquad f \in \mathscr{S}(M)$

(3) $\theta(X)(\Phi \wedge \Psi) = \theta(X)\Phi \wedge \Psi + \Phi \wedge \theta(X)\Psi$ $\qquad X, Y \in \mathscr{X}(M)$

(4) $\theta([X, Y]) = \theta(X)\,\theta(Y) - \theta(Y)\,\theta(X)$ $\qquad \Phi, \Psi \in A(M)$.

(5) $\theta(f \cdot X) = f \cdot \theta(X) + \mu(\delta f)\, i(X)$

Here μ denotes the multiplication operator in $A(M)$,

$$\mu(\Phi)\Psi = \Phi \wedge \Psi.$$

Remark: Property (3) states that for every vector field X on M $\theta(X)$ is a derivation in $A(M)$. Property (4) shows that the map $\mathscr{X}(M) \to \operatorname{Der} A(M)$ given by $X \to \theta(X)$ is a homomorphism of Lie algebras.

Proof: (1) We have,

$$\begin{aligned}\langle \theta(X)\,\delta f, Y\rangle &= X(Y(f)) - [X, Y](f) \\ &= Y(X(f)) = \langle \delta(X(f)), Y\rangle, \qquad Y \in \mathscr{X}(M),\end{aligned}$$

whence the result.

(2) Clear.

(3) We may assume that $\Phi \in A^p(M)$, $\Psi \in A^q(M)$ and induct on $p + q$. If $p + q = 0$, then (3) reduces to the derivation property of X

on functions. If (3) is true for $p + q < k$, then, for $p + q = k$, $X, Y \in \mathscr{X}(M)$, (2) gives

$$\begin{aligned}
&i(Y)\,\theta(X)(\Phi \wedge \Psi) \\
&\quad = \theta(X)\,i(Y)(\Phi \wedge \Psi) - i([X, Y])(\Phi \wedge \Psi) \\
&\quad = \theta(X)[i(Y)\Phi \wedge \Psi + (-1)^p \Phi \wedge i(Y)\Psi] - i([X, Y])\Phi \wedge \Psi \\
&\qquad -(-1)^p \Phi \wedge i([X, Y])\Psi \\
&\quad = \theta(X)\,i(Y)\Phi \wedge \Psi + i(Y)\Phi \wedge \theta(X)\Psi + (-1)^p\,\theta(X)\Phi \wedge i(Y)\Psi \\
&\qquad +(-1)^p \Phi \wedge \theta(X)\,i(Y)\Psi - i([X, Y])\Phi \wedge \Psi - (-1)^p \Phi \wedge i([X, Y])\Psi,
\end{aligned}$$

the last equality following from the inductive hypothesis.

Now apply (2) and the antiderivation rule for $i(Y)$ to this relation and obtain

$$i(Y)\,\theta(X)(\Phi \wedge \Psi) = i(Y)[\theta(X)\Phi \wedge \Psi + \Phi \wedge \theta(X)\Psi], \qquad Y \in \mathscr{X}(M).$$

Thus Lemma I implies that

$$\theta(X)(\Phi \wedge \Psi) = \theta(X)\Phi \wedge \Psi + \Phi \wedge \theta(X)\Psi$$

and the induction is closed.

(4) Both sides of (4) are derivations in $A(M)$. Since $A(M)$ is generated (as an algebra over $\mathbb{R}$) by functions and gradicnts (cf. sec. 3.19) it is sufficient to show that the effect of both sides of (4) on functions and gradients is the same. But (4), applied to functions, is the definition of the Lie product, while (1) yields

$$\begin{aligned}
\theta([X, Y])\,\delta f &= \delta([X, Y]f) = \delta(X(Y(f)) - Y(X(f))) \\
&= [\theta(X)\,\theta(Y) - \theta(Y)\,\theta(X)](\delta f).
\end{aligned}$$

(5) Both sides of (5) are derivations in $A(M)$. But each side, applied to $g \in \mathscr{S}(M)$ yields $f \cdot X(g)$; and applied to δg, yields

$$\delta(f \cdot X(g)) = f \cdot \delta(X(g)) + \delta f \wedge X(g).$$

Q.E.D.

Definition: A differential form Φ is called *invariant with respect to* $X \in \mathscr{X}(M)$ if $\theta(X)\Phi = 0$. The set of forms invariant with respect to X is a subalgebra of $A(M)$ because $\theta(X)$ is a derivation. (Recall that $A(M)$ is considered as an algebra over $\mathbb{R}$.)

4.3. The exterior derivative. Let Φ be a p-form ($p \geqslant 1$) on a manifold M and consider the map

$$\underbrace{\mathscr{X}(M) \times \cdots \times \mathscr{X}(M)}_{p+1 \text{ factors}} \to \mathscr{S}(M)$$

given by

$$(X_0, ..., X_p) \mapsto \sum_{j=0}^{p} (-1)^j X_j(\Phi(X_0, ..., \hat{X}_j, ..., X_p))$$
$$+ \sum_{0 \leqslant i < j \leqslant p} (-1)^{i+j} \Phi([X_i, X_j], ..., \hat{X}_i, ..., \hat{X}_j, ..., X_p)$$

(the notation $\hat{X}_j$ means that the argument X_j is deleted). The relations

$$X(f \cdot g) = X(f) \cdot g + f \cdot X(g), \qquad f, g \in \mathscr{S}(M), \quad X, Y \in \mathscr{X}(M)$$

and

$$[X, f \cdot Y] = f \cdot [X, Y] + X(f) \cdot Y$$

imply that this map is $(p + 1)$-linear over $\mathscr{S}(M)$. Since it is obviously skew-symmetric, it determines a $(p + 1)$-form on M.

Definition: The *exterior derivative* is the $\mathbb{R}$-linear map δ: $A(M) \to A(M)$, homogeneous of degree 1, defined by

$$\delta\Phi(X_0, ..., X_p) = \sum_{j=0}^{p} (-1)^j X_j(\Phi(X_0, ..., \hat{X}_j, ..., X_p))$$
$$+ \sum_{0 \leqslant i < j \leqslant p} (-1)^{i+j} \Phi([X_i, X_j], ..., \hat{X}_i, ..., \hat{X}_j, ..., X_p)$$
$$\Phi \in A^p(M), \quad p \geqslant 1, \quad X_j \in \mathscr{X}(M), \tag{4.1}$$

and

$$(\delta f)(X) = X(f), \qquad f \in \mathscr{S}(M), \quad X \in \mathscr{X}(M).$$

The differential form $\delta\Phi$ is called the *exterior derivative* of Φ. Observe that δf is the gradient of f (cf. sec. 3.17).

Combining the definition of the exterior derivative with that of the Lie derivative (cf. sec. 4.2) we obtain a second expression for $\delta\Phi$,

$$\delta\Phi(X_0, ..., X_p) = \sum_{j=0}^{p} (-1)^j (\theta(X_j))\Phi(X_0, ..., \hat{X}_j, ..., X_p)$$
$$- \sum_{i<j} (-1)^{i+j} \Phi([X_i, X_j], X_0, ..., \hat{X}_i, ..., \hat{X}_j, ..., X_p). \tag{4.2}$$

In particular, for a one-form these equations read

$$\delta\omega(X, Y) = X(\langle\omega, Y\rangle) - Y(\langle\omega, X\rangle) - \langle\omega, [X, Y]\rangle, \qquad X, Y \in \mathscr{X}(M)$$

and

$$\delta\omega(X, Y) = i(Y)\,\theta(X)\omega - i(X)\,\theta(Y)\omega + i([X, Y])\omega.$$

Proposition II: The exterior derivative has the following properties:

(1) $\theta(X) = i(X)\,\delta + \delta\, i(X), \qquad X \in \mathscr{X}(M)$
(2) $\delta(\Phi \wedge \Psi) = \delta\Phi \wedge \Psi + (-1)^p \Phi \wedge \delta\Psi, \qquad \Phi \in A^p(M), \Psi \in A(M)$
(3) $\delta^2 = 0$
(4) $\delta\, \theta(X) = \theta(X)\, \delta.$

Remark: (2) states that δ is an antiderivation in $A(M)$.

Proof: (1) This is an immediate consequence of the definitions.

(2) This identity may be proved by induction in essentially the same way that property (3) of Proposition I was proved. We omit the details except to remark that property (1) plays the same role in this proof as property (2) of Proposition I did in the earlier proof.

(3) Since δ is an antiderivation, δ^2 is a derivation. Since $A(M)$, as an $\mathbb{R}$-algebra, is generated by functions and gradients, it is sufficient to show that

$$\delta^2 f = 0, \quad \delta^2(\delta f) = 0, \qquad f \in \mathscr{S}(M).$$

But

$$\begin{aligned}(\delta^2 f)(X, Y) &= X(\langle\delta f, Y\rangle) - Y(\langle\delta f, X\rangle) - \langle\delta f, [X, Y]\rangle \\ &= X(Y(f)) - Y(X(f)) - [X, Y]f = 0, \qquad X, Y \in \mathscr{X}(M);\end{aligned}$$

i.e. $\delta^2 f = 0$. It follows that $\delta^2(\delta f) = 0$.

(4) Apply δ to both sides of (1) and use (3).

Q.E.D.

Example: Let O be an open subset of a vector space E and recall from sec. 3.19 that a p-form on O can be regarded as a smooth map $O \to \wedge^p E^*$. For any smooth function $f \in \mathscr{S}(O)$, the gradient is given by

$$\langle \delta f(x), h\rangle = f'(x; h), \qquad x \in O, \quad h \in E.$$

More generally, for $\Phi \in A^p(O)$, we have

$$\langle \delta\Phi(x), h_0 \wedge \cdots \wedge h_p\rangle = \sum_{j=0}^{p} (-1)^j \langle \Phi'(x; h_j), h_0, \ldots, \hat{h}_j, \ldots, h_p\rangle.$$

If e_ν, $e^{*\nu}$ ($\nu = 1, ..., n$) is a pair of dual bases for E and E^* and $\Phi \in A^p(O)$ is given by

$$\Phi = \sum_{i_1 < \cdots < i_p} f_{i_1 \cdots i_p} e^{*i_1} \wedge \cdots \wedge e^{*i_p}, \qquad f_{i_1 \cdots i_p} \in \mathscr{S}(O),$$

then

$$(\delta\Phi)(x) = \sum_{j=1}^{n} \sum_{i_1 < \cdots < i_p} f'_{i_1 \cdots i_p}(x; e_j)\, e^{*j} \wedge e^{*i_1} \wedge \cdots \wedge e^{*i_p}.$$

4.4. Smooth maps. Proposition III: Suppose that $\varphi: M \to N$ is a smooth map and that $X \in \mathscr{X}(M)$, $Y \in \mathscr{X}(N)$ are φ-related. Then

(1) $\varphi^* \circ i(Y) = i(X) \circ \varphi^*$
(2) $\varphi^* \circ \theta(Y) = \theta(X) \circ \varphi^*$
(3) $\varphi^* \circ \delta = \delta \circ \varphi^*$.

Proof: Observe that all the operators in (1), (2), and (3) are φ^*-derivations or φ^*-antiderivations. Hence it is sufficient to show that both sides agree on functions and gradients. This is immediate from Lemma V, sec. 3.13, and Equation (3.5) of sec. 3.17.

Q.E.D.

4.5. Carriers. Let $\Phi \in A(M)$ be a differential form on M. Recall from sec. 2.13 that the carrier of Φ is the closure of the set

$$\{x \in M \mid \Phi(x) \neq 0\}.$$

It is denoted by carr Φ.

Definition: A differential form Φ on M is said to have *compact carrier*, if carr Φ is compact. The set of differential forms on M with compact carrier is denoted by $A_c(M)$.

Proposition IV: Let X and Y be vector fields on M and let $\Phi, \Psi \in A(M)$. Then

(1) $\operatorname{carr}(\Phi + \Psi) \subset \operatorname{carr} \Phi \cup \operatorname{carr} \Psi$
(2) $\operatorname{carr}(\Phi \wedge \Psi) \subset \operatorname{carr} \Phi \cap \operatorname{carr} \Psi$
(3) $\operatorname{carr}[X, Y] \subset \operatorname{carr} X \cap \operatorname{carr} Y$
(4) $\operatorname{carr} i(X)\Phi \subset \operatorname{carr} X \cap \operatorname{carr} \Phi$
$\operatorname{carr} \theta(X)\Phi \subset \operatorname{carr} X \cap \operatorname{carr} \Phi$
(5) $\operatorname{carr} \delta\Phi \subset \operatorname{carr} \Phi$.

Corollary: $A_c(M)$ is a graded ideal in $A(M)$ and is stable under the operators $i(X)$, $\theta(X)$ ($X \in \mathscr{X}(M)$), and δ.

4.6. Product manifolds. Let M, N be manifolds. For $X \in \mathscr{X}(M)$, $Y \in \mathscr{X}(N)$ let the vector fields $i_M X$, $i_N Y \in \mathscr{X}(M \times N)$ (cf. sec. 3.14) be denoted simply by X and Y. Then for $\Phi \in A^p(M)$, $\Psi \in A(N)$ the relations

$$i(X)(\Phi \times \Psi) = i(X)\Phi \times \Psi, \qquad i(Y)(\Phi \times \Psi) = (-1)^p \Phi \times i(Y)\Psi$$

$$\theta(X)(\Phi \times \Psi) = \theta(X)\Phi \times \Psi, \qquad \theta(Y)(\Phi \times \Psi) = \Phi \times \theta(Y)\Psi$$

and

$$\delta(\Phi \times \Psi) = \delta\Phi \times \Psi + (-1)^p \Phi \times \delta\Psi$$

(cf. Example 2, sec. 3.20) follow from the formulae of sec. 3.14, together with the antiderivation or derivation properties of $i(X)$, $\theta(X)$, δ.

Now consider a differential form $\Omega \in A^r(M \times N)$. Lemma VII, (1) and (2), sec. 3.14, implies that the map

$$\underbrace{\mathscr{X}(M \times N) \times \cdots \times \mathscr{X}(M \times N)}_{r+1 \text{ factors}} \to \mathscr{S}(M \times N)$$

given by

$$(Z_0, ..., Z_r) \mapsto \sum_{j=0}^{r} (-1)^j Z_j^M(\Omega(Z_0, ..., \hat{Z}_j, ..., Z_r))$$
$$+ \sum_{0 \leqslant i < j \leqslant r} (-1)^{i+j} \Omega((Z_i, Z_j)_M, Z_0, ..., \hat{Z}_i, ..., \hat{Z}_j, ..., Z_r)$$

is skew-symmetric and $(r + 1)$-linear over $\mathscr{S}(M \times N)$. Thus it defines an $(r + 1)$-form on $M \times N$.

Definition: *The partial exterior derivative with respect to* M is the linear map, homogeneous of degree 1, $\delta_M : A(M \times N) \to A(M \times N)$ given by

$$(\delta_M \Omega)(Z_0, ..., Z_r) = \sum_{j=0}^{r} (-1)^j Z_j^M(\Omega(Z_0, ..., \hat{Z}_j, ..., Z_r))$$
$$+ \sum_{0 \leqslant i < j \leqslant r} (-1)^{i+j} \Omega((Z_i, Z_j)_M, Z_0, ..., \hat{Z}_i, ..., \hat{Z}_j, ..., Z_r).$$

The partial exterior derivative with respect to N is given by

$$(\delta_N \Omega)(Z_0, ..., Z_r) = \sum_{j=0}^{r} (-1)^j Z_j^N(\Omega(Z_0, ..., \hat{Z}_j, ..., Z_r))$$
$$+ \sum_{0 \leqslant i < j \leqslant r} (-1)^{i+j} \Omega((Z_i, Z_j)_N, Z_0, ..., \hat{Z}_i, ..., \hat{Z}_j, ..., Z_r),$$
$$\Omega \in A^r(M \times N).$$

As an immediate consequence of the definition we have the formulae

$$(\delta_M\Omega)(X_0, \ldots, X_p, Y_1, \ldots, Y_q)$$

$$= \sum_{\nu=0}^{p} (-1)^\nu X_\nu(\Omega(X_0, \ldots, \hat{X}_\nu, \ldots, X_p, Y_1, \ldots, Y_q)$$

$$+ \sum_{0\leqslant\nu<\mu\leqslant p} (-1)^{\nu+\mu} \Omega([X_\nu, X_\mu], X_0, \ldots, \hat{X}_\nu, \ldots, \hat{X}_\mu, \ldots, X_p, Y_1, \ldots, Y_q) \tag{4.3}$$

and

$$(\delta_N\Omega)(X_1, \ldots, X_p, Y_0, \ldots, Y_q)$$

$$= (-1)^p \sum_{\nu=0}^{q} (-1)^\nu Y_\nu(\Omega(X_1, \ldots, X_p, Y_0, \ldots, \hat{Y}_\nu, \ldots, Y_q))$$

$$+ (-1)^p \sum_{0\leqslant\nu<\mu\leqslant q} (-1)^{\nu+\mu} \Omega(X_1, \ldots, X_p, [Y_\nu, Y_\mu], Y_0, \ldots, \hat{Y}_\nu, \ldots, \hat{Y}_\mu, \ldots, Y_q), \tag{4.4}$$

where $\Omega \in A^{p+q}(M \times N)$, $X_i \in \mathscr{X}(M)$, and $Y_j \in \mathscr{X}(N)$.

Proposition V: The partial exterior derivatives have the following properties:

(1) $\delta = \delta_M + \delta_N$, $\delta_M^2 = 0$, $\delta_N^2 = 0$, $\delta_M\delta_N + \delta_N\delta_M = 0$

(2) δ_M and δ_N are antiderivations in $A(M \times N)$, homogeneous of bidegrees (1, 0) and (0, 1), respectively.

(3) $i(X)\delta_M + \delta_M i(X) = \theta(X)$, $\quad X \in \mathscr{X}(M)$,
$i(X)\delta_N + \delta_N i(X) = 0$

(4) $i(Y)\delta_N + \delta_N i(Y) = \theta(Y)$, $\quad Y \in \mathscr{X}(N)$,
$i(Y)\delta_M + \delta_M i(Y) = 0$

(5) $\delta_M\theta(Z) = \theta(Z)\delta_M$, $\quad Z \in \mathscr{X}(M)$ or $\mathscr{X}(N)$,
$\delta_N\theta(Z) = \theta(Z)\delta_N$

(6) $\delta_M(\Phi \times \Psi) = \delta\Phi \times \Psi$, $\quad \Phi \in A^p(M)$, $\Psi \in A(N)$,
$\delta_N(\Phi \times \Psi) = (-1)^p\Phi \times \delta\Psi$.

Proof: Use Lemma VII, sec. 3.14, and sec. 4.3 together with elementary arguments on bidegrees.

Q.E.D.

4.7. Vector-valued differential forms. Differential forms generalize as follows: Let M be a manifold and let E be a finite dimensional vector space. Consider the bundle $L(\wedge^p \tau_M; M \times E)$ (cf. sec. 2.10) whose fibre at $x \in M$ consists of the p-linear skew-symmetric maps

$$T_x(M) \times \cdots \times T_x(M) \to E.$$

Definition: *A p-form on M with values in E* is a cross-section in the vector bundle $L(\wedge^p \tau_M; M \times E)$. In other words, an E-valued p-form, Ω, on M is a smooth assignment to the points of M of skew-symmetric p-linear maps

$$\Omega_x: T_x(M) \times \cdots \times T_x(M) \to E.$$

The E-valued p-forms on M form a module over $\mathscr{S}(M)$, which will be denoted by $A^p(M; E)$. The direct sum of the modules $A^p(M; E)$ is denoted by $A(M; E)$

$$A(M; E) = \sum_p A^p(M; E).$$

In particular, we have

$$A^p(M; \mathbb{R}) = A^p(M).$$

The following lemma is trivial:

Lemma II: An $\mathscr{S}(M)$-module isomorphism

$$A(M) \otimes E \to A(M; E)$$

is given by $\Phi \otimes a \mapsto \Omega$, where

$$\Omega(x; \xi_1, \ldots, \xi_p) = \Phi(x; \xi_1, \ldots, \xi_p) \cdot a, \qquad x \in M, \quad \xi_i \in T_x(M).$$

The operators $i(X)$, $\theta(X)$ $(X \in \mathscr{X}(M))$ and δ extend to operators $i(X) \otimes \iota_E$, $\theta(X) \otimes \iota_E$ and $\delta \otimes \iota_E$ in $A(M; E)$. We denote them also by $i(X)$, $\theta(X)$, and δ.

Proposition VI: In $A(M; E)$ the following relations hold:

(1) $i([X, Y]) = \theta(X)\, i(Y) - i(Y)\, \theta(X)$
(2) $\theta([X, Y]) = \theta(X)\, \theta(Y) - \theta(Y)\, \theta(X)$
(3) δ satisfies formulae (4.1) and (4.2)
(4) $\theta(X) = i(X)\, \delta + \delta\, i(X)$
(5) $\delta^2 = 0$
(6) $\delta\, \theta(X) = \theta(X)\, \delta$, $\qquad X, Y \in \mathscr{X}(M)$.

Proof: Apply Proposition I, sec. 4.2, and Proposition II, sec. 4.3. Q.E.D.

Next consider the $\mathscr{S}(M)$-bilinear map

$$A(M) \times A(M; E) \to A(M; E)$$

given by

$$(\Phi, \Psi \otimes a) \mapsto (\Phi \wedge \Psi) \otimes a, \qquad \Phi, \Psi \in A(M), \quad a \in E.$$

We shall write

$$(\Phi, \Omega) \mapsto \Phi \wedge \Omega, \qquad \Phi \in A(M), \quad \Omega \in A(M; E).$$

This map makes $A(M; E)$ into a graded module over the graded algebra $A(M)$. The following relations are straightforward consequences of Proposition I, sec. 4.2, and Proposition II, sec. 4.3:

$$i(X)(\Phi \wedge \Omega) = i(X)\Phi \wedge \Omega + (-1)^p \Phi \wedge i(X)\Omega$$

$$\theta(X)(\Phi \wedge \Omega) = \theta(X)\Phi \wedge \Omega + \Phi \wedge \theta(X)\Omega$$

$$\delta(\Phi \wedge \Omega) = \delta\Phi \wedge \Omega + (-1)^p \Phi \wedge \delta\Omega$$

$$X \in \mathscr{X}(M), \quad \Phi \in A^p(M), \quad \Omega \in A(M; E).$$

A smooth map $\varphi: M \to N$ induces an $\mathbb{R}$-linear map

$$\varphi^*: A(M; E) \leftarrow A(N; E)$$

given by

$$(\varphi^*\Omega)(x; \xi_1, \ldots, \xi_p) = \Omega(\varphi(x); d\varphi\xi_1, \ldots, d\varphi\xi_p),$$

$$\Omega \in A^p(N; E), \quad x \in M, \quad \xi_i \in T_x(M)$$

or, equivalently,

$$\varphi^*(\Phi \otimes a) = \varphi^*\Phi \otimes a, \qquad \Phi \in A(N), \quad a \in E.$$

Proposition III of sec. 4.4 generalizes in an obvious way.

Every linear map $\alpha: E \to F$ induces a map

$$\alpha_*: A(M; E) \to A(M; F)$$

given by

$$(\alpha_*\Omega)(x; \xi_1, \ldots, \xi_p) = \alpha(\Omega(x; \xi_1, \ldots, \xi_p)),$$

$$\Omega \in A^p(M; E), \quad x \in M, \quad \xi_i \in T_x(M).$$

Evidently

$$\alpha_* i(X) = i(X)\,\alpha_*, \qquad \alpha_*\theta(X) = \theta(X)\,\alpha_*, \qquad \text{and} \qquad \alpha_*\delta = \delta\alpha_*.$$

Proposition IV, sec. 4.5, generalizes to vector-valued forms. The set of E-valued forms with compact support is denoted by $A_c(M; E)$; it is a module over $A(M)$.

Finally, assume that E is a (not necessarily associative or commutative) algebra. Then the multiplication in E induces a multiplication in the space $A(M; E)$, the product being given by

$$(\Phi \cdot \Psi)(x; \xi_1, \dots, \xi_{p+q})$$
$$= \frac{1}{p!q!} \sum_\sigma \epsilon_\sigma \Phi(x; \xi_{\sigma(1)}, \dots, \xi_{\sigma(p)}) \cdot \Psi(x; \xi_{\sigma(p+1)}, \dots, \xi_{\sigma(p+q)})$$
$$\Phi \in A^p(M; E), \quad \Psi \in A^q(M; E), \quad x \in M, \ \xi_i \in T_x(M).$$

The algebra $A(M; E)$ so obtained is isomorphic to the algebra $A(M) \otimes E$. The following special cases are of particular importance:

(1) if E is commutative,

$$\Phi \cdot \Psi = (-1)^{pq} \Psi \cdot \Phi$$

(2) if E is skew-commutative,

$$\Phi \cdot \Psi = (-1)^{pq+1} \Psi \cdot \Phi$$

(3) if E is a Lie algebra

$$(-1)^{pq}(\Phi \cdot \Psi) \cdot X + (-1)^{rp}(X \cdot \Phi) \cdot \Psi + (-1)^{qr}(\Psi \cdot X) \cdot \Phi = 0$$
$$\Phi \in A^p(M; E), \quad \Psi \in A^q(M; E), \quad X \in A^r(M; E).$$

Relation (3), a consequence of the Jacobi identity, implies that

$$(\Phi \cdot \Phi) \cdot \Phi = 0, \qquad \Phi \in A^p(M; E).$$

§2. Smooth families of differential forms

4.8. Smooth families of cross-sections. Let $\xi = (E, \pi, M, F)$ be a vector bundle. Suppose that

$$\sigma \colon \mathbb{R} \to \operatorname{Sec} \xi$$

is a set map; i.e., σ assigns to every real number $t \in \mathbb{R}$ a cross-section σ_t of ξ. Such a map will be called a *smooth family of cross-sections*, if the map $\mathbb{R} \times M \to E$ (also denoted by σ) given by $\sigma(t, x) = \sigma_t(x)$ is smooth. The set of smooth families of cross-sections in ξ will be denoted by $\{\operatorname{Sec}_t \xi\}_{t\in\mathbb{R}}$.

Each such family determines, for each fixed $x \in M$, a smooth map $\sigma_x : \mathbb{R} \to F_x$ given by $\sigma_x(t) = \sigma(t, x)$.

Definition: Let σ be a smooth family of cross-sections in ξ. The *derivative* of σ is the smooth family $\dot{\sigma}$ given by

$$\dot{\sigma}(t, x) = \lim_{s\to 0} \frac{\sigma(t + s, x) - \sigma(t, x)}{s} = \frac{d}{ds}\sigma_x \bigg|_{s=t}.$$

The *integral* (from $a \in \mathbb{R}$) of σ is the smooth family $\int_a \sigma$ given by

$$\left(\int_a \sigma\right)(t, x) = \int_a^t \sigma_x(s)\, ds.$$

The *definite integral* $\int_a^b \sigma$ is the cross-section in ξ given by

$$\left(\int_a^b \sigma\right)(x) = \int_a^b \sigma_x(t)\, dt.$$

It is often written $\int_a^b \sigma_t\, dt$.

The fundamental theorem of calculus yields the relations

$$\int_a^b \dot{\sigma}_t\, dt = \sigma_b - \sigma_a, \qquad a, b \in \mathbb{R}$$

and (4.5)

$$\left(\int_a \sigma\right)^{\cdot} (t, x) = \sigma(t, x), \qquad t \in \mathbb{R}, \quad x \in M,$$

for a smooth family σ.

4.9. Smooth families of differential forms. *A smooth family of p-forms on a manifold M* is a smooth family of cross-sections in the vector bundle $\wedge^p \tau_M^*$. The set of smooth families of p-forms on M will be denoted by $\{A_t^p(M)\}_{t\in\mathbb{R}}$. Evidently, (cf. Example 2, sec. 3.20)

$$\{A_t^p(M)\}_{t\in\mathbb{R}} = A^{0,p}(\mathbb{R} \times M).$$

Thus a smooth family of p-forms on M is a differential form on $\mathbb{R} \times M$, homogeneous of bidegree $(0, p)$.

In particular, if $X \in \mathscr{X}(M)$ and if we consider X as a vector field on $\mathbb{R} \times M$ (cf. sec. 3.14 and 4.6) then for a smooth family Φ of p-forms on M,

$$i(X)\Phi \in A^{0,p-1}(\mathbb{R} \times M) = \{A_t^{p-1}(M)\}_{t\in\mathbb{R}},$$

$$\theta(X)\Phi \in A^{0,p}(\mathbb{R} \times M) = \{A_t^p(M)\}_{t\in\mathbb{R}},$$

and

$$\delta_M\Phi \in A^{0,p+1}(\mathbb{R} \times M) = \{A_t^{p+1}(M)\}_{t\in\mathbb{R}}$$

are again smooth families of differential forms on M.

Let $j_t : M \to \mathbb{R} \times M$ be the inclusion map: $j_t(x) = (t, x)$. If $\Phi \in A^{0,p}(\mathbb{R} \times M)$ is a smooth family of p-forms, then the p-forms Φ_t on M are given by

$$\Phi_t = j_t^*\Phi, \qquad t \in \mathbb{R}.$$

Thus the smooth family $i(X)\Phi$ $(X \in \mathscr{X}(M))$ is given by

$$(i(X)\Phi)_t = j_t^* i(X)\Phi = i(X) j_t^*\Phi = i(X)\,\Phi_t .$$

Similarly,

$$(\theta(X)\Phi)_t = \theta(X)\,\Phi_t \qquad \text{and} \qquad (\delta_M\Phi)_t = \delta\Phi_t .$$

Now consider a smooth map $\varphi: M \to N$. Then

$$(\iota \times \varphi)^*: A(\mathbb{R} \times M) \leftarrow A(\mathbb{R} \times N)$$

restricts to linear maps

$$(\iota \times \varphi)^*: A^{0,p}(\mathbb{R} \times M) \leftarrow A^{0,p}(\mathbb{R} \times N).$$

Thus φ induces a map $(\iota \times \varphi)^*$ of smooth families of p-forms. If $\Phi \in A^{0,p}(\mathbb{R} \times N)$ is a smooth family, then

$$((\iota \times \varphi)^*\Phi)_t = j_t^*(\iota \times \varphi)^*\Phi = \varphi^* j_t^*\Phi = \varphi^*\Phi_t .$$

Proposition VII: Let $\varphi: M \to N$ be a smooth map, and let Φ be a smooth family of p-forms on N. Then

$$((\iota \times \varphi)^*\Phi)^{\cdot} = (\iota \times \varphi)^*\dot{\Phi} \tag{4.6}$$

and

$$\int_a^b \varphi^*\Phi_t \, dt = \varphi^* \int_a^b \Phi_t \, dt. \tag{4.7}$$

Proof: Let $\Phi_y : \mathbb{R} \to \wedge^p T_y(N)^*$ be the smooth map given by

$$\Phi_y(t) = \Phi(t, y), \qquad t \in \mathbb{R}, \quad y \in N.$$

Then

$$((\iota \times \varphi)^*\Phi)_x = \wedge^p(d\varphi)_x^* \circ \Phi_{\varphi(x)} : \mathbb{R} \to \wedge^p T_x(M)^*, \qquad x \in M.$$

Since $\wedge^p(d\varphi)_x^*$ is a *linear* map, it follows that

$$\frac{d}{ds}[(\iota \times \varphi)^*\Phi]_x = \wedge^p(d\varphi)_x^* \circ \frac{d}{ds}\Phi_{\varphi(x)};$$

i.e.,

$$[(\iota \times \varphi)^*\Phi]_x^{\cdot} = \wedge^p(d\varphi)_x^* \circ \dot{\Phi}_{\varphi(x)} = [(\iota \times \varphi^*)]\dot{\Phi}_x .$$

This proves (4.6).

Formula (4.7) is proved in the same way.

Q.E.D.

Proposition VIII: Let Φ be a smooth family of p-forms on M. Then

(1) $i(X) \int_a^b \Phi_t \, dt = \int_a^b i(X)\Phi_t \, dt$

(2) $\theta(X) \int_a^b \Phi_t \, dt = \int_a^b \theta(X)\Phi_t \, dt, \qquad X \in \mathscr{X}(M),$

(3) $\delta \int_a^b \Phi_t \, dt = \int_a^b \delta\Phi_t \, dt.$

Proof: (1) is clear. Next we verify (3). Using an atlas on M, reduce to the case M is a vector space E. In this case

$$A^{0,p}(\mathbb{R} \times E) = \mathscr{S}(\mathbb{R} \times E; \wedge^p E^*)$$

(cf. the example of sec. 3.19). Since both sides of (3) are linear we may restrict to the case $\Phi(t, x) = f(t, x)a$ where $a \in \wedge^p E^*, f \in \mathscr{S}(\mathbb{R} \times E)$. In this case (3) is equivalent to (cf. the example of sec. 4.3)

$$\sum_{\nu=1}^{n} \left[\frac{\partial}{\partial e^\nu} \int_a^b f(t, x) \, dt\right] e^{*\nu} \wedge a = \sum_\nu \left[\int_a^b \frac{\partial f}{\partial e^\nu}(t, x) \, dt\right] e^{*\nu} \wedge a,$$

where $e^{*\nu}$, e_ν is a pair of dual bases for E^*, E and $\partial/\partial e_\nu$ denotes the partial derivative in the e_ν direction. But this is standard calculus.

Finally, (2) follows from (1), (3), and the relation

$$\theta(X) = i(X)\delta + \delta i(X).$$

Q.E.D.

4.10. The operator I_a^b. Let M be a manifold and let $\Omega \in A^p(\mathbb{R} \times M)$. Then Ω can be uniquely decomposed in the form

$$\Omega = \Phi + \Psi, \qquad \Phi \in A^{0,p}(\mathbb{R} \times M), \quad \Psi \in A^{1,p-1}(\mathbb{R} \times M).$$

Now consider the smooth family of p-forms Φ. It satisfies

$$\Phi_t = j_t^*\Phi = j_t^*\Omega.$$

We shall abuse notation, and denote this smooth family by $j^*\Omega$, $(j^*\Omega)_t = j_t^*\Omega$.

Integrating this family yields a differential form

$$I_a^b\Omega = \int_a^b (j_t^*\Omega)\, dt$$

on M. The assignment $\Omega \mapsto I_a^b\Omega$ defines a linear map

$$I_a^b : A(\mathbb{R} \times M) \to A(M)$$

homogeneous of degree zero.

Lemma III: Let T denote the vector field d/dt on $\mathbb{R}$; consider it as a vector field on $\mathbb{R} \times M$. Then

$$(j^*\Omega)_s^{\cdot} = j_s^*\theta(T)\Omega, \qquad \Omega \in A(\mathbb{R} \times M).$$

Proof: We may assume that M is a vector space E, and that $\Omega \in A^{0,p}(\mathbb{R} \times E)$ is of the form

$$\Omega(t, x) = f(t, x)a$$

for $f \in \mathscr{S}(\mathbb{R} \times E)$ and $a \in \wedge^p E^*$. Then

$$\begin{aligned}(j^*\Omega)_s^{\cdot}(x) &= f'(s, x; d/dt)a = (\theta(T)\Omega)(s, x)\\ &= (j_s^*\theta(T)\Omega)(x).\end{aligned}$$

Q.E.D.

Proposition IX: The operator I_a^b satisfies

(1) $I_a^b \circ \delta = \delta \circ I_a^b$
(2) $j_b^* - j_a^* = I_a^b \circ \theta(T) = \delta \circ I_a^b \circ i(T) + I_a^b \circ i(T) \circ \delta$.

Proof: (1) Apply Proposition VIII, sec. 4.9.

(2) Formula (4.5) and Lemma III yield

$$j_b^*\Omega - j_a^*\Omega = \int_a^b (j^*\Omega)_s^{\cdot}\, ds = I_a^b\theta(T)\Omega.$$

Now (2) follows from (1).

Q.E.D.

4.11. Orbits. Let X be a vector field on M which generates a one-parameter group of diffeomorphisms

$$\varphi\colon \mathbb{R} \times M \to M$$

(cf. sec. 3.15). Then, for $\Phi \in A(M)$, $\varphi^*\Phi$, $\varphi^*\theta(X)\Phi \in A(\mathbb{R} \times M)$.

The corresponding smooth families of differential forms on M are given by

$$(j^*\varphi^*\Phi)_t = \varphi_t^*\Phi \qquad \text{and} \qquad (j^*\varphi^*\theta(X)\Phi)_t = \varphi_t^*\,\theta(X)\Phi,$$

where $\varphi_t : M \to M$ is the map, $\varphi_t(x) = \varphi(t, x)$.

Proposition X: Let $\Phi \in A^p(M)$. Then the family $\varphi_t^*\Phi$ satisfies the relation:

$$\varphi_t^*\Phi - \Phi = \int_0^t (\varphi_s^*\theta(X)\Phi)\, ds.$$

In particular,

$$(\varphi_t^*\Phi)_0^{\cdot} = \theta(X)\Phi.$$

Proof: Observe first that $T \underset{\varphi}{\sim} X$. It follows that $\varphi^*\theta(X) = \theta(T)\varphi^*$. Hence

$$\int_0^t \varphi_s^*\theta(X)\Phi\, ds = \int_0^t j_s^*\theta(T)\,\varphi^*\Phi\, ds = I_0^t\theta(T)\,\varphi^*\Phi$$
$$= j_t^*\varphi^*\Phi - j_0^*\varphi^*\Phi = \varphi_t^*\Phi - \Phi.$$

Now (4.5) yields

$$\varphi_s^*\theta(X)\Phi = (\varphi_t^*\Phi - \Phi)_s^{\cdot} = (\varphi_t^*\Phi)_s^{\cdot},$$

whence

$$\theta(X)\Phi = \varphi_0^*\theta(X)\Phi = (\varphi_t^*\Phi)_0^{\cdot}\,.$$

This completes the proof.

Q.E.D.

Corollary: A differential form Φ is invariant with respect to X if and only if it satisfies

$$\varphi_t^*\Phi = \Phi, \qquad t \in \mathbb{R}$$

(cf. sec. 4.2).

§3. Integration of n-forms

4.12. Integration in a vector space. Let O be an open subset of an oriented n-dimensional vector space E. We shall define a linear map

$$\int_O : A^n_c(E) \to \mathbb{R}$$

which depends only on the orientation of E.

First, let Δ be a positive determinant function in E and let $e_1, ..., e_n$ be a basis of E such that $\Delta(e_1, ..., e_n) = 1$. Each $\Phi \in A^n_c(E)$ can be written $\Phi = f \cdot \Delta$ some $f \in \mathscr{S}_c(E)$. We *define* $\int_O$ by

$$\int_O \Phi = \int_O f(x)\, dx^1 \cdots dx^n,$$

where $x^1, ..., x^n$ are the coordinate functions associated with the basis $e_1, ..., e_n$.

Then $\int_O$ is a linear map; it has to be shown that it is independent of the choice of Δ and of the basis. But if $\bar{\Delta}$ is a second positive determinant function, and $\bar{\Delta}(\bar{e}_1, ..., \bar{e}_n) = 1$, we write $\Phi = \bar{f} \cdot \bar{\Delta}$ and note that

$$\bar{f}(x) = \Phi(x; \bar{e}_1, ..., \bar{e}_n) = f(x)\, \Delta(\bar{e}_1, ..., \bar{e}_n), \qquad x \in E.$$

Set $\bar{e}_i = \sum_j \alpha_i^j e_j$ and rewrite this relation as

$$\bar{f}(x) = \det(\alpha_i^j) f(x), \qquad x \in E.$$

Since $\Delta, \bar{\Delta}$ are positive, so are the bases $\{e_i\}, \{\bar{e}_i\}$; hence

$$\det(\alpha_i^j) > 0.$$

Now the transformation formula for Riemannian integrals yields

$$\int_O \bar{f}(x)\, d\bar{x}^1 \cdots d\bar{x}^n = \int_O f(x) \det(\alpha_i^j)\, d\bar{x}^1 \cdots d\bar{x}^n$$
$$= \int_O f(x)\, dx^1 \cdots dx^n.$$

Hence $\int_O$ depends only on the orientation of E.

Remark: An intrinsic definition of the integral is given in [9] and [12].

Lemma IV: If carr $\Phi \subset W$ (W an open set), then

$$\int_O \Phi = \int_{O \cap W} \Phi, \qquad \Phi \in A_c^n(E).$$

Lemma V: Let $U \subset F$, $V \subset E$ be open subsets of oriented n-dimensional vector spaces. Suppose $\varphi\colon U \xrightarrow{\cong} V$ is a diffeomorphism, either orientation preserving or reversing. Then, for $\Phi \in A_c^n(V) \subset A_c^n(E)$, $\varphi^*\Phi \in A_c^n(U)$, and

$$\int_O \varphi^*\Phi = \epsilon \int_{\varphi(O)} \Phi$$

(O an open subset of U). Here $\epsilon = +1$ (resp. -1) if φ is orientation preserving (resp. reversing).

Proof: This is a straight translation of the "transformation of variables" law for Riemannian integrals.

Q.E.D.

4.13. Integration in manifolds. Let M be an oriented n-manifold and let O be an open subset of M. We shall define a linear map

$$\int_O : A_c^n(M) \to \mathbb{R}.$$

Let $\Phi \in A_c^n(M)$ and suppose first that carr $\Phi \subset U$ for some chart $(U, u, \mathbb{R}^n)$ on M. Give $\mathbb{R}^n$ the orientation induced from M via u, and set

$$\int_O \Phi = \int_{u(O \cap U)} (u^{-1})^*\Phi. \tag{4.8}$$

If $(V, v, \mathbb{R}^n)$ is a second such chart (with $\mathbb{R}^n$ given the orientation induced from V), set $W = U \cap V$. Then Lemmas IV and V of sec. 4.12 give

$$\int_{u(O\cap U)} (u^{-1})^*\Phi = \int_{u(O\cap W)} (u^{-1})^*\Phi = \int_{v(O\cap W)} (u \circ v^{-1})^*(u^{-1})^*\Phi$$
$$= \int_{v(O\cap V)} (v^{-1})^*\Phi.$$

Thus the definition (4.8) is independent of the choice of $(U, u, \mathbb{R}^n)$.

Finally, let $\Phi \in A_c^n(M)$ be arbitrary. Let $(U_i, u_i, \mathbb{R}^n)$ $(i = 1, \ldots, r)$ be a family of charts such that

$$\text{carr}\, \Phi \subset \bigcup_1^r U_i\,.$$

Set $U_0 = M - \operatorname{carr} \Phi$ and let $\{p_i\}_{i=0,\dots,r}$ be a partition of unity subordinate to the open cover $\{U_i\}$. Then

$$\Phi = \sum_{i=1}^{r} p_i \cdot \Phi \qquad \text{and} \qquad p_i \cdot \Phi \in A_c^n(U_i).$$

We *define*

$$\int_O \Phi = \sum_{i=1}^{r} \int_O p_i \cdot \Phi$$

where $\int_O p_i \cdot \Phi$ is given by (4.8) ($i = 1, \dots, r$).

It has to be shown that this definition is independent of the choice of the U_i, u_i, and p_i. Let $(V_j, v_j, \mathbb{R}^n)_{j=1,\dots,s}$ be a second family of charts on M such that $\operatorname{carr} \Phi \subset \bigcup_j V_j$. Set $V_0 = M - \operatorname{carr} \Phi$ and let $\{q_j\}_{j=0,\dots,s}$ be a partition of unity subordinate to the open cover $\{V_j\}$. We must prove that

$$\sum_{i=1}^{r} \int_O p_i \cdot \Phi = \sum_{j=1}^{s} \int_O q_j \cdot \Phi.$$

Since $q_0 \cdot \Phi = 0$, $p_0 \cdot \Phi = 0$, we have

$$p_i \cdot \Phi = \sum_{j=1}^{s} q_j p_i \cdot \Phi, \qquad i = 1, \dots, r,$$

whence

$$\sum_{i=1}^{r} \int_O p_i \cdot \Phi = \sum_{i=1}^{r} \sum_{j=1}^{s} \int_O q_j p_i \cdot \Phi = \sum_{j=1}^{s} \int_O q_j \cdot \Phi.$$

The elementary properties of the integral of an n-form are listed in the following

Proposition XI: Let M be an oriented n-manifold and $\Phi \in A_c^n(M)$. Then

(1) $\int_\emptyset \Phi = 0$.

(2) If $\{x \in U \mid \Phi(x) \neq 0\} = \{x \in V \mid \Phi(x) \neq 0\}$, where U, V are open subsets of M, then

$$\int_U \Phi = \int_V \Phi.$$

In particular, if $\Phi|_U = 0$, then $\int_U \Phi = 0$.

(3) If U, V are disjoint open subsets of M, then

$$\int_{U \cup V} \Phi = \int_U \Phi + \int_V \Phi.$$

(4) Let $\Phi = f \cdot \Delta$, where $f \in \mathscr{S}_c(M)$ and Δ represents the orientation. Assume $f(x) \geqslant 0$, $x \in M$, (Φ is called *nonnegative*), and $\Phi \neq 0$. Then

$$\int_M \Phi > 0.$$

Proof: With the aid of a suitable partition of unity, it is easy to reduce to the case $M = \mathbb{R}^n$. In this case these properties restate standard properties of the Riemannian integral (of functions).

Q.E.D.

Proposition XII: Let M, N be oriented n-manifolds and let $\varphi: M \to N$ be a diffeomorphism (either orientation preserving or reversing). Then

$$\int_O \varphi^*\Phi = \epsilon \cdot \int_{\varphi(O)} \Phi, \qquad \Phi \in A_c^n(N), \quad O \text{ open}, \quad O \subset M,$$

where $\epsilon = 1$ (resp. -1), if φ is orientation preserving (resp. reversing).

Proof: Write $\Phi = \sum_{i=1}^r \Phi_i$ with $\operatorname{carr} \Phi_i \subset U_i$ where U_i is diffeomorphic to $\mathbb{R}^n$. It is clearly sufficient to prove the proposition for each Φ_i ; i.e., we may assume $\operatorname{carr} \Phi \subset U$, $U \cong \mathbb{R}^n$. Since φ is a diffeomorphism, the proposition follows from Lemma V, sec. 4.12.

Q.E.D.

Proposition XIII (Fubini): Let M, N be oriented m- and n-manifolds and give $M \times N$ the product orientation. Then

$$\int_{M\times N} \Phi \times \Psi = \int_M \Phi \cdot \int_N \Psi, \qquad \Phi \in A_c^m(M), \quad \Psi \in A_c^n(N).$$

Remark: $\operatorname{carr}(\Phi \times \Psi) \subset \operatorname{carr} \Phi \times \operatorname{carr} \Psi$ is compact.

Proof: Use partitions of unity to reduce to the case that M and N are vector spaces. But in this case the proposition is a restatement of the formula

$$\int_{M\times N} f(x)\, g(y)\, dx^1 \cdots dx^m\, dy^1 \cdots dy^n$$
$$= \left(\int_M f(x)\, dx^1 \cdots dx^m\right)\left(\int_N g(y)\, dy^1 \cdots dy^n\right),$$

for Riemannian integrals of functions f, g with compact support.

Q.E.D.

Proposition XIV: Let M be an oriented n-manifold and assume $\Phi \in A_c^{n-1}(M)$. Then

$$\int_M \delta\Phi = 0.$$

Proof: Choose finitely many charts $(U_i, u_i, \mathbb{R}^n)$ $(i = 1, ..., r)$ so that the U_i cover carr Φ. Use a partition of unity to write $\Phi = \Phi_1 + \cdots + \Phi_r$ with $\Phi_i \in A_c^{n-1}(M)$ having carrier in U_i. It is sufficient to prove that

$$\int_M \delta\Phi_i = \int_{U_i} \delta\Phi_i = 0, \qquad i = 1, ..., r;$$

we are thus reduced to the case $M = \mathbb{R}^n$.

Choose a positive basis $e_1, ..., e_n$ of $\mathbb{R}^n$ with dual basis $e^{*1}, ..., e^{*n}$. Then $\Phi \in A_c^{n-1}(\mathbb{R}^n)$ is a sum of terms of the form

$$f_i \cdot e^{*1} \wedge \cdots \widehat{e^{*i}} \cdots \wedge e^{*n}, \qquad f_i \in \mathscr{S}_c(\mathbb{R}^n).$$

Hence it is sufficient to consider the case

$$\Phi = f \cdot e^{*2} \wedge \cdots \wedge e^{*n}, \qquad f \in \mathscr{S}_c(\mathbb{R}^n).$$

But then (cf. the example, sec. 4.3)

$$(\delta\Phi)(x) = f'(x; e_1)\, e^{*1} \wedge \cdots \wedge e^{*n}.$$

Since $e^{*1} \wedge \cdots \wedge e^{*n}$ is a determinant function in $\mathbb{R}^n$, and $\langle e^{*1} \wedge \cdots \wedge e^{*n}, e_1 \wedge ... \wedge e_n \rangle = 1$, we obtain

$$\int_{\mathbb{R}^n} \delta\Phi = \int_{\mathbb{R}^n} f'(x; e_1)\, dx^1 \cdots dx^n = \int_{\mathbb{R}^{n-1}} \left[\int_{-\infty}^{\infty} \frac{\partial f}{\partial e_1} dx^1\right] dx^2 \cdots dx^n$$
$$= 0.$$

Q.E.D.

4.14. Vector-valued forms. Let M be an oriented n-manifold and E be a finite-dimensional vector space. Recall the definition of the E-valued differential forms on M, $A(M; E)$, and the relation

$$A(M; E) = A(M) \otimes E$$

(cf. sec. 4.7). The space of E-valued forms with compact carrier is denoted $A_c(M; E)$. Evidently

$$A_c(M; E) = A_c(M) \otimes E.$$

Now let $\Omega \in A_c^n(M; E)$ and write

$$\Omega = \sum_{i=1}^{r} \Phi_i \otimes a_i$$

where $a_1, \ldots, a_r$ is a basis of E, and $\Phi_i \in A_c^n(M)$. It is easy to see that the vector in E, given by

$$\sum_{i=1}^{r} \left(\int_O \Phi_i\right) a_i$$

(O an open subset of M), is independent of the choice of basis $\{a_i\}$. We define *the integral of* Ω to be this vector,

$$\int_O \Omega = \sum_{i=1}^{r} \left(\int_O \Phi_i\right) a_i .$$

Let $\alpha: E \to F$ be a linear map of finite-dimensional vector spaces. $\alpha_*: A(M; E) \to A(M; F)$ restricts to a linear map

$$\alpha_*: A_c(M; E) \to A_c(M; F)$$

(cf. sec. 4.7). Evidently

$$\int_O \alpha_* \Phi = \alpha \left(\int_O \Phi\right), \qquad \Phi \in A_c^n(M; E). \tag{4.9}$$

Finally, observe that Propositions XII and XIV continue to hold for vector-valued forms.

4.15. Forms with noncompact carrier. Let U be an open subset of an oriented n-manifold M. Let $\Phi \in A^n(M)$ satisfy $\overline{\text{carr}\, \Phi \cap U} = K$ is compact. Choose $f \in \mathscr{S}_c(M)$ with $f = 1$ in K. Then $\Psi = f \cdot \Phi \in A_c^n(M)$ and satisfies

$$\Psi(x) = \Phi(x), \qquad x \in U. \tag{4.10}$$

Moreover, if $X \in A_c^n(M)$ also satisfies this equation, then it follows from Proposition XI, part 2, applied to $X - \Psi$, that

$$\int_U X = \int_U \Psi.$$

Thus we can define the *integral of* Φ *over* U by

$$\int_U \Phi = \int_U \Psi,$$

where $\Psi \in A_c^n(M)$ satisfies (4.10).

In particular, if $\bar{U}$ is compact *every* n-form on M can be integrated over U.

Examples: 1. Let S^1 be the unit circle in the complex plane $\mathbb{C}$,

$$S^1 = \{z \in \mathbb{C} \mid |z| = 1\}.$$

A determinant function Δ in $\mathbb{C}$ is given by

$$\Delta(z_1, z_2) = \operatorname{Im}(\bar{z}_1 \cdot z_2), \qquad z_i \in \mathbb{C}.$$

Hence a one-form $\Omega \in A^1(S^1)$ is given by

$$\Omega(z; \eta) = \operatorname{Im}(\bar{z} \cdot \eta) = \operatorname{Im}(\eta/z), \qquad z \in S^1, \quad \eta \in T_z(S^1),$$

and Ω orients S^1.

Now consider the smooth map $\varphi: \mathbb{R} \to S^1$ given by $\varphi(t) = \exp(2\pi i t)$. Then

$$\varphi^*\Omega(s; d/dt) = 2\pi \operatorname{Im}(i\, e^{-2\pi i s + 2\pi i s}) = 2\pi.$$

In particular, φ is orientation preserving.

Clearly φ restricts to a diffeomorphism

$$\varphi: (0, 1) \xrightarrow{\cong} S^1 - \{1\}.$$

Moreover, if $f \in \mathscr{S}(S^1)$, then

$$\int_{S^1} f \cdot \Omega = \int_{S^1 - \{1\}} f \cdot \Omega.$$

Hence

$$\int_{S^1} f \cdot \Omega = \int_{(0,1)} \varphi^* f \cdot \varphi^*\Omega = 2\pi \int_0^1 (\varphi^* f)(t)\, dt;$$

i.e.,

$$\int_0^1 (\varphi^* f)(t)\, dt = \frac{1}{2\pi} \int_{S^1} f \cdot \Omega.$$

In particular, $\int_{S^1} \Omega = 2\pi$.

2. Let E be an oriented $(n + 1)$-dimensional Euclidean space and let S^n denote the unit sphere in E. Denote by Δ_E the positive normed determinant function in E. Orient S^n by the form $\Omega \in A^n(S^n)$ given by

$$\Omega(y; \eta_1, \ldots, \eta_n) = \Delta_E(y, \eta_1, \ldots, \eta_n), \qquad y \in S^n, \quad \eta_i \in T_y(S^n)$$

(cf. Example 2, sec. 3.21). We shall compute $\int_{S^n} \Omega$. Choose

$$0 < a < b < \infty.$$

Let $\mathscr{I} = (a, b)$ and let $A \subset E$ be the annulus

$$A = \{x \mid a < |x| < b\}.$$

The orientation preserving diffeomorphism $\alpha: \mathbb{R}^+ \times S^n \to \dot{E}$ given by

$$(t, x) \mapsto tx$$

restricts to an orientation preserving diffeomorphism

$$\mathscr{I} \times S^n \xrightarrow{\cong} A$$

(cf. Example 3, sec. 3.22). A simple computation shows that

$$\alpha^* \Delta_E = (t^n \cdot \delta t) \times \Omega$$

(t: $\mathbb{R} \to \mathbb{R}$ is the identity map).

Next consider the $(n+1)$-form Φ in E given by

$$\Phi(x) = e^{-\langle x,x \rangle} \cdot \Delta_E, \qquad x \in E.$$

Then

$$(\alpha^*\Phi)(t, y) = (e^{-t^2} t^n \cdot \delta t) \times \Omega(y).$$

Since $\bar{A}$ and $\bar{\mathscr{I}}$ are compact, any $(n+1)$-form defined in $\dot{E}$ (resp. in $\mathbb{R}^+ \times S^n$) can be integrated over A (resp. over $\mathscr{I}$). Thus Proposition XII and Proposition XIII, sec. 4.13, give

$$\begin{aligned}\int_A \Phi = \int_{\mathscr{I} \times S^n} \alpha^*\Phi &= \int_{\mathscr{I}} e^{-t^2} t^n \, \delta t \cdot \int_{S^n} \Omega \\ &= \int_a^b e^{-t^2} t^n \, dt \cdot \int_{S^n} \Omega.\end{aligned}$$

On the other hand,

$$\int_A \Phi = \int_A e^{-\langle x,x \rangle} \, dx_1 \cdots dx_{n+1},$$

where $x_1, \ldots, x_{n+1}$ are the coordinate functions corresponding to an orthonormal basis. Taking limits as $a \to 0$, $b \to \infty$ gives

$$\begin{aligned}\int_0^\infty e^{-t^2} t^n \, dt \cdot \int_{S^n} \Omega = \int_E e^{-(x_1^2 + \cdots + x_{n+1}^2)} \, dx_1 \cdots dx_{n+1} \\ = \left(\int_{-\infty}^{\infty} e^{-x^2} \, dx \right)^{n+1}\end{aligned}$$

It follows that

$$\int_{S^n} \Omega = \begin{cases} \dfrac{2^{m+1}}{1 \cdot 3 \cdots (2m-1)} \pi^m, & n = 2m, \quad m \geq 1 \\ \dfrac{2}{m!} \pi^{m+1}, & n = 2m+1, \quad m \geq 0. \end{cases}$$

§4. Stokes' theorem

4.16. The annulus. Let M be an oriented n-manifold. A graded δ-stable ideal $A_M(\mathbb{R} \times M) \subset A(\mathbb{R} \times M)$ is defined as follows: $\Phi \in A_M(\mathbb{R} \times M)$ if, for all closed, finite intervals K

$$\operatorname{carr} \Phi \cap (K \times M)$$

is compact. Next, let $\mathbb{R}$ be oriented by the one-form δt (cf. Example 7, sec. 3.21) and give $\mathbb{R} \times M$ the product orientation. Let $\mathscr{I}$ denote a finite open interval $(a, b) \subset \mathbb{R}$ and let $j_a, j_b\colon M \to \mathbb{R} \times M$ be the inclusions opposite a and b.

Then, for $\Omega \in A_M^{n+1}(\mathbb{R} \times M)$, $\overline{\operatorname{carr} \Omega \cap (\mathscr{I} \times M)}$ is compact. Thus (cf. sec. 4.15) we can form the integral

$$\int_{\mathscr{I}\times M} \Omega.$$

In this section we prove

Theorem I (Stokes): Let M be an oriented n-manifold. Then, for $\Phi \in A_M^n(\mathbb{R} \times M)$,

$$\int_{\mathscr{I}\times M} \delta\Phi = \int_M j_b^*\Phi - \int_M j_a^*\Phi.$$

Remark: Since $\Phi \in A_M^n(\mathbb{R} \times M)$, $j_b^*\Phi$ and $j_a^*\Phi \in A_c^n(M)$.

Proof: First, consider the vector field T on $\mathbb{R} \times M$ given by

$$T(s, x) = (d/dt, 0), \qquad s \in \mathbb{R}, \quad x \in M.$$

T determines an operator (cf. sec. 4.10)

$$I_a^b \circ i(T) : A^p(\mathbb{R} \times M) \to A^{p-1}(M),$$

which clearly restricts to an operator

$$I_a^b \circ i(T) : A_M^p(\mathbb{R} \times M) \to A_c^{p-1}(M).$$

Lemma VI:

$$\int_{\mathscr{I}\times M} \Omega = \int_M I_a^b i(T)\Omega, \qquad \Omega \in A_M^{n+1}(\mathbb{R} \times M).$$

Proof of Lemma VI: Use a finite partition of unity in M, to reduce to the case $M = \mathbb{R}^n$ and

$$\operatorname{carr} \Omega \subset \mathbb{R} \times L$$

where L is a compact subset of $\mathbb{R}^n$.

Let $e_1, ..., e_n$ be a positive basis of $\mathbb{R}^n$. Then $\delta t \wedge e^{*1} \wedge \cdots \wedge e^{*n}$ is a positive $(n + 1)$-form in $\mathbb{R} \times \mathbb{R}^n$. Write

$$\Omega = f \cdot \delta t \wedge e^{*1} \wedge \cdots \wedge e^{*n}, \qquad f \in \mathscr{S}(\mathbb{R}^{n+1}), \quad \operatorname{carr} f \subset \mathbb{R} \times L.$$

Then $i(T)\Omega = f \cdot e^{*1} \wedge \cdots \wedge e^{*n}$.

It follows that

$$(I_a^b i(T)\Omega)(x) = \left(\int_a^b f(t, x)\, dt\right) e^{*1} \wedge \cdots \wedge e^{*n}$$

and so

$$\int_{\mathbb{R}^n} I_a^b i(T)\Omega = \int_{\mathbb{R}^n} \int_a^b f(t, x)\, dt\, dx^1 \cdots dx^n = \int_{\mathscr{I}\times\mathbb{R}^n} \Omega.$$

Q.E.D.

We return to the proof of Theorem I. Lemma VI yields

$$\int_{\mathscr{I}\times M} \delta\Phi = \int_M (I_a^b \circ i(T))\, \delta\Phi.$$

According to Proposition IX, sec. 4.10,

$$I_a^b i(T)\, \delta\Phi = j_b^*\Phi - j_a^*\Phi - \delta I_a^b i(T)\Phi.$$

Since $\Phi \in A_M^n(\mathbb{R} \times M)$, $I_a^b i(T)\Phi \in A_c^{n-1}(M)$. Thus Proposition XIV, sec. 4.13, implies that

$$\int_M \delta I_a^b i(T)\Phi = 0.$$

Hence

$$\int_M I_a^b i(T)\, \delta\Phi = \int_M j_b^*\Phi - \int_M j_a^*\Phi.$$

Q.E.D.

4.17. Stokes' theorem for the ball. Let B be the open unit ball in an oriented Euclidean $(n + 1)$-space E, and let S be the unit n-sphere with the induced orientation (cf. Example 2, sec. 3.21). Let $i: S \to E$ be the inclusion.

Theorem II: Let $\Phi \in A^n(U)$ where U is a neighbourhood of the closed unit-ball $\bar{B}$. Then

$$\int_B \delta\Phi = \int_S i^*\Phi \tag{4.11}$$

Proof: Let p be a smooth function in E such that

$$p(x) = 1, \quad |x| \leqslant 1 \qquad \text{and} \qquad \operatorname{carr} p \subset U.$$

Then neither side of (4.11) is changed if we replace Φ by $p \cdot \Phi$. But $p \cdot \Phi \in A^n_c(E)$ and thus we may assume that $\Phi \in A^n_c(E)$.

Next let q be a smooth function in E such that

$$q(x) = 1, \quad |x| \leqslant \tfrac{1}{4}; \qquad q(x) = 0, \quad |x| \geqslant \tfrac{1}{2}.$$

Then

$$i^*(1 - q)\Phi = i^*\Phi.$$

Moreover, since $q \cdot \Phi \in A^n_c(B)$, Proposition XIV, sec. 4.13 gives

$$\int_B \delta\Phi = \int_B \delta[(1 - q) \cdot \Phi] + \int_B \delta[q \cdot \Phi] = \int_B \delta[(1 - q) \cdot \Phi].$$

Thus both sides of (4.11) are unchanged if we replace Φ by $(1 - q) \cdot \Phi$; in other words, it is sufficient to consider the case $\Phi(x) = 0$, $|x| \leqslant \frac{1}{4}$. Then we have

$$\int_B \delta\Phi = \int_A \delta\Phi, \qquad A = \{x \mid \tfrac{1}{4} < |x| < 1\}.$$

Next consider the diffeomorphism $\alpha: \mathbb{R}^+ \times S \to E - \{0\}$ given by $\alpha(t, x) = tx$ ($t \in \mathbb{R}^+$, $x \in S$). Then α preserves orientations (cf. Example 3, sec. 3.22). Hence, setting $\mathscr{I} = (\frac{1}{4}, 1)$, we find

$$\int_B \delta\Phi = \int_A \delta\Phi = \int_{\mathscr{I}\times S} \alpha^*\delta\Phi = \int_{\mathscr{I}\times S} \delta(\alpha^*\Phi).$$

Applying Theorem I of the preceding section, we obtain

$$\int_B \delta\Phi = \int_S j_1^*\alpha^*\Phi - \int_S j_{1/4}^*\alpha^*\Phi = \int_S i^*\Phi,$$

because $i = \alpha \circ j_1$ and $j_{1/4}^*\alpha^*\Phi = 0$.

Q.E.D.

Problems

1. Fields of n-frames. *A field of n-frames* over an open set $O \subset M$ (M a manifold) is an n-tuple $e_1, \ldots, e_n$ of vector fields in O such that for each $x \in O$ the vectors $e_1(x), \ldots, e_n(x)$ form a basis of $T_x(M)$. Then $e^{*1}, \ldots, e^{*n}$ is the n-tuple of dual 1-forms.

(i) Show that

$$\delta = \sum_{\alpha} \mu(e^{*\alpha})\, \theta(e_\alpha) + \tfrac{1}{2} \sum_{\alpha,\beta} \mu(e^{*\alpha} \wedge e^{*\beta})\, i([e_\alpha, e_\beta])$$

and use this to prove that δ is an antiderivation.

(ii) Define functions $C^{\gamma}_{\alpha\beta}$ in O by $[e_\alpha, e_\beta] = \sum_\gamma C^{\gamma}_{\alpha\beta} e_\gamma$. Verify the relations

$$\theta(e_\alpha)\, e^{*\rho} = -\sum_{\beta} C^{\rho}_{\alpha\beta} e^{*\beta} \qquad \text{and} \qquad \delta e^{*\rho} = -\tfrac{1}{2} \sum_{\alpha,\beta} C^{\rho}_{\alpha\beta} e^{*\alpha} \wedge e^{*\beta}.$$

2. Given manifolds M and N, regard $\mathscr{S}(M \times N)$ as an $\mathscr{S}(N)$-module and show that the $\mathscr{S}(M \times N)$-modules $A^{0,p}(M \times N)$ and $\mathscr{S}(M \times N) \otimes_N A^p(N)$ are isomorphic.

3. Let $X \in \mathscr{X}(M)$, $\Phi \in A^n(M)$, where M is a compact oriented n-manifold. Show that $\int_M \theta(X)\Phi = 0$.

4. Let U be a domain in $\mathbb{R}^n$, star-shaped with respect to 0. Define

$$h: A^p(U) \to A^{p-1}(U) \qquad (p \geqslant 1)$$

by

$$(h\Phi)(x; \xi_1, \ldots, \xi_{p-1}) = \int_0^1 \Phi(tx; x, t\xi_1, \ldots, t\xi_{p-1})\, dt.$$

(i) Show that $\delta \circ h + h \circ \delta = \iota$.

(ii) If $f \in \mathscr{S}(U)$ and Δ is a determinant function show that $f \cdot \Delta = \delta(g \cdot \Phi)$, where

$$\Phi(x; \xi_1, \ldots, \xi_{n-1}) = \Delta(x, \xi_1, \ldots, \xi_{n-1}) \qquad \text{and} \qquad g(x) = \int_0^1 f(tx)\, t^{n-1}\, dt.$$

5. Define the integral of an n-form over an oriented manifold-with-boundary. Establish Stokes' theorem for compact oriented manifolds-with-boundary.

6. Solid angle. Define $\Omega \in A^{n-1}(\dot{E})$ (E an oriented Euclidean n-space, $\dot{E} = E - \{0\}$) by

$$\Omega(x; \xi_1, \ldots, \xi_{n-1}) = \frac{1}{|x|^n} \Delta(x, \xi_1, \ldots, \xi_{n-1}),$$

where Δ is the positive normed determinant function.

(i) Show that $\delta\Omega = 0$.

(ii) Fix a unit vector a, and let $U = E - \{ta \mid t \geqslant 0\}$. Construct an $(n-2)$-form Ψ in U such that $\Omega = \delta\Psi$ in U.

(iii) Let M be a compact oriented $(n-2)$-manifold and let $\alpha: M \to U$ be smooth. Show that $\int_M \alpha^*\Psi$ (Ψ defined in (ii)) is independent of the choice of Ψ. (*Hint*: Compare problem 4.) $\int_M \alpha^*\Psi$ is called the *solid angle* enclosed by $\alpha(M)$. If $\beta: M \to U$ is a second map such that $\beta = f \cdot \alpha$, $f \in \mathscr{S}(M)$, show that $\alpha(M)$ and $\beta(M)$ enclose the same angle.

(iv) Let N be a compact oriented $(n-1)$-manifold with boundary ∂N and let $\varphi: N \to E$ be a smooth map. Then $\int_N \varphi^*\Omega$ (Ω defined in (i)) is called the *solid angle subtended by* $\varphi(N)$. If $\varphi(N) \subset U$, show that this coincides with the solid angle enclosed by $\varphi(\partial N)$. Assume $\psi: N \to U$ is a second smooth map such that $\varphi(x) = \lambda_x \psi(x)$ for $x \in \partial N$. Show that $\varphi(N)$ and $\psi(N)$ subtend the same angle.

7. Show that a p-form Φ ($p \geqslant 1$) which satisfies $\theta(X)\Phi = 0$ for every vector field X must be zero.

8. Densities. A *p-density* on an n-manifold M is a cross-section in the bundle $\wedge^n \tau_M^* \otimes \wedge^p \tau_M$. The module of p-densities is denoted by $D_p(M)$.

(i) Express densities in terms of components with respect to a local coordinate system and find the transformation formula in an overlap of two coordinate systems.

(ii) Let F be an n-dimensional vector space. Show that

$$\Phi \otimes (x_1 \wedge \cdots \wedge x_p) \mapsto i(x_p) \cdots i(x_1)\Phi \qquad \Phi \in \wedge^n F^*, \quad x_\nu \in F$$

defines a canonical isomorphism $\wedge^n F^* \otimes \wedge^p F \to \wedge^{n-p} F^*$. Obtain the *Poincaré isomorphism* (of $\mathscr{S}(M)$-modules)

$$D: A^p(M) \xrightarrow{\cong} D_{n-p}(M) \qquad (0 \leqslant p \leqslant n).$$

(iii) Define the divergence operator $\partial: D_p(M) \to D_{p-1}(M)$ by $\partial = (-1)^p D \circ \delta \circ D^{-1}$. Show that $\partial^2 = 0$ and express ∂u in terms of the components of u with respect to a local coordinate system.

(iv) Suppose M is compact and oriented. Set

$$(\Phi, u) = \int_M \Phi \wedge D^{-1}u, \qquad \Phi \in A^p(M), \quad u \in D_p(M).$$

Prove the formula

$$(\delta\Phi, u) = (\Phi, \partial u) \qquad \Phi \in A^{p-1}(M), \quad u \in D_p(M).$$

9. Laplacian. Let M be an oriented n-manifold and let $\langle\,,\rangle$ be a Riemannian metric in τ_M.

(i) Use $\langle\,,\rangle$ and the orientation to identify Sec $\wedge^p\tau_M$ with $A^p(M)$ and hence obtain $\mathscr{S}(M)$-isomorphisms

$$A^p(M) \xrightarrow{\cong} D_p(M).$$

(ii) Use these isomorphisms to obtain, from the divergence, an operator $\delta^*\colon A^p(M) \to A^{p-1}(M)$. Express $\delta^*\Phi$ in terms of the components of Φ with respect to a local coordinate system.

(iii) Assume M to be compact and let Δ be the positive normed determinant function on M. Set

$$(\Phi, \Psi) = \int_M \langle\Phi, \Psi\rangle\Delta.$$

Show that

$$(\delta\Phi, \Psi) = (\Phi, \delta^*\psi) \qquad \Phi \in A^p(M), \quad \Psi \in A^{p+1}(M).$$

(iv) The Laplace operator $\underline{\Delta}\colon A^p(M) \to A^p(M)$ is defined by $\underline{\Delta} = \delta \circ \delta^* + \delta^* \circ \delta$. Establish Green's formula

$$(\delta\Phi, \delta\Psi) + (\delta^*\Phi, \delta^*\Psi) = (\underline{\Delta}\Phi, \Psi), \qquad \Phi, \Psi \in A^p(M),$$

and conclude that $\underline{\Delta}\Phi = 0$ if and only if $\delta\Phi = 0$ and $\delta^*\Phi = 0$.

(v) Find a square root of $\underline{\Delta}$.

(vi) Show that δ, ∂, δ^*, $\underline{\Delta}$ are differential operators in the sense of problem 7, Chap. III. Compute their symbols and decide which are elliptic.

10. Let M be a compact oriented n-manifold. Let Φ and Ψ represent the orientation of M. Show that $\int_M \Phi = \int_M \Psi$ if and only if there is an orientation preserving diffeomorphism φ of M such that $\Psi = \varphi^*\Phi$. (*Hint*: If $\int_M X = 0$, then $X = \delta\Omega$; cf. Theorem II, sec. 5.13.)

11. Symplectic manifolds. A *symplectic manifold* is a manifold together with a closed 2-form ω such that each $\omega(x)$ is a nondegenerate bilinear function in $T_x(M)$.

(i) Show that a symplectic manifold is even dimensional and orientable.

(ii) Show that if M is compact and $p \leqslant \frac{1}{2} \dim M$, then $\omega \wedge \cdots \wedge \omega$ (p factors) is not of the form $\delta\Phi$ (See hint of problem 10.)

(iii) Show that $\tau: X \mapsto i(X)\omega$ is an $\mathscr{S}(M)$-isomorphism from $\mathscr{X}(M)$ to $A^1(M)$.

(iv) Set $X_f = \tau^{-1}(\delta f)$ ($f \in \mathscr{S}(M)$) and define the *Poisson bracket* by

$$[f, g] = \omega(X_f, X_g), \qquad f, g \in \mathscr{S}(M).$$

Show that $X_{[f,g]} = [X_f, X_g]$ and conclude that the map $(f, g) \mapsto [f, g]$ makes $\mathscr{S}(M)$ into a Lie algebra.

(v) Suppose X_f generates the 1-parameter group φ_t. Prove that $\varphi_t^*\omega = \omega$.

(vi) Assume $\dim M = 4k$ and the class represented by ω generates the algebra $H(M)$. If M is compact prove that M is irreversible. Is this true if M is not compact? ($H(M)$ is defined in sec. 5.1.)

12. **Cotangent bundle.** Let $\tau_M^* = (T_M^*, \pi, M, \mathbb{R}^n)$ be the cotangent bundle of an n-manifold.

(i) Show that a 1-form θ is defined on T_M^* by

$$\theta(z; \zeta) = \langle z, (d\pi)\zeta \rangle, \qquad \zeta \in T_z(T_M^*).$$

(ii) Suppose U is open in $\mathbb{R}^n$ and write $T_U^* = U \times (\mathbb{R}^n)^*$. A basis e_i of $\mathbb{R}^n$ determines coordinate functions x^i in U. Coordinate functions in $(\mathbb{R}^n)^*$ are given by $p_i: e^* \mapsto \langle e^*, e_i \rangle$ and the x^i together with the p_i are coordinate functions in T_U^*. Show that, in T_U^*,

$$\theta = \sum_i p_i\, \delta x^i, \qquad \delta\theta = \sum_i \delta p_i \wedge \delta x^i.$$

(iii) Show that $(T_M^*, \delta\theta)$ is a symplectic manifold. Use $\delta\theta$ and θ to obtain a canonical vector field Z on T_M^* and express it in local coordinates. Show that if $z \in T_x(M)^*$, then $Z(z)$ is tangent to $T_x(M)^*$. Thus interpret $Z(z)$ as a vector in $T_x(M)^*$ and show that $Z(z) = z$.

(iv) A diffeomorphism $\varphi: T_M^* \to T_N^*$ which preserves the symplectic structures is called a *canonical transformation*. Prove that each diffeomorphism $M \to N$ induces a canonical transformation $T_M^* \to T_N^*$.

(v) Let τ denote the isomorphism of problem 11, iii, for the symplectic manifold $(T_M^*, \delta\theta)$. Show that a vector field X on T_M^* generates a local 1-parameter family φ_t of canonical transformations if and only if the 1-form τX is closed.

(vi) Suppose $f \in \mathscr{S}(T_M^*)$ and let $X_f = \tau^{-1}(\delta f)$. Show that f is constant along the orbits of X_f. Show that in local coordinates the differential equation for an orbit X_f is the classical Hamilton–Jacobi equation

$$\frac{dx^i}{dt} = -\frac{\partial f}{\partial p^i}, \qquad \frac{dp_i}{dt} = \frac{\partial f}{\partial x^i}.$$

(vii) If $X \in \mathscr{X}(T_M^*)$ interpret the condition $\theta(X)\theta = 0$ geometrically.

13. Integration. Let M be an n-manifold oriented by Δ. A *continuous n-form* on M is a continuous map $\Phi: M \to \wedge^n T_M^*$ such that $\pi \circ \Phi = \iota$.

(i) Define the integral of continuous compactly supported n-forms and show that the basic properties continue to hold.

(ii) Let Φ be a positive continuous n-form (i.e., $\Phi(x) = \lambda(x)\,\Delta(x)$, $\lambda(x) \geqslant 0$). Let U_i be an open covering of M such that $\bar{U}_i \subset U_{i+1}$, and $\bar{U}_i$ is compact. Show that $a_i = \int_{U_i} \Phi$ is an increasing sequence and that $a = \lim_{i\to\infty} a_i \leqslant \infty$ depends only on Φ. Show that a coincides with the integral whenever it is defined as in (i). Set $\int_M \Phi = a$ in any case.

(iii) Let Φ be any continuous n-form on M. Construct a continuous positive n-form, Φ^+, such that $\Phi^+(x) = \Phi(x)$ or $\Phi^+(x) = -\Phi(x)$. Assuming that $\int_M \Phi^+ < \infty$, define $\int_M \Phi$. Show that $\int_M \Phi \leqslant \int_M \Phi^+$.

14. Parallelizable manifolds. Let ζ be the vector bundle over $M \times M$ whose fibre at (x, y) is the space $L(T_x(M); T_y(M))$. A *parallelism* on M is a cross-section P in ζ such that

$$P(z, y) \circ P(x, z) = P(x, y) \qquad \text{and} \qquad P(x, x) = \iota, \qquad x, y, z \in M.$$

A vector field X is called *parallel* (with respect to P) if

$$P(x, y)\, X(x) = X(y), \qquad x, y \in M.$$

A manifold which admits a parallelism is called *parallelizable*.

Let (M, P) and $(\hat{M}, \hat{P})$ be manifolds with parallelisms. A diffeomorphism $\varphi: M \to \hat{M}$ is called *parallelism preserving* if $d\varphi \circ P(x, y) = \hat{P}(\varphi x, \varphi y) \circ (d\varphi)_x$.

(i) Given a parallelism P, fix a point $a \in M$. Show that $(x, \xi) \mapsto P(a, x)\xi$ defines a strong bundle isomorphism $M \times T_a(M) \xrightarrow{\cong} \tau_M$. In this way obtain a bijection between parallelisms on M and trivializations of τ_M. Show that the parallel vector fields correspond to the constant cross-sections under this bijection.

(ii) Let P be a parallelism. Fix a point a and set $T_a(M) = F$. Define $\theta_a \in A^1(M; F)$ by $\theta_a(x; \xi) = P(a, x)^{-1}\xi$. Show that the relation

$$S(x; \xi, \eta) = -P(a, x)\, \delta\theta_a(x; \xi, \eta)$$

defines a tensor field, S, of type (2, 1) on M and that this tensor field is independent of a. It is called the *torsion* of P. Regard S as a map $\mathscr{X}(M) \times \mathscr{X}(M) \to \mathscr{X}(M)$ and show that if X and Y are parallel, then $S(X, Y) = [X, Y]$.

(iii) Show that if φ is parallelism preserving, then

$$S(\varphi(x); (d\varphi)\xi, (d\varphi)\eta) = (d\varphi)\, S(x; \xi, \eta).$$

(iv) Assume that P is a parallelism such that $S = 0$. Show that for every point $a \in M$ there exists a neighbourhood U and a parallelism preserving diffeomorphism of U onto an open subset $\hat{U}$ of $\mathbb{R}^n$, where $\hat{U}$ is given the parallelism induced by that of $\mathbb{R}^n$.

(v) Use the Cayley numbers (problem 5, Chap. III) to define a parallelism on S^7. Compute its torsion.

(vi) Two parallelisms, P, $\hat{P}$ on M are called *conjugate*, if whenever X is P-parallel and Y is $\hat{P}$-parallel, then $[X, Y] = 0$. Show that if M is connected then a parallelism has at most one conjugate parallelism. Show that the torsions of conjugate parallelisms are connected by $\hat{S} = -S$.

(vii) Show that if P admits a conjugate parallelism, then S satisfies

$$P(x, y)\, S(x; \xi, \eta) = S(y; P(x, y)\xi, P(x, y)\eta).$$

Conversely, if this relation holds, show that every point has a neighbourhood in which a conjugate parallelism exists.

(viii) Show that the parallelism of S^7 (part v) does not admit a conjugate parallelism.

15. Legendre transformation. Let $L \in \mathscr{S}(T_M)$ and let

$$j_x\colon T_x(M) \to T_M$$

be the inclusion map. Regard $(dj_x)_h$ as a linear map from $T_x(M)$.

(i) Show that $\langle \mathscr{L}(\xi), \eta\rangle = \delta L(\xi; dj_x(\eta))$, $\xi, \eta \in T_x(M)$, defines a strong bundle map $\mathscr{L}\colon \tau_M \to \tau_M^*$. It is called the *Legendre transformation* induced by L. When is $\mathscr{L}$ an isomorphism?

(ii) Suppose $\langle\,,\rangle$ is a Riemannian metric on M and let $f \in \mathscr{S}(M)$. Define T, $V \in \mathscr{S}(T_M)$ by $T(\xi) = \frac{1}{2}\langle \xi, \xi\rangle$, $V = \pi^* f$. Show that the function $L = T - V$ induces an invertible Legendre transformation.

(iii) If $\mathscr{L}$ is an isomorphism define $H \in \mathscr{S}(T_M^*)$ by

$$H(\xi^*) = \langle \xi^*, \mathscr{L}^{-1}\xi^*\rangle - L(\mathscr{L}^{-1}\xi^*).$$

If L is defined as in (ii), show that $\mathscr{L}^* H = T + V$.

Chapter V

De Rham Cohomology

§1. The axioms

5.1. Cohomology algebra of a manifold. Given an n-manifold M consider the graded algebra

$$A(M) = \sum_{p=0}^{n} A^p(M)$$

of differential forms on M. It follows from Proposition II of sec. 4.3 that the exterior derivative makes $A(M)$ into a graded differential algebra. The *cocycles* in this differential algebra consist of the differential forms Φ which satisfy the condition $\delta\Phi = 0$. Such a differential form is called *closed*. Since δ is an antiderivation, the closed forms are a graded subalgebra $Z(M)$ of $A(M)$.

The subset $B(M) = \delta A(M)$ is a graded ideal in $Z(M)$. The differential forms in $B(M)$ are called *exact*, or *coboundaries*. The corresponding (graded) cohomology algebra is given by

$$H(M) = Z(M)/B(M).$$

It is called the *de Rham cohomology algebra* of M.

Suppose $\varphi: M \to N$ is a smooth map. Then

$$\varphi^*: A(M) \leftarrow A(N)$$

is a homomorphism of graded differential algebras ($A(M)$ and $A(N)$ are considered as *real* algebras) as was shown in sec. 4.4 (Proposition III). Thus φ^* induces a homomorphism of cohomology algebras, homogeneous of degree zero, denoted by

$$\varphi^\#: H(M) \leftarrow H(N).$$

If $\psi: N \to Q$ is another smooth map, then $(\psi \circ \varphi)^* = \varphi^* \circ \psi^*$ and so

$$(\psi \circ \varphi)^\# = \varphi^\# \circ \psi^\#.$$

Moreover,

$$(\iota_M)^\# = \iota_{H(M)}\,.$$

In particular, if φ and ψ are inverse diffeomorphisms, then $\varphi^\#$ and $\psi^\#$ are inverse isomorphisms.

The gradation of $H(M)$ is given by

$$H(M) = \sum_{p=0}^{n} H^p(M),$$

where

$$H^p(M) = Z^p(M)/B^p(M).$$

Since $A^p(M) = 0$ for $p > n$ it follows that

$$H^p(M) = 0, \qquad p > n$$

and

$$H^n(M) = A^n(M)/B^n(M).$$

On the other hand, $B^0(M) = 0$, so that

$$H^0(M) = Z^0(M).$$

Now $Z^0(M)$ consists of the smooth functions f on M which satisfy $\delta f = 0$, and hence Proposition XI of sec. 3.17 can be restated in the form:

If M is connected, then $H^0(M) \cong \mathbb{R}$;

i.e., the cohomology algebra of a connected manifold is connected (cf. sec. 0.3).

In any case the constant functions represent elements of $H^0(M)$. In particular, the function $1: M \to 1$ represents the identity element, 1, of the algebra $H(M)$. If M is connected, the map $\lambda \mapsto \lambda \cdot 1$ provides a canonical isomorphism $\mathbb{R} \xrightarrow{\cong} H^0(M)$.

Example: If M consists of a single point, then

$$H^p(M) = 0 \quad (p \geq 1) \qquad \text{and} \qquad H^0(M) = \mathbb{R}.$$

If the spaces $H^p(M)$ have finite dimension (it will be shown in sec. 5.15 and, independently, in sec. 5.22 that this is the case for a compact manifold), then the number

$$b_p = \dim H^p(M)$$

is called the pth *Betti number of* M, and the polynomial

$$f_M(t) = \sum_{p=0}^{n} b_p t^p$$

is called the *Poincaré polynomial of* M. The alternating sum

$$\chi_M = \sum_{p=0}^{n} (-1)^p \, b_p = f_M(-1)$$

is called the *Euler–Poincaré characteristic of* M.

It is the purpose of this article to establish the following axioms for de Rham cohomology.

A1: $H(\text{point}) = \mathbb{R}$

A2: (homotopy axiom) If $\varphi \sim \psi \colon M \to N$, then $\varphi^\# = \psi^\#$.

A3: (disjoint union) If M is the disjoint union of open submanifolds M_α, then

$$H(M) \cong \prod_\alpha H(M_\alpha).$$

A4: (Mayer–Vietoris) If $M = U \cup V$ (U, V open) there is an exact triangle

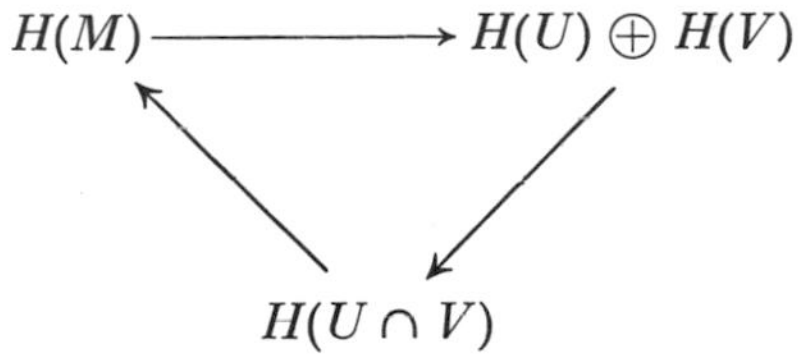

5.2. Homotopy. Recall from sec. 1.10 that two smooth maps $\varphi, \psi \colon M \to N$ are homotopic if there is a smooth map $H \colon \mathbb{R} \times M \to N$ such that

$$H(0, x) = \varphi(x) \qquad \text{and} \qquad H(1, x) = \psi(x).$$

Given such a homotopy H, define a linear map

$$h \colon A(M) \leftarrow A(N)$$

homogeneous of degree -1, by

$$h = I_0^1 \circ i(T) \circ H^*.$$

($T = (d/dt, 0)$, and I_0^1 is defined in sec. 4.10.) h is called the *homotopy operator induced from* H.

Remark: Let $\Phi \in A(N)$ and assume

$$(H^{-1}(\operatorname{carr} \Phi)) \cap ([0, 1] \times M) \subset [0, 1] \times C$$

for some closed set $C \subset M$. Then

$$\operatorname{carr}(i(T) H^*\Phi) \cap ([0, 1] \times M) \subset [0, 1] \times C$$

and so

$$\operatorname{carr} h\Phi \subset C.$$

Proposition I: The homotopy operator h satisfies

$$\psi^* - \varphi^* = h\delta + \delta h.$$

Proof: Let $j_t: M \to \mathbb{R} \times M$ denote inclusion opposite t. Apply Proposition IX, sec. 4.10, to obtain

$$(j_1^* - j_0^*)H^* = \delta h + h\delta.$$

Then observe that $H \circ j_1 = \psi$ and $H \circ j_0 = \varphi$.

Q.E.D.

Corollary: If $\varphi \sim \psi: M \to N$, then

$$\varphi^\# = \psi^\#: H(M) \leftarrow H(N).$$

5.3. Disjoint unions. Let M be a manifold which is the disjoint union of open submanifolds M_α,

$$M = \bigcup_\alpha M_\alpha .$$

The inclusion map $i_\alpha: M_\alpha \to M$ induces a homomorphism

$$i_\alpha^*: A(M_\alpha) \leftarrow A(M).$$

Denoting the direct product of the algebras $A(M_\alpha)$ by $\prod_\alpha A(M_\alpha)$ we obtain a homomorphism

$$i^*: A(M) \to \prod_\alpha A(M_\alpha)$$

given by

$$(i^*\Phi)_\alpha = i_\alpha^*\Phi, \qquad \Phi \in A(M).$$

Clearly, i^* is an isomorphism of graded differential algebras if $\prod_\alpha A(M_\alpha)$ is given the differential operator $\prod_\alpha \delta_\alpha$ (δ_α denotes the exterior derivative in $A(M_\alpha)$). Hence i^* induces an isomorphism

$$i^\#: H(M) \xrightarrow{\cong} \prod_\alpha H(M_\alpha)$$

given by

$$(i^\# \gamma)_\alpha = i_\alpha^\# (\gamma), \qquad \gamma \in H(M).$$

5.4. Mayer–Vietoris sequence. Let M be a manifold and let U_1, U_2 be open subsets such that $U_1 \cup U_2 = M$. Consider the inclusion maps

$$j_1: U_1 \cap U_2 \to U_1, \qquad j_2: U_1 \cap U_2 \to U_2$$
$$i_1: U_1 \to M, \qquad i_2: U_2 \to M.$$

They induce a sequence of linear mappings

$$0 \longrightarrow A(M) \xrightarrow{\alpha} A(U_1) \oplus A(U_2) \xrightarrow{\beta} A(U_1 \cap U_2) \longrightarrow 0 \tag{5.1}$$

given by

$$\alpha\Phi = (i_1^*\Phi, i_2^*\Phi), \qquad \Phi \in A(M)$$

and

$$\beta(\Phi_1, \Phi_2) = j_1^*\Phi_1 - j_2^*\Phi_2, \qquad \Phi_i \in A(U_i), \quad i = 1, 2.$$

Denote the exterior derivatives in $A(U_1)$, $A(U_2)$, $A(U_1 \cap U_2)$, and $A(M)$ by δ_1, δ_2, δ_{12}, and δ respectively. Then

$$\alpha \circ \delta = (\delta_1 \oplus \delta_2) \circ \alpha \qquad \text{and} \qquad \beta \circ (\delta_1 \oplus \delta_2) = \delta_{12} \circ \beta$$

and so α, β induce linear maps

$$\alpha_\#: H(M) \to H(U_1) \oplus H(U_2), \qquad \beta_\#: H(U_1) \oplus H(U_2) \to H(U_1 \cap U_2).$$

Lemma I: The sequence (5.1) is exact.

Proof: (1) β *is surjective*: Let p_1, p_2 be a partition of unity for M subordinate to the covering U_1, U_2. Then

$$\operatorname{carr} i_1^* p_2, \ \operatorname{carr} i_2^* p_1 \subset U_1 \cap U_2.$$

Now let $\Phi \in A(U_1 \cap U_2)$. Define

$$\Phi_1 = i_1^* p_2 \cdot \Phi \in A(U_1), \qquad \Phi_2 = i_2^* p_1 \cdot \Phi \in A(U_2).$$

Then

$$\Phi = \beta(\Phi_1, -\Phi_2).$$

(2) $\ker\beta = \operatorname{Im}\alpha$: Clearly $\beta\circ\alpha = 0$ so that $\ker\beta \supset \operatorname{Im}\alpha$. To prove equality, let $(\Phi_1, \Phi_2) \in \ker\beta$. Then, if $x \in U_1 \cap U_2$, $\Phi_1(x) = \Phi_2(x)$. Thus a differential form $\Phi \in A(M)$ is given by

$$\Phi(x) = \begin{cases} \Phi_1(x), & x \in U_1 \\ \Phi_2(x), & x \in U_2 . \end{cases}$$

Clearly $\alpha\Phi = (\Phi_1, \Phi_2)$; thus $\operatorname{Im}\alpha \supset \ker\beta$.

(3) α *is injective*: If $\alpha\Phi = 0$ (some $\Phi \in A(M)$), then $\Phi(x) = 0$, for $x \in U_1 \cup U_2 = M$.

Q.E.D.

The short exact sequence (5.1) induces an exact triangle

$$\begin{array}{ccc} H(M) & \xrightarrow{\alpha_\#} & H(U_1) \oplus H(U_2) \\ & {\scriptstyle\partial}\nwarrow \quad \swarrow{\scriptstyle\beta_\#} & \\ & H(U_1 \cap U_2) & \end{array}$$

where ∂ is homogeneous of degree $+1$ (cf. sec. 0.7). In other words, we have a long exact sequence

$$\cdots \to H^p(M) \xrightarrow{\alpha_\#} H^p(U_1) \oplus H^p(U_2) \xrightarrow{\beta_\#} H^p(U_1 \cap U_2) \xrightarrow{\partial} H^{p+1}(M) \to \cdots .$$

This sequence is called the *Mayer–Vietoris sequence* of the triple (M, U_1, U_2). ∂ is called the *connecting homomorphism*: However, *it is not an algebra homomorphism.* If $\alpha \in H(U_1 \cap U_2)$ is represented by X and $\Psi_i \in A(U_i)$ satisfy $\Psi_1|_{U_1\cap U_2} - \Psi_2|_{U_1\cap U_2} = X$, then $\partial\alpha$ is represented by Ω, where $\Omega|_{U_i} = \delta\Psi_i$.

Now let N be a second manifold and let $V_1, V_2 \subset N$ be open sets such that $N = V_1 \cup V_2$. Let $\varphi: M \to N$ be a smooth map which restricts to maps

$$\varphi_1: U_1 \to V_1 \qquad \text{and} \qquad \varphi_2: U_2 \to V_2 .$$

Then φ restricts to a map $\varphi_{12}: U_1 \cap U_2 \to V_1 \cap V_2$.

The commutative diagram

$$\begin{array}{ccccccccc} 0 & \longrightarrow & A(M) & \xrightarrow{\alpha_M} & A(U_1) \oplus A(U_2) & \xrightarrow{\beta_M} & A(U_1 \cap U_2) & \longrightarrow & 0 \\ & & \uparrow{\scriptstyle\varphi^*} & & \uparrow{\scriptstyle\varphi_1^*\oplus\varphi_2^*} & & \uparrow{\scriptstyle\varphi_{12}^*} & & \\ 0 & \longrightarrow & A(N) & \xrightarrow{\alpha_N} & A(V_1) \oplus A(V_2) & \xrightarrow{\beta_N} & A(V_1 \cap V_2) & \longrightarrow & 0 \end{array} \tag{5.2}$$

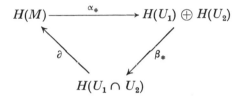

induces the commutative diagram

$$\begin{array}{ccccccccc}
\cdots \to H^p(M) & \to & H^p(U_1) \oplus H^p(U_2) & \to & H^p(U_1 \cap U_2) & \xrightarrow{\partial_M} & H^{p+1}(M) \to \cdots \\
\uparrow \varphi^{\#} & & \uparrow \varphi_1^{\#} \oplus \varphi_2^{\#} & & \uparrow \varphi_{12}^{\#} & & \uparrow \varphi^{\#} \\
\cdots \to H^p(N) & \to & H^p(V_1) \oplus H^p(V_2) & \to & H^p(V_1 \cap V_2) & \xrightarrow{\partial_N} & H^{p+1}(N) \to \cdots
\end{array} \tag{5.3}$$

between the Mayer–Vietoris sequences of (M, U_1, U_2) and (N, V_1, V_2).

§2. Examples

5.5. Retracts. Let N be a retract of M with inclusion i and retraction ρ (cf. Example 3, sec. 3.10). Since $\rho \circ i = \iota_N$ it follows that

$$i^\# \circ \rho^\# = \iota_{H(N)} .$$

Thus $\rho^\#$ is injective, $i^\#$ is surjective, and

$$H(M) = \operatorname{Im} \rho^\# \oplus \ker i^\#.$$

If in addition $i \circ \rho: M \to M$ is homotopic to the identity map of M, then N is called a *deformation retract of* N. In this case

$$\rho^\# \circ i^\# = \iota_{H(M)}$$

and so $\rho^\#$ and $i^\#$ are inverse isomorphisms.

Examples: 1. A manifold M is called *contractible* if it contains a point $a \in M$ as deformation retract (equivalently, the constant map $\gamma: M \to a$ is homotopic to the identity). The homotopy connecting ι_M and γ is called a *contraction*, or a *contracting* homotopy.

If M is contractible, then

$$H^p(M) \cong H^p(\text{point}) = \begin{cases} 0, & p > 0 \\ \mathbb{R}, & p = 0. \end{cases}$$

As a special case we have the

Poincaré lemma: If M is a star-shaped domain of a vector space, then $H^p(M) = 0$ ($p > 0$) and $H^0(M) = \mathbb{R}$.

2. Let M be a manifold, and let N be a manifold contractible to a point b. Then

$$\pi_M: M \times N \to M, \qquad j_b: M \to M \times N.$$

(π_M the projection, j_b the inclusion opposite b) make M into a deformation retract of $M \times N$.

In fact, $\pi_M \circ j_b = \iota_M$. Moreover, if H is a contracting homotopy for N, then

$$\psi(t, x, y) = (x, H(t, y)), \qquad t \in \mathbb{R}, \quad x \in M, \quad y \in N$$

defines a homotopy connecting $\iota_{M\times N}$ and $j_b \circ \pi_M$. Hence

$$\pi_M^{\#}: H(M) \xrightarrow{\cong} H(M \times N).$$

3. Let $\xi = (E, \pi, B, F)$ be a vector bundle, and let $o: B \to E$ be the zero cross-section. Then $\pi \circ o = \iota$, and so B is a retract of E. Moreover, the homotopy $H: \mathbb{R} \times E \to E$ given by $H(t, z) = tz$ ($t \in \mathbb{R}$, $z \in E$) connects ι_E and $o \circ \pi$. Thus B is a deformation retract of E. In particular

$$\pi^{\#}: H(B) \to H(E) \qquad \text{and} \qquad o^{\#}: H(E) \to H(B)$$

are inverse isomorphisms.

4. Assign a Riemannian metric to the bundle of Example 3. Let $\dot{\xi} = (\dot{E}, \dot{\pi}, B, \dot{F})$ and $\xi_S = (E_S, \pi_S, B, S)$ be the deleted bundle and the associated sphere bundle (cf. sec. 3.10, Examples 5, 6). Recall that $i: E_S \to \dot{E}$, $\rho: \dot{E} \to E_S$ ($\rho(z) = z/|z|$) make E_S into a retract of $\dot{E}$.

On the other hand, a homotopy $H: \mathbb{R} \times \dot{E} \to \dot{E}$ connecting $i \circ \rho$ and $\iota_{\dot{E}}$ is given by

$$H(t, z) = \frac{(t^2 + (1-t)^2 |z|)z}{|z|}.$$

Thus E_S is a deformation retract of $\dot{E}$; in particular

$$\rho^{\#}: H(E_S) \to H(\dot{E}) \qquad \text{and} \qquad i^{\#}: H(\dot{E}) \to H(E_S)$$

are inverse isomorphisms.

5. Consider the special case of Example 4 in which

$$E = F, \qquad B = (\text{point}).$$

In this case Example 4 states that the unit sphere, S, of F is a deformation retract of $\dot{F} = F - \{0\}$, and so $H(S) \cong H(\dot{F})$.

6. Let $a \in S^n$ ($n \geqslant 1$) be a fixed point. Then

$$S^{n-1} = \{x \in S^n \mid \langle a, x\rangle = 0\}$$

is the unit sphere of the Euclidean space $F = a^\perp \subset E$. Let $U \subset S^n$ be the open set given by

$$U = \{x \in S^n \mid -\epsilon < \langle a, x\rangle < \epsilon\}$$

for some fixed ϵ with $0 < \epsilon < 1$. Then S^{n-1} is a deformation retract of U.

In fact, let $i: S^{n-1} \to U$ be the inclusion map, and define

$$\rho: U \to S^{n-1}$$

by

$$\rho(x) = \frac{x - \langle x, a\rangle a}{|x - \langle x, a\rangle a|}, \qquad x \in U$$

(observe that $x - \langle x, a\rangle a \neq 0$, for $x \in U$). Clearly $\rho \circ i = \iota$.

Finally, define $H: \mathbb{R} \times U \to U$ by

$$H(t, x) = \frac{x - t\langle x, a\rangle a}{|x - t\langle x, a\rangle a|}, \qquad x \in U, \quad t \in \mathbb{R}.$$

Then H connects $i \circ \rho$ and the identity map of U.

7. Let $\varphi, \psi: M \to S^n$ $(n \geqslant 1)$ be smooth maps such that $\varphi(x) \neq -\psi(x)$, for $x \in M$. Then $\varphi \sim \psi$ (cf. Example 3, sec. 1.10). Thus $\varphi^\# = \psi^\#$.

5.6. The cohomology of S^n. As an application of the Mayer–Vietoris sequence we shall determine the cohomology of S^n.

Proposition II: $H(S^n)$ $(n \geqslant 1)$ is given by

$$H^0(S^n) \cong H^n(S^n) \cong \mathbb{R}$$

and

$$H^p(S^n) = 0 \qquad (1 \leqslant p \leqslant n - 1).$$

Proof: Consider S^n as embedded in an $(n + 1)$-dimensional Euclidean space E. Since S^n is connected, $H^0(S^n) = \mathbb{R}$. Now let $a \in S^n$ and fix $\epsilon \in (0, 1)$. Define open sets $U, V \subset S^n$ by

$$U = \{x \in S^n \mid \langle x, a\rangle > -\epsilon\}, \qquad V = \{x \in S^n \mid \langle x, a\rangle < \epsilon\}.$$

Then $S^n = U \cup V$, and so there is a long exact Mayer–Vietoris sequence

$$\cdots \to H^p(S^n) \to H^p(U) \oplus H^p(V) \to H^p(U \cap V) \to H^{p+1}(S^n) \to \cdots.$$

Next observe that U and V are contractible, while $U \cap V$ contains S^{n-1} as a deformation retract (Example 6, sec. 5.5). Thus we may rewrite the Mayer–Vietoris sequence as the exact sequence

$$\cdots \to H^p(S^n) \to H^p(\text{point}) \oplus H^p(\text{point}) \to H^p(S^{n-1}) \to H^{p+1}(S^n) \to \cdots.$$

Hence it splits into the exact sequences

$$0 \to H^0(S^n) \to H^0(\text{point}) \oplus H^0(\text{point}) \to H^0(S^{n-1}) \to H^1(S^n) \to 0$$

and

$$0 \longrightarrow H^p(S^{n-1}) \xrightarrow{\cong} H^{p+1}(S^n) \longrightarrow 0, \qquad p \geqslant 1.$$

From the first of these we obtain

$$0 = \dim H^1(S^n) - \dim H^0(S^{n-1}) + 2 \dim H^0(\text{point}) - \dim H^0(S^n).$$

Since S^{n-1} is connected if $n \geqslant 2$, while S^0 consists of two points, this equation yields

$$H^1(S^n) \cong \begin{cases} \mathbb{R}, & n = 1 \\ 0, & n > 1. \end{cases}$$

Finally, the second of the exact sequences shows that $H^p(S^n) \cong H^1(S^{n-p+1})$ $(1 \leqslant p \leqslant n)$ and the proposition follows.

Q.E.D.

Corollary I: The algebra $H(S^n)$ is the exterior algebra over the one-dimensional graded space $H^n(S^n)$.

Corollary II: The Poincaré polynomial of S^n is given by

$$f(t) = 1 + t^n \qquad (n \geqslant 1).$$

The Euler–Poincaré characteristic of S^n is given by

$$\chi_{S^n} = \begin{cases} 0, & n \text{ odd} \\ 2, & n \text{ even.} \end{cases}$$

Remark: Consider S^n as a submanifold of Euclidean space E. Orient S^n by the n-form Ω given by $\Omega(x; h_1, \ldots, h_n) = \Delta(x, h_1, \ldots, h_n)$, where Δ is a determinant function in E. Then

$$\int_{S^n} \Omega > 0.$$

Hence Proposition XIV, sec. 4.13, shows that Ω is *not* exact; i.e., it represents a nontrivial element $\alpha \in H^n(S^n)$.

Since $H^n(S^n) \cong \mathbb{R}$, α is a basis of $H^n(S^n)$.

5.7. Free involutions. Let ω be an involution of a manifold M

($\omega^2 = \iota$) without fixed points. Assume $\pi: M \to N$ is a surjective local diffeomorphism with

$$\pi^{-1}(x) = \{z, \omega(z)\}, \qquad x \in N.$$

Recall from Example 4, sec. 3.20, that

$$A(M) = A_+(M) \oplus A_-(M),$$

where $A_+(M) = \{\Phi \mid \omega^*\Phi = \Phi\}$ and $A_-(M) = \{\Phi \mid \omega^*\Phi = -\Phi\}$. Moreover, π^* may be considered as an isomorphism $A(N) \xrightarrow{\cong} A_+(M)$.

Since $\omega^*\delta = \delta\omega^*$, it follows from the definitions that $A_+(M)$, $A_-(M)$ are stable under δ. Hence

$$H(M) = H(A_+(M), \delta) \oplus H(A_-(M), \delta) = H_+(M) \oplus H_-(M),$$

where

$$H_+(M) = \{\alpha \mid \omega^\#(\alpha) = \alpha\} \qquad \text{and} \qquad H_-(M) = \{\alpha \mid \omega^\#(\alpha) = -\alpha\}.$$

Moreover, $\pi^\#$ may be considered as an isomorphism $H(N) \xrightarrow{\cong} H_+(M)$.

Example: The cohomology of $\mathbb{R}P^n$ ($n \geqslant 1$) is given by

$$H^0(\mathbb{R}P^n) = \mathbb{R}, \quad H^p(\mathbb{R}P^n) = 0, \qquad 1 \leqslant p < n$$

and

$$H^n(\mathbb{R}P^n) = \begin{cases} 0, & n \text{ even} \\ \mathbb{R}, & n \text{ odd.} \end{cases}$$

In fact, applying the discussion above to the projection $\pi: S^n \to \mathbb{R}P^n$ and the involution $\omega: z \mapsto -z$ of S^n, we find that

$$H(\mathbb{R}P^n) = H_+(S^n).$$

It follows that

$$H^0(\mathbb{R}P^n) = \mathbb{R} \qquad \text{and} \qquad H^p(\mathbb{R}P^n) = 0 \qquad (1 \leqslant p < n).$$

Finally the positive n-form $\Omega \in A^n(S^n)$ of Example 3, sec. 3.21, satisfies $\omega^*\Omega = (-1)^{n+1}\Omega$. By the remark at the end of sec. 5.6, Ω represents a basis of $H^n(S^n)$. Thus

$$H^n(S^n) = \begin{cases} H^n_-(S^n), & n \text{ even} \\ H^n_+(S^n), & n \text{ odd} \end{cases}$$

and so

$$H^n(\mathbb{R}P^n) = H^n_+(S^n) = \begin{cases} 0, & n \text{ even} \\ \mathbb{R}, & n \text{ odd.} \end{cases}$$

5.8. Germs of forms. Let M be a manifold and let $b \in M$. Consider the set $\mathscr{I}_b$ of differential forms Φ which are zero in some neighbourhood of b (possibly dependent on Φ). Then $\Phi \in \mathscr{I}_b$ if and only if $b \notin \operatorname{carr} \Phi$. Proposition IV, sec. 4.5, shows that $\mathscr{I}_b$ is a graded ideal in $A(M)$ stable under δ.

Denote the factor algebra $A(M)/\mathscr{I}_b$ by $A_b(M)$. δ induces an operator δ_b in $A_b(M)$ which makes $A_b(M)$ into a graded differential algebra. $A_b(M)$ is called the *algebra of germs of forms at b*.

Proposition III: Let M be any manifold. Then

$$H^0(A_b(M), \delta_b) = \mathbb{R} \qquad \text{and} \qquad H^p(A_b(M), \delta_b) = 0, \qquad p > 0.$$

Proof: Let $\pi\colon A(M) \to A_b(M)$ be the projection. Assume first that $\delta_b \pi f = 0$, for some $f \in \mathscr{S}(M) = A^0(M)$. Then $\pi \delta f = \delta_b \pi f = 0$ and hence δf is zero in some neighbourhood of b.

It follows (cf. Proposition XI, sec. 3.17) that f is constant in some neighbourhood of b, whence, for some $\lambda \in \mathbb{R}$, $\pi f = \pi \lambda$.
Thus

$$H^0(A_b(M)) = Z^0(A_b(M)) = \pi(\mathbb{R} \cdot 1) = \mathbb{R}.$$

Now let $\pi\Phi \in A_b^p(M)$ ($p \geqslant 1$) satisfy $\delta_b \pi \Phi = 0$. Then the restriction of Φ to some neighbourhood U of b is closed. Choose a contractible neighbourhood V of b such that $V \subset U$. Then according to Example 1, sec. 5.5, there is a ($p - 1$)-form $\Psi \in A^{p-1}(V)$ such that $(\delta\Psi)(x) = \Phi(x)$ ($x \in V$).

Choose $\Omega \in A^{p-1}(M)$ so that Ω equals Ψ in some neighbourhood of b. Then

$$\delta_b \pi \Omega = \pi \delta \Omega = \pi \Phi$$

and hence $H^p(A_b(M)) = 0$, $p \geqslant 1$.

Q.E.D.

Corollary: Let $\alpha \in H^p(M)$ ($p \geqslant 1$). Then there is a representing cocycle Φ such that Φ is zero in a neighbourhood of b.

Proof: Let Φ_1 be any cocycle representing α. Then

$$\pi \Phi_1 = \delta_b \pi \Psi = \pi \delta \Psi,$$

for some $\Psi \in A^{p-1}(M)$. Set $\Phi = \Phi_1 - \delta \Psi$.

Q.E.D.

§3. Cohomology with compact supports

5.9. Let M be a manifold. Recall (sec. 4.5) that $A_c(M) \subset A(M)$ is the graded ideal of forms with compact carrier. It is stable under $i(X)$, $\theta(X)$ $(X \in \mathscr{X}(M))$ and δ.

Definition: The graded algebra (possibly without unit) $H(A_c(M), \delta)$ is called the *cohomology of M with compact supports*, and is denoted by $H_c(M)$,

$$H_c(M) = \sum_{p=0}^{n} H_c^p(M), \qquad n = \dim M.$$

Since $A_c(M)$ is an ideal, multiplication in $A(M)$ restricts to a real bilinear map $A(M) \times A_c(M) \to A_c(M)$ which makes $A_c(M)$ into a left graded $A(M)$-module. This map induces a bilinear map

$$H(M) \times H_c(M) \to H_c(M)$$

written

$$(\alpha, \beta) \mapsto \alpha * \beta, \qquad \alpha \in H(M), \quad \beta \in H_c(M),$$

which makes $H_c(M)$ into a left graded $H(M)$-module. Similarly, $H_c(M)$ is made into a right graded $H(M)$-module, and we write $\beta * \alpha$, $\beta \in H_c(M)$, $\alpha \in H(M)$.

The inclusion map γ_M: $A_c(M) \to A(M)$ induces an algebra homomorphism

$$(\gamma_M)_{\#}: H_c(M) \to H(M),$$

which converts the module structures above to ordinary multiplication.

Example: Let M be a manifold with no compact component. Then $H_c^0(M) = 0$. In particular $H_c^0(\mathbb{R}^n) = 0$.

In fact,

$$H_c^0(M) = \{f \in \mathscr{S}_c(M) \mid \delta f = 0\}.$$

But $\delta f = 0$ if and only if f is constant on each component of M. Moreover, if f has compact support, and is constant on each component, it can be different from zero only on compact components. Thus, if M has no compact components, $f = 0$; i.e., $H_c^0(M) = 0$.

Remark: If M is compact, $A_c(M) = A(M)$, $H_c(M) = H(M)$.

A smooth map $\varphi\colon M \to N$ is called *proper* if the inverse image under φ of every compact subset of N is compact. For any $\Phi \in A(N)$,

$$\operatorname{carr} \varphi^*\Phi \subset \varphi^{-1}(\operatorname{carr} \Phi).$$

Thus if φ is proper, φ^* restricts to a homomorphism of graded differential algebras $\varphi_c^*\colon A_c(N) \to A_c(M)$ which in turn induces a homomorphism

$$\varphi_c^{\#}\colon H_c(M) \leftarrow H_c(N).$$

In particular, a diffeomorphism $\varphi\colon M \xrightarrow{\cong} N$ induces isomorphisms, φ_c^*, $\varphi_c^{\#}$.

Next, let $\varphi\colon M \to N$ be a diffeomorphism onto an open subset U of N. If $\Phi \in A_c(M)$, we can form

$$(\varphi^{-1})_c^* \Phi \in A_c(U).$$

We extend this to a differential form $(\varphi_c)_*\Phi \in A_c(N)$ by setting

$$(\varphi_c)_* \Phi(x) = 0, \qquad x \notin \operatorname{carr}(\varphi^{-1})_c^* \Phi.$$

(Since $\operatorname{carr}(\varphi^{-1})_c^*\Phi$ is compact, it is closed *in* N).

In this way we obtain a homomorphism

$$(\varphi_c)_*\colon A_c(M) \to A_c(N)$$

which commutes with δ. Thus it induces a homomorphism

$$(\varphi_c)_{\#}\colon H_c(M) \to H_c(N).$$

5.10. Axioms for $H_c(M)$. In this section we establish axioms for cohomology with compact supports, analogous to those given at the end of sec. 5.1.

Proposition IV: The cohomology of $\mathbb{R}^n$ with compact supports is given by

$$H_c^p(\mathbb{R}^n) = \begin{cases} 0, & p < n \\ \mathbb{R}, & p = n. \end{cases}$$

Proof: If $n = 0$, the proposition is trivial. For $n > 0$ consider S^n as the one-point compactification of $\mathbb{R}^n$ (cf. Example 10, sec. 1.5) and let $b \in S^n$ be the compactifying point; i.e. $S^n - \{b\} = \mathbb{R}^n$. Recall from sec. 5.8 the short exact sequence

$$0 \to \mathscr{I}_b \to A(S^n) \to A_b(S^n) \to 0,$$

where $\mathscr{I}_b$ denotes the ideal of differential forms on S^n which are zero in a neighbourhood of b. Clearly then, $\mathscr{I}_b = A_c(\mathbb{R}^n)$.

The short exact sequence above gives rise to a long exact sequence in cohomology. Since (sec. 5.8) $H(A_b(S^n)) = H(\text{point})$, this sequence splits into exact sequences

$$0 \to H_c^0(\mathbb{R}^n) \to H^0(S^n) \to \mathbb{R} \to H_c^1(\mathbb{R}^n) \to H^1(S^n) \to 0$$

and

$$0 \longrightarrow H_c^p(\mathbb{R}^n) \xrightarrow{\cong} H^p(S^n) \longrightarrow 0, \qquad p \geqslant 2.$$

Since $H_c^0(\mathbb{R}^n) = 0$ (Example of sec. 5.9), while $H^0(S^n) = \mathbb{R}$, the first sequence gives the exact sequence

$$0 \longrightarrow H_c^1(\mathbb{R}^n) \xrightarrow{\cong} H^1(S^n) \longrightarrow 0.$$

In view of the second sequence the proposition follows from Proposition II, sec. 5.6.

Q.E.D.

Next a homotopy axiom is established. Let $H: \mathbb{R} \times M \to N$ be a homotopy connecting smooth maps $\varphi, \psi: M \to N$. H will be called a *proper homotopy* if, for all compact sets $K \subset N$, $H^{-1}(K) \cap ([0, 1] \times M)$ is compact (equivalently, the restriction of H to $[0, 1] \times M$ is a proper map). If H is a proper homotopy φ and ψ are proper maps, because they are the restrictions of H to $0 \times M$ and $1 \times M$.

Proposition V: Assume $\varphi, \psi: M \to N$ are connected by a proper homotopy $H: \mathbb{R} \times M \to N$. Then the induced homotopy operator h restricts to a linear map

$$h_c: A_c(N) \to A_c(M)$$

and

$$\psi_c^* - \varphi_c^* = h_c\delta + \delta h_c .$$

In particular $\psi_c^\# = \varphi_c^\#$.

Proof: Apply the remark of sec. 5.2, together with Proposition I of that section.

Q.E.D.

Next suppose $M = \bigcup_\alpha M_\alpha$ is the disjoint union of open submanifolds M_α. The inclusion maps $i^\alpha\colon M_\alpha \to M$ induce homomorphisms

$$(i_c^\alpha)_*\colon A_c(M_\alpha) \to A_c(M).$$

Moreover, the linear map

$$i_*\colon \bigoplus_\alpha A_c(M_\alpha) \to A_c(M)$$

defined by

$$i_*\left(\sum_{\nu=1}^{p} \Phi_{\alpha_\nu}\right) = \sum_{\nu=1}^{p} (i_c^{\alpha_\nu})_* \Phi_{\alpha_\nu}, \qquad \Phi_{\alpha_\nu} \in A_c(M_{\alpha_\nu}),$$

is an isomorphism of graded differential algebras. Here $\bigoplus_\alpha A_c(M_\alpha)$ is given the differential operator $\bigoplus_\alpha \delta_\alpha$ (δ_α denotes the exterior derivative in M_α). Observe that this is the direct *sum*; in sec. 5.3 we used the direct *product*.

It follows that i_* induces an isomorphism

$$i_\#\colon \bigoplus_\alpha H_c(M_\alpha) \xrightarrow{\cong} H_c(M),$$

given by

$$i_\#\left(\sum_{\nu=1}^{p} \gamma_{\alpha_\nu}\right) = \sum_{\nu=1}^{p} (i_c^{\alpha_\nu})_\# \gamma_{\alpha_\nu}, \qquad \gamma_{\alpha_\nu} \in H_c(M_{\alpha_\nu}).$$

Finally, suppose $M = U_1 \cup U_2$ (U_ν open). Let

$$j^\nu\colon U_{12} \to U_\nu, \qquad i^\nu\colon U_\nu \to M, \qquad \nu = 1, 2$$

denote the inclusion maps. Define a sequence

$$0 \longrightarrow A_c(U_{12}) \xrightarrow{\beta_c} A_c(U_1) \oplus A_c(U_2) \xrightarrow{\alpha_c} A_c(M) \longrightarrow 0$$

of linear maps by

$$\beta_c \Phi = ((j_c^1)_*\Phi, -(j_c^2)_*\Phi), \qquad \Phi \in A_c(U_{12})$$

and

$$\alpha_c(\Phi_1, \Phi_2) = (i_c^1)_* \Phi_1 + (i_c^2)_* \Phi_2, \qquad \Phi_\nu \in A_c(U_\nu).$$

(This should be contrasted with the situation of sec. 5.4.)

An argument similar to that of sec. 5.4 shows that this sequence is short exact. The induced exact triangle of cohomology reads

$$\begin{array}{ccc} H_c(U_{12}) & \xrightarrow{(\beta_c)_\#} & H_c(U_1) \oplus H_c(U_2) \\ & {\scriptstyle \partial_c} \nwarrow \quad \swarrow {\scriptstyle (\alpha_c)_\#} & \\ & H_c(M) & \end{array}$$

where ∂_c is the connecting homomorphism, homogeneous of degree $+1$. The corresponding long exact sequence

$$\cdots \to H_c^p(U_{12}) \to H_c^p(U_1) \oplus H_c^p(U_2) \to H_c^p(M) \xrightarrow{\partial_c} H_c^{p+1}(U_{12}) \to \cdots$$

is called the *Mayer–Vietoris sequence for the triple* (M, U_1, U_2) *with respect to compact carriers*. If $\alpha \in H_c(M)$ is represented by Φ and $\Psi_i \in A_c(U_i)$ satisfy $\Psi_1 + \Psi_2 = \Phi$, then $\partial_c\alpha$ is represented by $\delta\Psi_1 = -\delta\Psi_2$.

If $N = V_1 \cup V_2$ is a second manifold and $\varphi: M \to N$ restricts to proper maps $\varphi_\nu: U_\nu \to V_\nu$ (φ is then proper), then the following diagrams commute:

$$\begin{array}{ccccccccc} 0 & \to & A_c(U_{12}) & \to & A_c(U_1) \oplus A_c(U_2) & \to & A_c(M) & \to & 0 \\ & & \uparrow{\scriptstyle (\varphi_{12})_c^*} & & \uparrow{\scriptstyle (\varphi_1)_c^* \oplus (\varphi_2)_c^*} & & \uparrow{\scriptstyle \varphi_c^*} & & \\ 0 & \to & A_c(V_{12}) & \to & A_c(V_1) \oplus A_c(V_2) & \to & A_c(N) & \to & 0 \end{array} \tag{5.4}$$

and

$$\begin{array}{ccccccccccc} \cdots & \to & H_c^p(U_{12}) & \to & H_c^p(U_1) \oplus H_c^p(U_2) & \to & H_c^p(M) & \to & H_c^{p+1}(U_{12}) & \to & \cdots \\ & & \uparrow{\scriptstyle (\varphi_{12})_c^\#} & & \uparrow{\scriptstyle (\varphi_1)_c^\# \oplus (\varphi_2)_c^\#} & & \uparrow{\scriptstyle \varphi_c^\#} & & \uparrow{\scriptstyle (\varphi_{12})_c^\#} & & \\ \cdots & \to & H_c^p(V_{12}) & \to & H_c^p(V_1) \oplus H_c^p(V_2) & \to & H_c^p(N) & \to & H_c^{p+1}(V_{12}) & \to & \cdots \end{array} \tag{5.5}$$

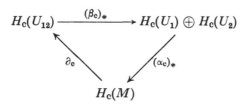

§4. Poincaré duality

5.11. Definition. Let M be an oriented n-manifold. Recall (Proposition XIV, sec. 4.13) that the surjective linear map

$$\int_M : A_c^n(M) \to \mathbb{R}$$

satisfies $\int_M \circ \delta = 0$. It follows that it induces a surjective linear map

$$\int_M^{\#} : H_c^n(M) \to \mathbb{R}.$$

Definition: The *Poincaré scalar product*

$$\mathscr{P}_M^p : H^p(M) \times H_c^{n-p}(M) \to \mathbb{R}$$

is the bilinear map given by

$$\mathscr{P}_M^p(\alpha, \beta) = \int_M^{\#} \alpha * \beta, \qquad \alpha \in H^p(M), \quad \beta \in H_c^{n-p}(M)$$

(cf. sec. 5.9).

Evidently,

$$\mathscr{P}_M^0(1, \beta) = \int_M^{\#} \beta, \qquad \beta \in H_c^n(M)$$

and

$$\mathscr{P}_M^{p+q}(\alpha \cdot \gamma, \beta) = \mathscr{P}_M^p(\alpha, \gamma * \beta), \tag{5.6}$$

$$\alpha \in H^p(M), \quad \gamma \in H^q(M), \quad \beta \in H_c^{n-p-q}(M).$$

Finally, combine the $\mathscr{P}_M^p$ into a single bilinear map

$$\mathscr{P}_M : H(M) \times H_c(M) \to \mathbb{R}$$

by setting

$$\mathscr{P}_M(H^p(M), H_c^q(M)) = 0, \qquad p + q \neq n.$$

Now denote by $H_c^p(M)^*$ the space $L(H_c^p(M); \mathbb{R})$ of linear functions in $H_c^p(M)$. Then $H_c(M)^* = \sum_p H_c^p(M)^*$ is the space of linear functions in $H_c(M)$. The Poincaré scalar products determine linear maps

$$D_M^p : H^p(M) \to H_c^{n-p}(M)^*$$

by the equation

$$\langle D_M^p \alpha, \beta \rangle = \mathscr{P}_M^p(\alpha, \beta), \qquad \alpha \in H^p(M), \quad \beta \in H_c^{n-p}(M).$$

Combining these maps yields a linear map

$$D_M: H(M) \to H_c(M)^*.$$

(It will be shown in sec. 5.12 that D_M is a linear isomorphism.)

Example: The linear function $\int_M^\# \in H_c^n(M)^*$ is given by

$$\int_M^\# = D_M(1).$$

Now consider an orientation preserving diffeomorphism $\varphi: M \to N$ of M onto an open subset of an oriented n-manifold N. φ induces homomorphisms

$$\varphi^\#: H(M) \leftarrow H(N) \qquad \text{and} \qquad (\varphi_c)_\#: H_c(M) \to H_c(N).$$

Let $(\varphi_c)_\#^*$ denote the linear map dual to $(\varphi_c)_\#$.

Proposition VI: The diagram

$$\begin{array}{ccc} H(M) & \xleftarrow{\varphi^\#} & H(N) \\ {\scriptstyle D_M}\downarrow & & \downarrow{\scriptstyle D_N} \\ H_c(M)^* & \xleftarrow[(\varphi_c)_\#^*]{} & H_c(N)^* \end{array}$$

commutes.

Proof: Let $\alpha \in H^p(N)$, $\beta \in H_c^{n-p}(M)$ be represented by $\Phi \in A^p(N)$, $\Psi \in A_c^{n-p}(M)$. Then $(\varphi_c)_\# \beta \in H_c^{n-p}(N)$ is represented by $(\varphi_c)_* \Psi$, and

$$\varphi^*(\Phi \wedge (\varphi_c)_* \Psi) = \varphi^* \Phi \wedge \Psi.$$

Hence Proposition XII, sec. 4.13, can be applied to give

$$\int_M^\# (\varphi^\# \alpha) * \beta = \int_M \varphi^* \Phi \wedge \Psi = \int_N \Phi \wedge (\varphi_c)_* \Psi = \int_N^\# \alpha * (\varphi_c)_\# \beta.$$

Thus

$$\mathscr{P}_M^p(\varphi^\# \alpha, \beta) = \mathscr{P}_N^p(\alpha, (\varphi_c)_\# \beta)$$

and the proposition follows.

Q.E.D.

Next, suppose $M = U \cup V$ (U, V open). Recall, from secs. 5.4 and 5.10, the Mayer–Vietoris sequences

$$\cdots \xrightarrow{\partial} H^p(M) \longrightarrow H^p(U) \oplus H^p(V) \longrightarrow H^p(U \cap V) \xrightarrow{\partial} H^{p+1}(M) \longrightarrow \cdots$$

and

$$\cdots \xleftarrow{\partial_c} H_c^{n-p}(M) \leftarrow H_c^{n-p}(U) \oplus H_c^{n-p}(V) \leftarrow H_c^{n-p}(U \cap V) \xleftarrow{\partial_c} H_c^{n-p-1}(M) \leftarrow \cdots.$$

Dualizing the second sequence and replacing ∂ by $(-1)^{p+1}\partial$ we obtain the row-exact diagram

$$\begin{array}{ccccccccc}
\cdots \xrightarrow{(-1)^p\partial} & H^p(M) & \rightarrow & H^p(U) \oplus H^p(V) & \rightarrow & H^p(U \cap V) & \xrightarrow{(-1)^{p+1}\partial} & H^{p+1}(M) & \rightarrow \cdots \\
& \downarrow D_M & ① & \downarrow D_U \oplus D_V & ② & \downarrow D_{U\cap V} & ③ & \downarrow D_M & \\
\cdots \xrightarrow{\partial_c^*} & H_c^{n-p}(M)^* & \rightarrow & H_c^{n-p}(U)^* \oplus H_c^{n-p}(V)^* & \rightarrow & H_c^{n-p}(U\cap V)^* & \xrightarrow{\partial_c^*} & H_c^{n-p-1}(M)^* & \rightarrow \cdots
\end{array} \qquad (5.7)$$

Proposition VII: The diagram (5.7) commutes.

Proof: It is immediate from Proposition VI that squares ① and ② commute. To show that ③ commutes, let

$$\alpha \in H^p(U \cap V) \qquad \text{and} \qquad \beta \in H_c^{n-p-1}(M)$$

be cohomology classes, and let $\Phi \in A^p(U \cap V)$, $\Psi \in A_c^{p-n-1}(M)$ be representing cocycles. We must prove that

$$\mathscr{P}_M^{p+1}((-1)^{p+1}\,\partial\alpha, \beta) = \mathscr{P}_{U\cap V}^p(\alpha, \partial_c\beta).$$

Choose $\Phi_1 \in A^p(U)$, $\Phi_2 \in A^p(V)$ so that

$$\Phi_1|_{U\cap V} - \Phi_2|_{U\cap V} = \Phi.$$

Then $\partial\alpha$ is represented by Ω, where

$$\Omega|_U = \delta\Phi_1 \qquad \text{and} \qquad \Omega|_V = \delta\Phi_2.$$

Next, choose $\Psi_1 \in A_c^{n-p-1}(U)$, $\Psi_2 \in A_c^{n-p-1}(V)$ so that

$$\Psi = \Psi_1 + \Psi_2.$$

Then $\delta\Psi_1 = -\delta\Psi_2 \in A_c^{n-p}(U \cap V)$ and $\partial_c\beta$ is represented by $\delta\Psi_1$.

It follows that

$$\mathscr{P}_M^{p+1}((-1)^{p+1}\,\partial\alpha,\beta)$$

$$= (-1)^{p+1}\int_M \Omega\wedge\Psi = (-1)^{p+1}\int_U \Omega\wedge\Psi_1 + (-1)^{p+1}\int_V \Omega\wedge\Psi_2$$

$$= (-1)^{p+1}\int_U \delta\Phi_1\wedge\Psi_1 + (-1)^{p+1}\int_V \delta\Phi_2\wedge\Psi_2 .$$

But $\Phi_1\wedge\Psi_1\in A_c^{n-1}(U)$, $\Phi_2\wedge\Psi_2\in A_c^{n-1}(V)$. Thus by Proposition XIV, sec. 4.13,

$$\int_U \delta(\Phi_1\wedge\Psi_1) = 0 = \int_V \delta(\Phi_2\wedge\Psi_2).$$

Hence

$$\mathscr{P}_M^{p+1}((-1)^{p+1}\,\partial\alpha,\beta) = \int_U \Phi_1\wedge\delta\Psi_1 + \int_V \Phi_2\wedge\delta\Psi_2$$

$$= \int_{U\cap V}(\Phi_1-\Phi_2)\wedge\delta\Psi_1 = \int_{U\cap V}\Phi\wedge\delta\Psi_1$$

$$= \mathscr{P}_{U\cap V}^p(\alpha,\partial_c\beta),$$

and so ③ commutes.

Q.E.D.

5.12. The main theorem. Theorem I: Let M be an oriented n-manifold. Then

$$D_M\colon H(M)\to H_c(M)^*$$

is a linear isomorphism.

Definition: D_M will be called the *Poincaré isomorphism.*

Before proving the theorem we establish three lemmas. Let $\mathcal{O}$ be an i-basis for the topology of M. Denote by $\mathcal{O}_f$ the open sets of M which can be expressed as finite unions of elements in $\mathcal{O}$. Denote by $\mathcal{O}_s$ the open subsets of M which can be expressed as (at most countable) disjoint unions of elements of $\mathcal{O}$. Then $\mathcal{O}_f$ and $\mathcal{O}_s$ are i-bases for the topology of M (cf. sec. 0.11 and sec. 1.1).

Lemma II: Let $\mathcal{O}$ be an i-basis for the topology of M. Assume D_O is an isomorphism for every $O\in\mathcal{O}$. Then D_O is an isomorphism for every $O\in\mathcal{O}_f$.

Proof: If $O \in \mathcal{O}_\mathrm{f}$, we can write

$$O = O_1 \cup \cdots \cup O_m, \qquad O_i \in \mathcal{O}.$$

Set $U = O_1$, $V = O_2 \cup \cdots \cup O_m$. Then

$$U \cap V = (O_1 \cap O_2) \cup \cdots \cup (O_1 \cap O_m).$$

Since $\mathcal{O}$ is an i-basis each $O_1 \cap O_i \in \mathcal{O}$. Thus we may assume by induction on m that D_U, D_V, and $D_{U\cap V}$ are all isomorphisms.

Now apply the five-lemma to the commutative diagram of Proposition VII, sec. 5.11, to obtain that D_O is an isomorphism.

Q.E.D.

Lemma III: Let $\mathcal{O}$ be a basis for the topology of M. Assume that D_O is an isomorphism for every $O \in \mathcal{O}$. Then D_O is an isomorphism for every $O \in \mathcal{O}_s$.

Proof: An element $O \in \mathcal{O}_s$ can be written

$$O = \bigcup_{i=1}^{\infty} O_i, \qquad O_i \in \mathcal{O}$$

where the O_i are disjoint. Now recall (cf. secs. 5.3 and 5.10) the canonical isomorphisms

$$H(O) \xrightarrow{\cong} \prod_{i=1}^{\infty} H(O_i)$$

and

$$H_c(O) \xleftarrow{\cong} \textstyle\bigoplus_{i=1}^{\infty} H_c(O_i).$$

The latter dualizes to yield an isomorphism

$$H_c(O)^* \xrightarrow{\cong} \prod_{i=1}^{\infty} H_c(O_i)^*.$$

Denote the linear isomorphisms D_{O_i} by D_i. An elementary computation shows that the diagram

$$\begin{array}{ccc} H(O) & \xrightarrow{D_O} & H_c(O)^* \\ \cong\downarrow & & \downarrow\cong \\ \prod_i H(O_i) & \xrightarrow[\Pi_i D_i]{\cong} & \prod_i H_c(O_i)^* \end{array}$$

commutes. It follows that D_O is an isomorphism.

Q.E.D.

Lemma IV: Let $\mathcal{O}$ be an i-basis for the topology of M. Assume D_O is an isomorphism for every $O \in \mathcal{O}$. Then D_O is an isomorphism for every open subset $O \subset M$.

Proof: According to Proposition II, sec. 1.1, the basis $((\mathcal{O}_{\rm f})_{\rm s})_{\rm f}$ contains all the open sets of M. Now apply Lemmas II and III.

Q.E.D.

Proof of Theorem I: We come to the actual proof of Theorem I, and proceed in three stages:

Case I: $M = \mathbb{R}^n$. Since (cf. Example 1, sec. 5.5, and Proposition IV, sec. 5.10)

$$H^p(\mathbb{R}^n) = \begin{cases} \mathbb{R}, & p = 0 \\ 0, & p \neq 0 \end{cases} \quad \text{and} \quad H_c^p(\mathbb{R}^n) = \begin{cases} \mathbb{R}, & p = n \\ 0, & p \neq n, \end{cases}$$

we need only show that

$$D\colon H^0(\mathbb{R}^n) \to H_c^n(\mathbb{R}^n)^*$$

is a linear isomorphism. Moreover, since

$$\dim H^0(\mathbb{R}^n) = 1 = \dim H_c^n(\mathbb{R}^n)^*,$$

it is sufficient to show that $D \neq 0$.

Let Δ be a positive determinant function in $\mathbb{R}^n$, and let $f \in \mathscr{S}_c(\mathbb{R}^n)$ be a nonnegative function which is not identically zero. Then, for a suitable basis of $\mathbb{R}^n$,

$$\int_{\mathbb{R}^n} f \cdot \Delta = \int_{\mathbb{R}^n} f(x)\, dx^1 \cdots dx^n > 0.$$

Thus, in view of Proposition XIV, sec. 4.13, $f \cdot \Delta$ represents an element $\beta \neq 0$ in $H_c^n(\mathbb{R}^n)$.

But it follows immediately from the definitions that

$$\langle D(1), \beta \rangle = \int_{\mathbb{R}^n} 1 \wedge (f \cdot \Delta) = \int_{\mathbb{R}^n} f \cdot \Delta \neq 0.$$

Hence $D(1) \neq 0$, and so $D \neq 0$.

Case II: *M is an open subset of $\mathbb{R}^n$.* Let $e_1, \ldots, e_n$ be a basis of $\mathbb{R}^n$. If $x \in \mathbb{R}^n$, write $x = \sum_{i=1}^n x^i e_i$. Then the open subsets of the form

$$O = \{x \in \mathbb{R}^n \mid a^i < x^i < b^i, \quad i = 1, \ldots, n\}$$

are an i-basis for the topology of $\mathbb{R}^n$. Each such O is diffeomorphic to $\mathbb{R}^n$. Thus Proposition VI, sec. 5.11, together with Case I imply that D_O is an isomorphism for each such O. Now it follows from Lemma IV above that D_M is an isomorphism for every open set $M \subset \mathbb{R}^n$.

Case III: M arbitrary. Let $\mathcal{O}$ be the collection of open subsets of M which are diffeomorphic to open subsets of $\mathbb{R}^n$. Clearly $\mathcal{O}$ is an i-basis for the topology of M. In view of Proposition VI, sec. 5.11, and Case II above, D_O is an isomorphism for every $O \in \mathcal{O}$. Thus Lemma IV above implies that D_U is an isomorphism for every open subset $U \subset M$. In particular, D_M is an isomorphism,

$$D_M: H(M) \xrightarrow{\cong} H_c(M)^*.$$

Q.E.D.

Corollary I: The bilinear maps $\mathscr{P}_M, \mathscr{P}_M^p$ are nondegenerate (i.e., scalar products in the sense of linear algebra).

Corollary II: Let M be an oriented manifold, and let $j: U \to M$ be the inclusion of an open subset. Then

$$j^\#: H(U) \leftarrow H(M)$$

is an isomorphism if and only if

$$(j_c)_\#: H_c(U) \to H_c(M)$$

is an isomorphism.

Proof: Apply Proposition VI, sec. 5.11 and Theorem I.

Q.E.D.

§5. Applications of Poincaré duality

5.13. Orientation class. **Theorem II:** Let M be a connected oriented n-manifold. Then

$$\int_M^{\#} : H_c^n(M) \to \mathbb{R}$$

is a linear isomorphism. Moreover,

$$\ker \int_M = \delta(A_c^{n-1}(M)).$$

Proof: The second statement is an immediate consequence of the first. To prove the first observe that Theorem 1, sec. 5.12, gives

$$\dim H_c^n(M) = \dim H^0(M) = 1$$

(since M is connected). Since $\int_M^{\#}$ is surjective (cf. sec. 5.11), it must be an isomorphism.

Q.E.D.

Definition: The unique cohomology class $\omega_M \in H_c^n(M)$ which satisfies

$$\int_M^{\#} \omega_M = 1$$

is called the *orientation class for* M.

Remark: If M is compact $\int_M^{\#}$ is an isomorphism from $H^n(M)$, and $\omega_M \in H^n(M)$.

5.14. Nonorientable manifolds. Let M be a connected nonorientable n-manifold, and let $\pi: \tilde{M} \to M$ be the induced double cover (cf. Example 9, sec. 3.21). Thus $\tilde{M}$ is a connected orientable manifold. Let $\tau: \tilde{M} \to \tilde{M}$ be the covering transformation.

Lemma V: τ is orientation reversing.

Proof: Suppose $\Delta \in A^n(\tilde{M})$ orients $\tilde{M}$. Then $\tau^*\Delta = f \cdot \Delta$ where either $f > 0$ or $f < 0$ (because $\tilde{M}$ is connected). If $f > 0$, the n-form

$$\Omega = \Delta + \tau^*\Delta = (1 + f) \cdot \Delta$$

would have no zeros.

On the other hand $\tau^*\Omega = \Omega$; hence by Example 4, sec. 3.20, $\Omega = \pi^*\Phi$ for some $\Phi \in A^n(M)$. If Ω has no zeros, neither does Φ, thus Φ would orient M. It follows that $f < 0$; i.e. τ reverses orientations.

Q.E.D.

Next, write (in analogy with Example 4, sec. 3.20)

$$A_c(\tilde{M}) = (A_c)_+(\tilde{M}) \oplus (A_c)_-(\tilde{M}),$$

where $\Phi \in (A_c)_+(\tilde{M})$ (resp. $(A_c)_-(\tilde{M})$) if $\tau^*\Phi = \Phi$ (resp. $\tau^*\Phi = -\Phi$). This leads to the decomposition

$$H_c(\tilde{M}) = (H_c)_+(\tilde{M}) \oplus (H_c)_-(\tilde{M}),$$

where $\alpha \in (H_c)_+(\tilde{M})$ (resp. $(H_c)_-(\tilde{M})$) if $\tau_c^\#\alpha = \alpha$ (resp. $\tau_c^\#\alpha = -\alpha$). Thus

$$(H_c)_+(\tilde{M}) = H((A_c)_+(\tilde{M}), \delta), \qquad (H_c)_-(\tilde{M}) = H((A_c)_-(\tilde{M}), \delta).$$

π_c^*, $\pi_c^\#$ can be considered as isomorphisms

$$A_c(M) \xrightarrow{\cong} (A_c)_+(\tilde{M}), \qquad H_c(M) \xrightarrow{\cong} (H_c)_+(\tilde{M}).$$

Lemma VI: $H_c^n(\tilde{M}) = (H_c^n)_-(\tilde{M}); \qquad (H_c^n)_+(\tilde{M}) = 0.$

Proof: Since $\tilde{M}$ is connected and orientable, $\dim H_c^n(\tilde{M}) = 1$ (by Theorem II, sec. 5.13). Since

$$H_c^n(\tilde{M}) = (H_c^n)_+(\tilde{M}) \oplus (H_c^n)_-(\tilde{M})$$

it is sufficient to prove that $(H_c^n)_-(\tilde{M}) \neq 0$.

Orient $\tilde{M}$ and let $\Omega \in A_c^n(\tilde{M})$ be positive. Since τ reverses orientations,

$$\Phi = \Omega - \tau^*\Omega$$

is again positive. Hence $\int_{\tilde{M}} \Phi > 0$; i.e. Φ represents a nontrivial cohomology class $\alpha \in H_c^n(\tilde{M})$. But $\tau^*\Phi = -\Phi$; thus

$$\alpha \in (H_c^n)_-(\tilde{M})$$

and so

$$(H_c^n)_-(\tilde{M}) \neq 0.$$

Q.E.D.

Corollary: $H_c^n(M) = 0$.

Proof: Recall $H_c^n(M) \cong (H_c^n)_+(\tilde{M})$.

Q.E.D.

Proposition VIII: If M is a nonorientable connected n-manifold with double cover $\tilde{M}$, then

(1) $\mathscr{P}_{\tilde{M}}(H_+^p(\tilde{M}),\ (H_c^{n-p})_+(\tilde{M})) = 0$

$\mathscr{P}_{\tilde{M}}(H_-^p(\tilde{M}),\ (H_c^{n-p})_-(\tilde{M})) = 0$

and

(2) $D_{\tilde{M}}$ restricts to linear isomorphisms

$$H_+(\tilde{M}) \xrightarrow{\cong} (H_c)_-(\tilde{M})^*, \qquad H_-(\tilde{M}) \xrightarrow{\cong} (H_c)_+(\tilde{M})^*.$$

Proof: If $\alpha \in H_+^p(\tilde{M})$, $\beta \in (H_c^{n-p})_+(\tilde{M})$, then

$$\alpha * \beta \in (H_c^n)_+(\tilde{M}) = 0$$

(cf. Lemma VI, above). Hence $\mathscr{P}_{\tilde{M}}^p(\alpha, \beta) = 0$ and the first relation of (1) is proved. The second equation in (1) follows in the same way. Finally, (2) is an immediate consequence of (1) and some elementary linear algebra.

Q.E.D.

Corollary I: Precomposing $D_{\tilde{M}}$ with $\pi^\#$ yields a linear isomorphism

$$H(M) \xrightarrow{\cong} (H_c)_-(\tilde{M})^*.$$

Corollary II: Composing $D_{\tilde{M}}$ with $(\pi_c^\#)^*$ yields a linear isomorphism

$$H_-(\tilde{M}) \xrightarrow{\cong} H_c(M)^*.$$

5.15. Compact manifolds. Let M be a compact oriented n-manifold. Then $H_c(M) = H(M)$ and so the Poincaré scalar products are bilinear maps,

$$\mathscr{P}_M^p : H^p(M) \times H^{n-p}(M) \to \mathbb{R}$$

while the Poincaré isomorphism is a linear map

$$D_M\colon H(M) \xrightarrow{\cong} H(M)^*.$$

In particular,

$$\mathscr{P}_M^p(\alpha, \beta) = \langle D_M(1), \alpha \cdot \beta \rangle = \int_M^{\#} \alpha \cdot \beta, \qquad \alpha \in H^p(M), \quad \beta \in H^{n-p}(M)$$

and hence

$$\mathscr{P}_M^p(\alpha, \beta) = (-1)^{p(n-p)} \mathscr{P}_M^{n-p}(\beta, \alpha), \qquad \alpha \in H^p(M), \quad \beta \in H^{n-p}(M). \quad (5.8)$$

If n is even, this yields

$$\mathscr{P}_M^p(\alpha, \beta) = (-1)^p \mathscr{P}_M^{n-p}(\beta, \alpha);$$

while if n is odd, we have

$$\mathscr{P}_M^p(\alpha, \beta) = \mathscr{P}_M^{n-p}(\beta, \alpha).$$

Formula (5.6), sec. 5.11, becomes

$$\mathscr{P}_M^{p+q}(\alpha \cdot \gamma, \beta) = \mathscr{P}_M^p(\alpha, \gamma \cdot \beta),$$

where

$$\alpha \in H^p(M), \quad \gamma \in H^q(M), \quad \beta \in H^{n-p-q}(M).$$

The duality theorem has the following further corollaries.

Corollary III: If $\alpha \in H^p(M)$, $\alpha \neq 0$ (M compact oriented), then for some $\beta \in H^{n-p}(M)$,

$$\alpha \cdot \beta = \omega_M .$$

Proof: Choose β so that $\mathscr{P}_M^p(\alpha, \beta) = 1$.

Q.E.D.

Corollary IV: Suppose $\varphi: Q \to M$ is smooth (M compact oriented). Assume $\varphi^{\#}\omega_M \neq 0$. Then $\varphi^{\#}$ is injective.

Proof: $\ker \varphi^{\#}$ is an ideal in $H(M)$ not containing ω_M. Corollary III implies that every nonzero ideal in $H(M)$ contains ω_M. Hence $\ker \varphi^{\#} = 0$.

Q.E.D.

Theorem III: Let M be any compact manifold. Then

$$\dim H(M) < \infty.$$

Proof: Assume first that M is orientable. In view of the duality theorem, formula (5.8) above shows that $\mathscr{P}^p_M$ induces *two* linear isomorphisms; namely

$$H^p(M) \xrightarrow{\cong} H^{n-p}(M)^* \qquad \text{and} \qquad H^{n-p}(M) \xrightarrow{\cong} H^p(M)^*.$$

It follows now from elementary linear algebra that each $H^p(M)$ has finite dimension; hence the theorem, in this case.

If M is nonorientable the double cover $\tilde{M}$ is orientable (and compact). In this case we have (cf. sec. 5.7)

$$\dim H(M) = \dim H_+(\tilde{M}) \leqslant \dim H(\tilde{M}) < \infty.$$

Q.E.D.

Corollary: If M is compact then the Betti numbers $b_p = \dim H^p(M)$ are defined. If, in addition, M is orientable, then

$$b_p = b_{n-p}, \qquad 0 \leqslant p \leqslant n.$$

Proposition IX: Let M be a connected n manifold. If M is compact and orientable, then

$$H^n(M) \cong \mathbb{R}.$$

Otherwise,

$$H^n(M) = 0.$$

Proof: Suppose first that M is compact. If M is orientable, Theorem II of sec. 5.13 implies $H^n(M) \cong \mathbb{R}$. If M is nonorientable, the Corollary to Lemma VI, sec. 5.14, gives $H^n(M) = 0$.

Next assume M is not compact. If M is orientable, the duality theorem gives

$$H^n(M) \cong H^0_c(M)^* = 0$$

(use the example of sec. 5.9). If M is nonorientable the double cover $\tilde{M}$ is connected, orientable, and noncompact. Hence

$$H^n(M) \cong H^n_+(\tilde{M}) \subset H^n(\tilde{M}) = 0.$$

Q.E.D.

5.16. Euler characteristic and signature. Let M be a compact n-manifold. Recall from sec. 5.1 that the Euler characteristic χ_M is defined by

$$\chi_M = \sum_{p=0}^{n} (-1)^p b_p$$

where b_p is the pth Betti number of M.

Next, consider a compact oriented $2m$-manifold M. According to formula (5.8), sec. 5.15, the nondegenerate scalar product

$$\mathscr{P}_M^m : H^m(M) \times H^m(M) \to \mathbb{R}$$

is skew-symmetric if m is odd and symmetric if m is even.

Recall (sec. 0.1) the definition of the signature of a symmetric scalar product.

Definition: If M is a compact oriented manifold of dimension $4k$ then the signature of the scalar product $\mathscr{P}_M^{2k}$ is called the *signature of* M. It will be denoted by $\operatorname{Sig}(M)$.

Theorem IV: Let M be a compact oriented n-manifold.

(1) If n is odd, then $\chi_M = 0$.
(2) If $n = 2m$, m odd, then $\chi_M \equiv b_m \equiv 0 \pmod 2$.
(3) If $n = 2m = 4k$, then $\operatorname{Sig}(M) \equiv b_m \equiv \chi_M \pmod 2$.
(4) If $n = 4k$ and $\operatorname{Sig}(M) \neq 0$, then M is irreversible.

Proof: The corollary to Theorem III, sec. 5.15, yields

$$\chi_M = \sum_p (-1)^p b_p = \sum_p (-1)^p b_{n-p} = (-1)^n \sum_p (-1)^{n-p} b_{n-p} = (-1)^n \chi_M .$$

This implies that $\chi_M = 0$ if n is odd and so (1) is proved.

Now assume that $n = 2m$. Then we have (again via the corollary to Theorem III)

$$\chi_M = 2 \sum_{p=0}^{m-1} (-1)^p b_p + (-1)^m b_m ,$$

whence $\chi_M \equiv b_m \pmod 2$. Since $\mathscr{P}_M^m$ is skew and nondegenerate if m is odd, $b_m \equiv 0 \pmod 2$, in this case.

Next, assume that $n = 2m = 4k$. It is evident that

$$b_m \equiv \operatorname{Sig}(M) \pmod 2$$

and (3) follows.

Finally, assume that $n = 2m = 4k$ and let $\varphi : M \to M$ be an orientation-reversing diffeomorphism. We must show that $\mathscr{P}_M^m$ has zero signature. Proposition XII, sec. 4.13, shows that

$$\int_M \varphi^* \Phi = - \int_M \Phi, \qquad \Phi \in A^n(M).$$

Hence it follows from Theorem II, sec. 5.13, that

$$\varphi^{\#}\gamma = -\gamma, \qquad \gamma \in H^n(M).$$

Now, for $\alpha, \beta \in H^m(M)$, we have

$$\mathscr{P}_M^m(\varphi^{\#}\alpha, \varphi^{\#}\beta) = \int_M^{\#} \varphi^{\#}(\alpha \cdot \beta) = -\int_M^{\#} \alpha \cdot \beta = -\mathscr{P}_M^m(\alpha, \beta).$$

Thus part (4) of the theorem follows from Lemma VII below:

Lemma VII: Let E be a finite-dimensional real vector space with a symmetric scalar product $\langle\ ,\ \rangle$. Assume $\varphi: E \to E$ is a linear map such that

$$\langle \varphi(x), \varphi(y) \rangle = -\langle x, y \rangle, \qquad x, y \in E.$$

Then $\langle\ ,\ \rangle$ has zero signature.

Proof: Let F be a subspace of maximum dimension, s, such that the restriction of $\langle\ ,\ \rangle$ to F is positive definite. Then the relation

$$\langle \varphi(x), \varphi(x) \rangle = -\langle x, x \rangle$$

shows that the restriction of $\langle\ ,\ \rangle$ to $\varphi(F)$ is negative definite. Moreover, $\varphi(F)$ is a subspace of maximum dimension with this property. Finally, our hypothesis implies that φ is injective. Hence

$$\dim F = \dim \varphi(F)$$

and so the signature of $\langle\ ,\ \rangle$ is zero.

Q.E.D.

§6. Künneth theorems

5.17. Künneth homomorphisms. Recall that the tensor product of two graded differential $\mathbb{R}$-algebras (E, δ_E) and (F, δ_F) is the graded algebra $E \otimes F$ (anticommutative tensor product) together with the differential operator $\delta_{E\otimes F}$ given by

$$\delta_{E\otimes F}(a \otimes b) = \delta_E(a) \otimes b + (-1)^p a \otimes \delta_F(b), \qquad a \in E^p, \quad b \in F.$$

If a, b are cocycles representing $\alpha \in H(E)$, $\beta \in H(F)$, then the cohomology class $\gamma \in H(E \otimes F)$ represented by $a \otimes b$ depends only on α and β. Thus $\alpha \otimes \beta \mapsto \gamma$ defines a linear map $H(E) \otimes H(F) \to H(E \otimes F)$.

Moreover (cf. [6, pp. 54–60]) this linear map is an isomorphism of graded algebras. Henceforth we shall identify $H(E) \otimes H(F)$ with $H(E \otimes F)$ under this isomorphism.

Now let M and N be manifolds. The linear map

$$\kappa\colon A(M) \otimes A(N) \to A(M \times N)$$

defined by

$$\kappa(\Phi \otimes \Psi) = \Phi \times \Psi$$

(cf. Example 2, sec. 3.20) is a homomorphism of graded differential algebras. Thus it induces a homomorphism

$$\kappa_\#\colon H(M) \otimes H(N) \to H(M \times N)$$

called the *Künneth homomorphism:*

$$\kappa_\#(\alpha \otimes \beta) = (\pi_M{}^\# \alpha) \cdot (\pi_N{}^\# \beta), \qquad \alpha \in H(M), \quad \beta \in H(N).$$

Suppose $\varphi\colon M \to M_1$ and $\psi\colon N \to N_1$ are smooth maps. Then the diagram

$$\begin{array}{ccc} H(M) \otimes H(N) & \xrightarrow{\kappa_\#} & H(M \times N) \\ \uparrow{\scriptstyle \varphi^\# \otimes \psi^\#} & & \uparrow{\scriptstyle (\varphi \times \psi)^\#} \\ H(M_1) \otimes H(N_1) & \xrightarrow[\kappa_\#]{} & H(M_1 \times N_1) \end{array}$$

commutes.

In sec. 5.20 it will be shown that $\kappa_\#$ is an isomorphism whenever $\dim H(M) < \infty$ or $\dim H(N) < \infty$.

Examples: 1. *The multiplication map:* If M is a manifold, the multiplication map is the homomorphism of graded differential algebras

$$\mu: A(M) \otimes A(M) \to A(M)$$

given by

$$\mu(\Phi \otimes \Psi) = \Phi \wedge \Psi.$$

On the other hand, the diagonal map $\Delta: M \to M \times M$ is the smooth map defined by

$$\Delta(x) = (x, x), \qquad x \in M.$$

Let $\pi_L: M \times M \to M$ be projection on the left factor. Since $\pi_L \circ \Delta = \iota_M$, we have

$$\Delta^*\kappa(\Phi \otimes 1) = \Delta^*\pi_L^*\Phi = \Phi = \mu(\Phi \otimes 1), \qquad \Phi \in A(M).$$

Similarly,

$$\Delta^*\kappa(1 \otimes \Psi) = \Psi = \mu(1 \otimes \Psi), \qquad \Psi \in A(M).$$

Now Δ^*, κ, and μ are algebra homomorphisms, and $A(M) \otimes 1$, $1 \otimes A(M)$ generate $A(M) \otimes A(M)$. Thus these relations imply that the diagram

$$\begin{array}{ccc} A(M) \otimes A(M) & \xrightarrow{\kappa} & A(M \times M) \\ & {\scriptstyle\mu} \searrow \quad \swarrow {\scriptstyle\Delta^*} & \\ & A(M) & \end{array}$$

commutes.

Passing to cohomology gives the commutative diagram

$$\begin{array}{ccc} H(M) \otimes H(M) & \xrightarrow{\kappa_\#} & H(M \times M) \\ & {\scriptstyle\mu_\#} \searrow \quad \swarrow {\scriptstyle\Delta^\#} & \\ & H(M) & \end{array}$$

Thus

$$\alpha \cdot \beta = \Delta^\#\kappa_\#(\alpha \otimes \beta), \qquad \alpha, \beta \in H(M).$$

2. *Fibre projection:* Let M and N be manifolds such that M is connected. Fix $a \in M$. The inclusion map opposite a, $j_a: N \to M \times N$, induces a homomorphism

$$j_a^\#: H(N) \leftarrow H(M \times N).$$

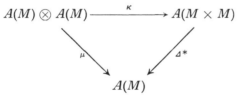

If $b \in M$ is a second point, there is a smooth path $f: \mathbb{R} \to M$ such that $f(0) = a$ and $f(1) = b$ (cf. sec. 1.11). f is a homotopy between j_a and j_b ; thus

$$j_a^{\#} = j_b^{\#}.$$

The homomorphism $j_a^{\#}$, which is independent of a, is called the *fibre projection.*

On the other hand, since M is connected, we have (cf. sec. 5.1) $H(M) = \mathbb{R} \oplus H^+(M)$ $(H^+(M) = \sum_1^n H^p(M))$. Hence a homomorphism $\rho_N: H(M) \otimes H(N) \to H(N)$ is defined by

$$\rho_N(1 \otimes \beta) = \beta, \quad \rho_N(\alpha \otimes \beta) = 0, \qquad \beta \in H(N), \quad \alpha \in H^+(M).$$

A trivial argument shows that the diagram

$$\begin{array}{ccc} H(M) \otimes H(N) & \xrightarrow{\kappa_{\#}} & H(M \times N) \\ & \searrow^{\rho_N} \quad {}^{j_a^{\#}}\swarrow & \\ & H(N) & \end{array}$$

commutes.

5.18. The homomorphism $(\kappa_c)_{\#}$. Let M and N be manifolds. κ restricts to a homomorphism

$$\kappa_c: A_c(M) \otimes A_c(M) \to A_c(M \times N)$$

and κ_c induces a homomorphism

$$(\kappa_c)_{\#}: H_c(M) \otimes H_c(N) \to H_c(M \times N).$$

$(\kappa_c)_{\#}$ is called the *compact Künneth homomorphism.*

The homomorphisms $\kappa_{\#}$ and $(\kappa_c)_{\#}$ are related as follows, via Poincaré duality. Let ϵ denote the linear isomorphism of $H(M) \otimes H(N)$ given by

$$\epsilon(\alpha \otimes \beta) = (-1)^{(m-p)q}\, \alpha \otimes \beta, \qquad \alpha \in H^p(M), \quad \beta \in H^q(N)$$

$(m = \dim M)$.

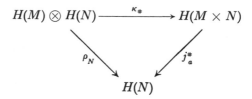

Proposition X: Suppose M, N are oriented manifolds and give $M \times N$ the product orientation. Then the diagram

$$\begin{array}{ccc} H(M) \otimes H(N) & \xrightarrow{\kappa_\#} & H(M \times N) \\ {\scriptstyle (D_M \otimes D_N) \circ \epsilon} \downarrow \cong & & \cong \downarrow {\scriptstyle D_{M \times N}} \\ H_c(M)^* \otimes H_c(N)^* & & H_c(M \times N)^* \\ & {\scriptstyle i} \searrow \quad \swarrow {\scriptstyle (\kappa_c)_\#^*} & \\ & [H_c(M) \otimes H_c(N)]^* & \end{array}$$

commutes, where i is the standard inclusion map.

Proof: Fix $\alpha \in H^p(M)$, $\beta \in H^q(N)$, $\gamma \in H_c^{m-p}(M)$, $\lambda \in H_c^{n-q}(N)$ ($m = \dim M$, $n = \dim N$). We must show that

$$(-1)^{(m-p)q} \langle D_M\alpha \otimes D_N\beta, \gamma \otimes \lambda \rangle = \langle D_{M\times N}\kappa_\#(\alpha \otimes \beta), (\kappa_c)_\#(\gamma \otimes \lambda) \rangle.$$

Let $\Phi_1 \in A^p(M)$, $\Psi_1 \in A^q(N)$, $\Phi_2 \in A_c^{m-p}(M)$, $\Psi_2 \in A_c^{n-q}(N)$ represent α, β, γ, λ. Then $\Phi_1 \times \Psi_1$ represents $\kappa_\#(\alpha \otimes \beta)$; $\Phi_2 \times \Psi_2$ represents $(\kappa_c)_\#(\gamma \otimes \lambda)$. Thus we must show

$$(-1)^{(m-p)q} \int_M \Phi_1 \wedge \Phi_2 \int_N \Psi_1 \wedge \Psi_2 = \int_{M \times N} (\Phi_1 \times \Psi_1) \wedge (\Phi_2 \times \Psi_2).$$

But this follows at once from Proposition XIII, sec. 4.13.

Q.E.D.

5.19. The Künneth theorem for $(\kappa_c)_\#$. Theorem V: The compact Künneth homomorphism is an isomorphism,

$$(\kappa_c)_\#: H_c(M) \otimes H_c(N) \xrightarrow{\cong} H_c(M \times N).$$

Corollary: If M and N are compact, then the Künneth homomorphism is an isomorphism,

$$\kappa_\#: H(M) \otimes H(N) \xrightarrow{\cong} H(M \times N).$$

In particular, the Poincaré polynomial of a product of two compact manifolds is given by

$$f_{M \times N}(t) = f_M(t) \cdot f_N(t).$$

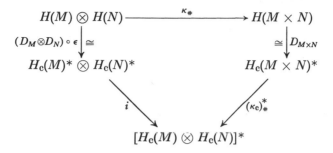

To prove Theorem V, we begin with preliminary results. Recall that if $\mathcal{O}$ is an i-basis for the topology of M, then $\mathcal{O}_{\mathrm{f}}$ is the an i-basis consisting of finite unions of open sets $O_\alpha \in \mathcal{O}$ while $\mathcal{O}_{\mathrm{s}}$ is the i-basis consisting of disjoint, at most countable, unions of open sets $O_\alpha \in \mathcal{O}$ (cf. sec. 1.1).

Lemma VIII: Suppose $\mathcal{O}$ is an i-basis for the topology of M and that

$$(\kappa_{\mathrm{c}})_{\#}: H_{\mathrm{c}}(O) \otimes H_{\mathrm{c}}(N) \to H_{\mathrm{c}}(O \times N) \tag{5.9}$$

is an isomorphism for each $O \in \mathcal{O}$. Then (5.9) is an isomorphism for every $O \in \mathcal{O}_{\mathrm{f}}$.

Proof: Let $O \in \mathcal{O}_{\mathrm{f}}$ and write

$$O = U_1 \cup \cdots \cup U_p, \qquad U_j \in \mathcal{O}.$$

Set $U = U_1$, $V = U_2 \cup \cdots \cup U_p$, $W = U \cap V$. Then

$$W = (U_1 \cap U_2) \cup \cdots \cup (U_1 \cap U_p)$$

and each $U_1 \cap U_j \in \mathcal{O}$. Thus by induction on p we may assume the lemma holds for U, V, and W.

Now consider (cf. sec. 5.10) the exact sequences

$$0 \to A_{\mathrm{c}}(W) \to A_{\mathrm{c}}(U) \oplus A_{\mathrm{c}}(V) \to A_{\mathrm{c}}(O) \to 0$$

and

$$0 \to A_{\mathrm{c}}(W \times N) \to A_{\mathrm{c}}(U \times N) \oplus A_{\mathrm{c}}(V \times N) \to A_{\mathrm{c}}(O \times N) \to 0.$$

Tensoring the first with $A_{\mathrm{c}}(N)$ yields the row-exact commutative diagram

$$\begin{array}{ccccccccc} 0 \to & A_{\mathrm{c}}(W)\otimes A_{\mathrm{c}}(N) & \to & [A_{\mathrm{c}}(U)\otimes A_{\mathrm{c}}(N)]\oplus[A_{\mathrm{c}}(V)\otimes A_{\mathrm{c}}(N)] & \to & A_{\mathrm{c}}(O)\otimes A_{\mathrm{c}}(N) & \to 0 \\ & \downarrow{\scriptstyle \kappa_{\mathrm{c}}} & & \downarrow{\scriptstyle \kappa_{\mathrm{c}}\oplus\kappa_{\mathrm{c}}} & & \downarrow{\scriptstyle \kappa_{\mathrm{c}}} & \\ 0 \longrightarrow & A_{\mathrm{c}}(W \times N) & \longrightarrow & A_{\mathrm{c}}(U \times N) \oplus A_{\mathrm{c}}(V \times N) & \longrightarrow & A_{\mathrm{c}}(O \times N) & \longrightarrow 0. \end{array} \tag{5.10}$$

By induction the maps

$$(\kappa_{\mathrm{c}})_{\#}: H_{\mathrm{c}}(W) \otimes H_{\mathrm{c}}(N) \to H_{\mathrm{c}}(W \times N)$$

$$(\kappa_{\mathrm{c}})_{\#}: H_{\mathrm{c}}(U) \otimes H_{\mathrm{c}}(N) \to H_{\mathrm{c}}(U \times N)$$

and

$$(\kappa_{\mathrm{c}})_{\#}: H_{\mathrm{c}}(V) \otimes H_{\mathrm{c}}(N) \to H_{\mathrm{c}}(V \times N)$$

are isomorphisms. Thus applying the five-lemma to the row-exact commutative diagram of cohomology induced by (5.10) we find that

$$(\kappa_c)_\#: H_c(O) \otimes H_c(N) \to H_c(O \times N)$$

is an isomorphism.

Q.E.D.

Lemma IX: Let $\mathcal{O}$ be a basis for the topology of M and assume that

$$(\kappa_c)_\#: H_c(O) \otimes H_c(N) \to H_c(O \times N) \tag{5.11}$$

is an isomorphism for every $O \in \mathcal{O}$. Then (5.11) is an isomorphism for every $O \in \mathcal{O}_s$.

Proof: If $O \in \mathcal{O}_s$ we can write O as the disjoint union

$$O = \bigcup_\alpha O_\alpha, \qquad O_\alpha \in \mathcal{O}.$$

Using sec. 5.10 we can construct a commutative diagram

$$\begin{array}{ccc} \bigoplus_\alpha (H_c(O_\alpha) \otimes H_c(N)) & \xrightarrow{\oplus_\alpha (\kappa_c)_\#} & \bigoplus_\alpha H_c(O_\alpha \times N) \\ \varphi \big\downarrow \cong & & \cong \big\downarrow \\ H_c(O) \otimes H_c(N) & \xrightarrow[(\kappa_c)_\#]{} & H_c(O \times N), \end{array}$$

where φ is the composite map

$$\bigoplus_\alpha (H_c(O_\alpha) \otimes H_c(N)) \xrightarrow{\cong} (\bigoplus_\alpha H_c(O_\alpha)) \otimes H_c(N) \xrightarrow{\cong} H_c(O) \otimes H_c(N).$$

By hypothesis the maps

$$(\kappa_c)_\#: H_c(O_\alpha) \otimes H_c(N) \to H_c(O_\alpha \times N)$$

are isomorphisms. The lemma follows.

Q.E.D.

Proposition XI: Suppose that for some i-basis $\mathcal{O}$ of the topology of M

$$(\kappa_c)_\#: H_c(O) \otimes H_c(N) \to H_c(O \times N) \tag{5.12}$$

is an isomorphism for every $O \in \mathcal{O}$. Then (5.12) is an isomorphism for every open subset of M.

Proof: Lemma VIII shows that (5.12) is an isomorphism for every $O \in \mathcal{O}_\mathrm{f}$. Thus Lemma IX implies that (5.12) is an isomorphism for every $O \in (\mathcal{O}_\mathrm{f})_\mathrm{s}$. Applying Lemma VIII again we see that (5.12) is an isomorphism for every $O \in ((\mathcal{O}_\mathrm{f})_\mathrm{s})_\mathrm{f}$. But according to Proposition II, sec. 1.1, every open subset of M is in $((\mathcal{O}_\mathrm{f})_\mathrm{s})_\mathrm{f}$.

Q.E.D.

Proof of Theorem V: *Case 1:* $M = \mathbb{R}^n$, $N = \mathbb{R}^r$: Let $f \in \mathscr{S}_\mathrm{c}(M)$ and $g \in \mathscr{S}_\mathrm{c}(N)$ be nonzero nonnegative functions. Let Δ_M and Δ_N be determinant functions in $\mathbb{R}^n$ and $\mathbb{R}^r$. Then (using Proposition XIII, sec. 4.13)

$$\int_M f \cdot \Delta_M > 0, \qquad \int_N g \cdot \Delta_N > 0, \qquad \text{and} \qquad \int_{M \times N} f \cdot \Delta_M \times g \cdot \Delta_N > 0.$$

Thus according to Proposition XIV, sec. 4.13, $f \cdot \Delta_M$, $g \cdot \Delta_N$ and $f \cdot \Delta_M \times g \cdot \Delta_N = \kappa_\mathrm{c}(f \cdot \Delta_M \otimes g \cdot \Delta_N)$ represent nonzero classes in $H_\mathrm{c}^n(M)$, $H_\mathrm{c}^r(N)$ and $H_\mathrm{c}^{n+r}(M \times N)$.

Hence

$$(\kappa_\mathrm{c})_\#: H_\mathrm{c}^n(\mathbb{R}^n) \otimes H_\mathrm{c}^r(\mathbb{R}^r) \to H_\mathrm{c}^{n+r}(\mathbb{R}^{n+r})$$

is nonzero. Now it follows immediately from Proposition IV, sec. 5.10, that

$$(\kappa_\mathrm{c})_\#: H_\mathrm{c}(\mathbb{R}^n) \otimes H_\mathrm{c}(\mathbb{R}^r) \to H_\mathrm{c}(\mathbb{R}^{n+r})$$

is an isomorphism.

Case 2: M is an open subset of $\mathbb{R}^n$, $N = \mathbb{R}^r$: Consider the *i*-basis of the topology of $\mathbb{R}^n$ consisting of the sets

$$U = \{(x^1, \ldots, x^n) \mid a^i < x^i < b^i, \quad i = 1, \ldots, n\}.$$

Each of these sets is diffeomorphic to $\mathbb{R}^n$. Thus (in view of Case 1) Proposition XI implies that

$$(\kappa_\mathrm{c})_\#: H_\mathrm{c}(M) \otimes H_\mathrm{c}(\mathbb{R}^r) \to H_\mathrm{c}(M \times \mathbb{R}^r)$$

is an isomorphism for any open subset M of $\mathbb{R}^n$.

Case 3: M arbitrary, $N = \mathbb{R}^r$: The fact that $(\kappa_\mathrm{c})_\#$ is an isomorphism in this case, follows from Proposition XI and Case 2, once we observe that M has an *i*-basis of open sets each diffeomorphic to an open subset of $\mathbb{R}^n$.

Case 4: M arbitrary, N an open subset of $\mathbb{R}^r$: Evidently Proposition XI continues to hold if M and N are interchanged. Applying it to an i-basis for the topology of $\mathbb{R}^r$ consisting of open sets diffeomorphic to $\mathbb{R}^r$ we see that $(\kappa_c)_\#$ is an isomorphism if N is an open subset of $\mathbb{R}^r$.

Case 5: M and N arbitrary: Apply Proposition XI to an i-basis of the topology of N consisting of open sets diffeomorphic to open subsets of $\mathbb{R}^r$.

Q.E.D.

Examples: 1. Let

$$M = S^{k_1} \times \cdots \times S^{k_r}$$

where S^{k_j} is a sphere of dimension k_j. Since the Poincaré polynomial of S^{k_j} is given by

$$f(t) = 1 + t^{k_j}$$

(cf. sec. 5.6), it follows that the Poincaré polynomial for M is given by

$$f_M(t) = (1 + t^{k_1})(1 + t^{k_2}) \cdots (1 + t^{k_r}).$$

2. Let T^n be the n-torus. Since T^n is diffeomorphic to $S^1 \times \cdots \times S^1$ (n factors) and the Poincaré polynomial of S^1 is given by $f(t) = 1 + t$ it follows that

$$f_{T^n}(t) = (1 + t)^n.$$

5.20. The Künneth theorem for $\kappa_\#$. Theorem VI: Let M and N be manifolds such that either $\dim H(M) < \infty$ or $\dim H(N) < \infty$. Then

$$\kappa_\#: H(M) \otimes H(N) \to H(M \times N)$$

is an isomorphism.

Proof: Assume first that M and N are orientable, and consider the diagram of Proposition X, sec. 5.18. Since $(\kappa_c)_\#$ (Theorem V above) is an isomorphism we need only show that the inclusion

$$i: H_c(M)^* \otimes H_c(N)^* \to [H_c(M) \otimes H_c(N)]^*$$

is surjective. But this is the case since either $H_c(M)^* \cong H(M)$ or $H_c(N)^* \cong H(N)$ has finite dimension (cf. Theorem I, sec. 5.12).

Next, assume M nonorientable, N orientable. Let $\rho: \tilde{M} \to M$ be the double cover, and consider $\tilde{M} \times N$ as a double cover of $M \times N$. $\tilde{M} \times N$ is orientable and the isomorphism $\kappa_\#$ for $\tilde{M} \times N$ satisfies

$$\kappa_\# \circ (\tau^\# \otimes \iota) = (\tau \times \iota)^\# \circ \kappa_\# ,$$

where τ is the covering transformation of $\tilde{M}$. Hence it restricts to an isomorphism

$$\kappa_\#: H_+(\tilde{M}) \otimes H(N) \xrightarrow{\cong} H_+(\tilde{M} \times N).$$

Now the commutative diagram

$$\begin{array}{ccc} H_+(\tilde{M}) \otimes H(N) & \xrightarrow[\cong]{\kappa_\#} & H_+(\tilde{M} \times N) \\ {\scriptstyle \rho^\# \otimes \iota} \uparrow {\scriptstyle \cong} & & {\scriptstyle \cong} \uparrow {\scriptstyle (\rho \times \iota)^\#} \\ H(M) \otimes H(N) & \xrightarrow[\kappa_\#]{} & H(M \times N) \end{array}$$

establishes the theorem.

The remaining two cases (M orientable and N nonorientable; M, N nonorientable) are proved in the same way.

Q.E.D.

§7. The De Rham theorem

5.21. The nerve of an open cover. Let $\mathscr{I}$ be a set. An *abstract simplicial complex* is a collection, K, of finite subsets of $\mathscr{I}$ subject to the following condition: if $\sigma \in K$, then every subset of σ is also in K. An element $\{i_0, ..., i_p\}$ of K is called a *p-simplex*, and the 0-simplices are called the *vertices* of K (cf. [11]).

Let $\mathscr{U} = \{U_i \mid i \in \mathscr{I}\}$ be an open cover of a manifold M. The *nerve* $\mathscr{N}$ of such a cover is the abstract simplicial complex whose vertices are the indices $i \in \mathscr{I}$, and which is defined as follows: A set $\{i_0, ..., i_q\}$ of distinct elements of $\mathscr{I}$ is a q-simplex of $\mathscr{N}$ if and only if

$$U_{i_0} \cap \cdots \cap U_{i_q} \neq \varnothing .$$

(If the covering is such that each U_i meets only a finite number of the U_j we call it *star-finite*. Then the corresponding nerve $\mathscr{N}$ is a *locally finite* simplicial complex.)

An *ordered q-simplex* of $\mathscr{N}$ is an ordered set $\sigma = (i_0, ..., i_q)$ of elements of $\mathscr{I}$ (not necessarily distinct) such that the distinct elements form a simplex of $\mathscr{N}$. If $\tau = (i_1, ..., i_q)$, we write $\sigma = (i_0, \tau)$. Every ordered q-simplex σ determines a nonempty open subset $U_\sigma = U_{i_0} \cap \cdots \cap U_{i_q}$ of M. If σ is an ordered q-simplex of $\mathscr{N}$ ($q \geqslant 1$), we define $\partial_j\sigma$ to be the ordered $(q - 1)$-simplex given by

$$\partial_j\sigma = (i_0 \cdots \hat{i_j} \cdots i_q)$$

($\hat{i_j}$ means the argument, i_j, is deleted).

We call $\partial_j\sigma$ the jth *face* of σ and we note that

$$\partial_i\partial_j = \partial_j\partial_{i+1}, \qquad j \leqslant i.$$

Denote the set of ordered q-simplices of $\mathscr{N}$ by $\mathscr{N}^q$. Then the set maps $\mathscr{N}^q \to \mathbb{R}$ form a real vector space, $C^q(\mathscr{N})$; the linear structure being given by

$$(\lambda f + \mu g)(\sigma) = \lambda \cdot f(\sigma) + \mu \cdot g(\sigma), \qquad \lambda, \mu \in \mathbb{R}, \quad f, g \in C^q(\mathscr{N}), \quad \sigma \in \mathscr{N}^q.$$

The graded space

$$C(\mathscr{N}) = \sum_q C^q(\mathscr{N})$$

is made into a graded algebra by the following multiplication map:

$$(f \cdot g)(\omega) = f(\sigma)\, g(\tau), \qquad f \in C^p(\mathcal{N}), \quad g \in C^q(\mathcal{N}), \quad \omega \in \mathcal{N}^{p+q},$$

where, if $\omega = (i_0, \ldots, i_{p+q}) \in \mathcal{N}^{p+q}$, then σ, τ are defined by

$$\sigma = (i_0, \ldots, i_p), \qquad \tau = (i_p, \ldots, i_{p+q}).$$

In $C(\mathcal{N})$ we define a linear operator, d, homogeneous of degree $+1$, by

$$(df)(\sigma) = \sum_{\nu=0}^{p} (-1)^\nu f(\partial_\nu \sigma), \qquad f \in C^{p-1}(\mathcal{N}), \quad \sigma \in \mathcal{N}^p.$$

It is easy to verify that d is an antiderivation of square zero. Thus $(C(\mathcal{N}), d)$ is a graded differential algebra. The corresponding cohomology algebra will be denoted by $H(\mathcal{N})$, and called *the cohomology algebra of* $\mathcal{N}$.

If $\mathscr{U}$ is finite then $C(\mathcal{N})$ (and hence $H(\mathcal{N})$) are finite dimensional.

5.22. Simple covers. An open cover $\mathscr{U} = \{U_i \mid i \in \mathscr{I}\}$ of a manifold M is called *simple* if all the nonvoid intersections $U_{i_1} \cap \cdots \cap U_{i_p}$ $(U_{i_\nu} \in \mathscr{U})$ are contractible. For a simple open cover $\mathscr{U}$ we have, then, that

$$H^+(U_{i_1} \cap \cdots \cap U_{i_p}) = 0, \qquad U_{i_\nu} \in \mathscr{U}.$$

It is easy to see that if, in the terminology of [7, p. 34] each $U_i \in \mathscr{U}$ is simple and convex, then $\mathscr{U}$ is a simple cover. Hence Lemma 6.4 of [7, p. 35] implies that every manifold admits a simple cover.

It is the purpose of this article to establish the following fundamental theorem.

Theorem VII (De Rham): The cohomology algebra of a manifold is isomorphic to the cohomology algebra of the nerve of a simple covering (as graded algebras).

Corollary: If M is compact, then $H(M)$ has finite dimension.

The proof of Theorem VII is carried out in the next five sections.

5.23. Cochains of differential forms. Let U be an open subset of M and let V be an open subset of U. The inclusion map $j: V \to U$ induces a homomorphism, $j^*: A(V) \leftarrow A(U)$, which, in this article, is denoted ρ_V^U

$$\rho_V^U : A(U) \to A(V).$$

Clearly, we have

$$\rho_W^U = \rho_W^V \circ \rho_V^U \qquad \text{if} \quad W \subset V \subset U$$

and

$$\rho_U^U = \iota_{A(U)} .$$

Now let $\mathscr{U}$ be an open cover of M with nerve $\mathscr{N}$. If σ is a face of a simplex $\tau \in \mathscr{N}$ we shall denote the restriction

$$\rho_{U_\tau}^{U_\sigma}: A(U_\sigma) \to A(U_\tau)$$

by ρ_τ^σ.

Definition: Let $\mathscr{U}$ be an open cover of a manifold M, with nerve $\mathscr{N}$. A *q-cochain of differential forms* for the covering $\mathscr{U}$ is a function f which assigns to each ordered q-simplex σ a differential form $f(\sigma)$ in U_σ.

In particular, a 0-cochain of differential forms assigns to each index i a differential form on U_i. If for each ordered q-simplex σ, $f(\sigma) \in A^p(U_\sigma)$, we call f a *q-cochains of p-forms*.

If f and g are q-cochains of forms we define $\lambda f + \mu g$ by

$$(\lambda f + \mu g)(\sigma) = \lambda f(\sigma) + \mu g(\sigma), \qquad \lambda, \mu \in \mathbb{R}, \quad \sigma \in \mathscr{N}^q.$$

With this definition the set of q-cochains of p-forms becomes a vector space, denoted by $C^{p,q}$. We put

$$C^{\cdot,q} = \sum_p C^{p,q} \qquad \text{and} \qquad C^{p,\cdot} = \sum_q C^{p,q}.$$

Finally, we define the space of cochains of differential forms to be the bigraded vector space

$$C = \sum_{p,q} C^{p,q}.$$

Next we introduce a multiplication in C as follows: Let $f \in C^{p,q}$ and $g \in C^{r,s}$. Given an ordered $(q + s)$-simplex $\omega = (i_0, ..., i_{q+s})$ of $\mathscr{N}$, write

$$\sigma = (i_0, ..., i_q) \qquad \text{and} \qquad \tau = (i_q, ..., i_{q+s}).$$

Then $f \wedge g$, defined by

$$(f \wedge g)(\omega) = (-1)^{qr} \rho_\omega^\sigma(f(\sigma)) \wedge \rho_\omega^\tau(g(\tau)),$$

is a $(q+s)$-cochain of $(p+r)$-forms. It is easily seen that this multiplication makes C into a bigraded associative algebra. The unit element is the 0-cochain which assigns to every index i the constant function

$$x \mapsto 1, \qquad x \in U_i .$$

5.24. The operator δ. Every q-cochain, f, of p-forms determines the q-cochain of $(p+1)$-forms, δf, given by

$$\delta f(\sigma) = \delta(f(\sigma)), \qquad \sigma \in \mathcal{N}^q.$$

The operator $\delta: f \mapsto \delta f$ so obtained is homogeneous of bidegree $(1, 0)$. Clearly

$$\delta(f \wedge g) = \delta f \wedge g + (-1)^{p+q} f \wedge \delta g, \qquad f \in C^{p,q}, \quad g \in C,$$

and $\delta^2 = 0$. Thus we can form the (bigraded) cohomology algebra of C with respect to δ,

$$H(C, \delta) = \ker \delta/\operatorname{Im} \delta; \qquad H(C, \delta) = \sum_{p,q} H^{p,q}(C, \delta).$$

Denote ker δ by Z_δ,

$$Z_\delta = \sum_{p,q \geqslant 0} Z_\delta^{p,q}.$$

Lemma X: If $\mathscr{U}$ is a simple cover, then the inclusion

$$Z_\delta^{0,\bullet} \to C$$

induces an isomorphism

$$Z_\delta^{0,\bullet} \xrightarrow{\cong} H(C, \delta).$$

Proof: Evidently

$$H^{0,q}(C, \delta) = Z_\delta^{0,q}, \qquad q \geqslant 0.$$

Thus we have only to show that

$$H^{+,q}(C, \delta) = 0, \qquad q \geqslant 0.$$

Consider the map

$$\gamma_q: C^{\bullet,q} \to \prod_{\sigma \in \mathcal{N}^q} A(U_\sigma)$$

given by

$$(\gamma_q(f))_\sigma = f(\sigma), \qquad f \in C^{\bullet,q}.$$

Evidently,

$$\gamma_q \circ \delta = \left(\prod_{\sigma \in \mathcal{N}^q} \delta_\sigma\right) \circ \gamma_q$$

where δ_σ is the exterior derivative in $A(U_\sigma)$. Moreover, γ_q is an isomorphism. Thus it induces isomorphisms

$$\gamma_q^{\#}: H^{p,q}(C, \delta) \xrightarrow{\cong} \prod_{\sigma \in \mathcal{N}^q} H^p(U_\sigma).$$

By hypothesis each U_σ is contractible. Hence

$$H^+(U_\sigma) = 0, \qquad \sigma \in \mathcal{N}^q, \quad q \geqslant 0$$

and so

$$H^{+,q}(C, \delta) = 0, \qquad q \geqslant 0.$$

Q.E.D.

5.25. The operator D. Consider the operator $D: C^{\cdot,q} \to C^{\cdot,q+1}$ given by

$$Df(\sigma) = (-1)^p \sum_{i=0}^{q+1} (-1)^i \rho_\sigma^{\partial_i \sigma} f(\partial_i \sigma), \qquad f \in C^{p,q},$$

where σ is an ordered $(q+1)$-simplex. In particular, for $q = 0$ and $q = 1$ we have

$$(-1)^p\, Df(i, j) = \rho_{ij}^{j} f(j) - \rho_{ij}^{i} f(i)$$

and

$$(-1)^p\, Df(i, j, k) = \rho_{ijk}^{jk} f(j, k) - \rho_{ijk}^{ik} f(i, k) + \rho_{ijk}^{ij} f(i, j).$$

The operator D is homogeneous of bidegree (0, 1). Moreover, it has the following properties:

$$D(f \wedge g) = Df \wedge g + (-1)^{p+q} f \wedge Dg, \qquad f \in C^{p,q}, \quad g \in C \tag{5.13}$$

and

$$D^2 = 0, \tag{5.14}$$

as follows directly from the relations $\partial_i \partial_j = \partial_j \partial_{i+1}$ of sec. 5.21.

In view of relations (5.13) and (5.14), we can form the bigraded cohomology algebra

$$H(C, D) = \ker D/\operatorname{Im} D; \qquad H(C, D) = \sum_{p,q} H^{p,q}(C, D).$$

Denote ker D by Z_D,

$$Z_D = \sum_{p,q\geqslant 0} Z_D^{p,q}$$

Lemma XI: The inclusion

$$Z_D^{\bullet,0} \to C$$

induces an isomorphism

$$Z_D^{\bullet,0} \xrightarrow{\cong} H(C, D).$$

Proof: As in Lemma X, we have only to show that

$$H^{p,+}(C, D) = 0, \qquad p \geqslant 0.$$

Choose a partition of unity $\{p_i\}$ subordinate to the covering $\{U_i\}$. Let τ be a face of σ and let i be an index such that $U_i \cap U_\sigma \neq \varnothing$. Then, for $\Phi \in A(U_\tau \cap U_i)$, $p_i \cdot \Phi \in A(U_\tau)$, and

$$\rho_\sigma^\tau(p_i \cdot \Phi) = p_i \cdot (\rho_{(i,\sigma)}^{(i,\tau)}\Phi).$$

Now consider the operator

$$k\colon C^{\bullet,q} \to C^{\bullet,q+1}, \qquad q \geqslant 1$$

given by

$$kf(i_1, ..., i_q) = \sum_{i\in\mathscr{I}} p_i \cdot f(i, i_1, ..., i_q), \qquad f \in C^{\bullet,q}.$$

(Observe that $p_i \cdot f(i, i_1, ..., i_q) = 0$, if $U_i \cap U_{i_1,...,i_q} = \varnothing$, and so this is a finite sum!) Then k is a homotopy operator for D in $C^{\bullet,+}$,

$$k \circ D + D \circ k = \iota\colon C^{\bullet,q} \to C^{\bullet,q}, \qquad q \geqslant 1.$$

In fact, let $f \in C^{p,q}$, $q \geqslant 1$. Then we have, for a q-simplex $\sigma = (i_0, ..., i_q)$

$$Dkf(i_0, ..., i_q) = \sum_{\nu=0}^{q} (-1)^\nu \rho_\sigma^{\partial_\nu\sigma}(k(f))(i_0, ... \hat{i}_\nu ..., i_q)$$

$$= \sum_{\nu=0}^{q} (-1)^\nu \rho_\sigma^{\partial_\nu\sigma}\left(\sum p_i \cdot f(i, i_0, ... \hat{i}_\nu ..., i_q)\right).$$

But since

$$\rho_\sigma^{\partial_\nu\sigma} p_i f(i, i_0, ... \hat{i}_\nu ..., i_q) = 0 \qquad \text{if} \quad U_i \cap U_\sigma = \varnothing,$$

the sum has only to be taken over the indices for which $U_i \cap U_\sigma \neq \varnothing$. Hence,

$$Dkf(i_0, \ldots, i_q) = \sum_{\nu=0}^{q} (-1)^\nu \rho_\sigma^{\partial_\nu \sigma} \Big(\sum_{U_i \cap U_\sigma \neq \varnothing} p_i \cdot f(i, i_0, \ldots \hat{i}_\nu \ldots, i_q) \Big)$$

$$= \sum_{U_i \cap U_\sigma \neq \varnothing} p_i \Big(\sum_{\nu=0}^{q} (-1)^\nu \rho_{(i,\sigma)}^{(i,\partial_\nu \sigma)} f(i, i_0, \ldots \hat{i}_\nu \ldots, i_q) \Big).$$

On the other hand, we have

$$kDf(i_0, \ldots, i_q) = \sum_{U_i \cap U_\sigma \neq \varnothing} p_i \cdot Df(i, i_0, \ldots, i_q)$$

$$= \sum_{U_i \cap U_\sigma \neq \varnothing} p_i \cdot \rho_{(i,\sigma)}^{\sigma} f(i_0, \ldots, i_q)$$

$$+ \sum_{U_i \cap U_\sigma \neq \varnothing} p_i \Big(\sum_{\nu=0}^{q} (-1)^{\nu+1} \rho_{(i,\sigma)}^{(i,\partial_\nu \sigma)} f(i, i_0, \ldots, \hat{i}_\nu, \ldots, i_q) \Big).$$

Adding these equations we find

$$((D \circ k + k \circ D)(f))(i_0, \ldots, i_q) = \sum_{U_i \cap U_\sigma \neq \varnothing} p_i \cdot \rho_{(i,\sigma)}^{\sigma} f(i_0, \ldots, i_q) = f(i_0, \ldots, i_q).$$

This completes the proof.

Q.E.D.

5.26. The operator ∇. Define an operator ∇ in C by

$$\nabla = \delta + D$$

Grade C by setting $C^r = \sum_{p+q=r} C^{p,q}$; then ∇ is an antiderivation. A simple computation shows that $D\delta + \delta D = 0$. It follows that $\nabla^2 = 0$ and so we can form the graded cohomology algebra

$$H(C, \nabla) = \ker \nabla / \operatorname{Im} \nabla.$$

Next, observe that since $D \circ \delta = -\delta \circ D$, Z_D is stable under δ while Z_δ is stable under D. In particular, we have the graded differential algebras

$$(Z_\delta^{0,\cdot}, D) \qquad \text{and} \qquad (Z_D^{\cdot,0}, \delta).$$

Moreover, the inclusions

$$\sigma \colon Z_\delta^{0,\cdot} \to C, \qquad \tau \colon Z_D^{\cdot,0} \to C$$

are homogeneous of degree zero, and satisfy

$$\nabla\sigma = \sigma D \qquad \text{and} \qquad \nabla\tau = \tau\delta.$$

Hence they induce homomorphisms

$$\sigma_\#\colon H(Z_\delta^{0,\cdot}, D) \to H(C, \nabla)$$

and

$$\tau_\#\colon H(Z_D^{\cdot,0}, \delta) \to H(C, \nabla).$$

of graded algebras.

Lemma XII: If the cover $\mathscr{U}$ is simple, then $\sigma_\#$ and $\tau_\#$ are isomorphisms of graded algebras.

Proof: To show that $\sigma_\#$ is an isomorphism we recall first from Lemma X, sec. 5.24, that $H(C, \delta) \cong Z_\delta^{0,\cdot}$.

It follows that there is a linear map, homogeneous of bidegree (0, 0),

$$\pi\colon C \to Z_\delta^{0,\cdot}$$

and a linear map, homogeneous of bidegree (−1, 0),

$$h\colon C \to C$$

such that $\pi\sigma = \iota$ and $\sigma\pi - \iota = h\delta + \delta h$.

Define $\alpha\colon C \to C$ by

$$\alpha = h\nabla + \nabla h + \iota.$$

Evidently,

$$\alpha = hD + Dh + \sigma\pi.$$

Since D is homogeneous of bidegree (0, 1), $hD + Dh$ is homogeneous of bidegree (−1, 1). It follows that

$$\alpha(C^{p,\cdot}) \subset C^{p-1,\cdot} + C^{0,\cdot}.$$

In particular, for every $f \in C$ there is an integer p such that

$$\alpha^p(f) \in C^{0,\cdot}.$$

Moreover, note that α is homogeneous of total degree zero, so that if f has degree q, then

$$\alpha^p(f) \in C^{0,q}.$$

Now we show that $\sigma_\#$ is surjective. In fact, let $z \in H^q(C, \nabla)$ and let $f \in \ker \nabla$ be a representing cocycle of degree q. Choose an integer p so that

$$g = \alpha^p(f) \in C^{0,q}.$$

Since $\alpha = h\nabla + \nabla h + \iota$, we have $\alpha\nabla = \nabla\alpha$ and

$$\alpha_\# = \iota : H(C, \nabla) \to H(C, \nabla).$$

Thus $g \in \ker \nabla$ and represents z.

But

$$\delta g \in C^{1,q} \qquad \text{and} \qquad Dg \in C^{0,q+1}.$$

Since

$$0 = \nabla g = \delta g + Dg,$$

it follows that $\delta g = 0 = Dg$. Thus $g \in Z_\delta^{0,q} \cap \ker D$; i.e., it represents an element w in $H(Z_\delta^{0,\cdot}, D)$. Because g represents z in $H(C, \nabla)$ we have $\sigma_\# w = z$. Hence $\sigma_\#$ is surjective.

Next we show that $\sigma_\#$ is injective. In fact, suppose

$$f \in Z_\delta^{0,q} \cap \ker D \qquad \text{and} \qquad f = \nabla g.$$

We must show that

$$f = Dg_1 \tag{5.15}$$

for some $g_1 \in Z_\delta^{0,q-1}$. We may assume g to be homogeneous of degree $q - 1$. Choose p so that

$$g_1 = \alpha^p(g) \in C^{0,q-1}.$$

Now observe that since $f \in \ker \delta \cap \ker D$, $\nabla f = 0$. Moreover, since $f \in C^{0,\cdot}$, $h(f) = 0$. Hence

$$\alpha(f) = (h\nabla + \nabla h)f + f = f,$$

and so

$$f = \alpha^p(f) = \alpha^p \nabla g = \nabla\alpha^p(g) = \nabla g_1 .$$

Since $g_1 \in C^{0,q-1}$ and $\delta g_1 = f - Dg_1$, we have

$$\delta g_1 \in C^{1,q-1} \cap C^{0,q} = 0,$$

whence $g_1 \in Z_\delta^{0,q-1}$. Thus

$$f = \nabla g_1 = Dg_1$$

and (5.15) is proved.

It has now been established that $\sigma_{\#}$ is an isomorphism. The identical argument, using Lemma XI, sec. 5.25, shows that $\tau_{\#}$ is an isomorphism.

Q.E.D.

5.27. Proof of De Rham's theorem. Define an inclusion map

$$\varphi: C(\mathcal{N}) \to C^{0,\bullet}$$

as follows: if $f \in C^q(\mathcal{N})$, let $\varphi(f)$ be the q-cochain which assigns to each $\sigma \in \mathcal{N}^q$ the constant function

$$U_\sigma \to f(\sigma).$$

Evidently φ is a homomorphism of graded algebras. Moreover, it is clear that $\delta \circ \varphi = 0$. Thus φ can be considered as a homomorphism

$$\varphi: C(\mathcal{N}) \to Z_\delta^{0,\bullet}$$

of graded algebras.

Lemma XIII: φ is an isomorphism of graded differential algebras,

Proof: It follows from the definitions that

$$\varphi \circ d = D \circ \varphi.$$

If $\varphi(f) = 0$, then $f(\sigma) = 0$ for all $\sigma \in \mathcal{N}$; whence $f = 0$. Thus φ is injective. If $\Phi \in Z_\delta^{0,q}$, then to each $\sigma \in \mathcal{N}^q$, Φ assigns a function $\Phi_\sigma \in \mathcal{S}(U_\sigma)$ such that

$$\delta\Phi_\sigma = 0.$$

Since U_σ is contractible, it is connected. It follows (cf. sec. 5.1) that Φ_σ is constant. Hence an element $f \in C^q(\mathcal{N})$ is defined by

$$f(\sigma) = \Phi_\sigma(x), \qquad x \in U_\sigma, \quad \sigma \in \mathcal{N}^q$$

and clearly $\varphi(f) = \Phi$. Thus φ is surjective.

Q.E.D.

Now we define an inclusion map

$$\psi: A(M) \to C^{\bullet,0} \subset C.$$

In fact, if $\Phi \in A^p(M)$, we define $\psi(\Phi) \in C^{p,0}$ by

$$\psi(\Phi)(i) = \rho_{U_i}^M(\Phi), \qquad i \in \mathcal{N}^0.$$

Evidently ψ is a homomorphism of graded algebras. Moreover, for $\Phi \in A(M)$,

$$D(\psi(\Phi))(i,j) = \rho_{ij}^{j}(\psi(\Phi))(i) - \rho_{ij}^{i}(\psi(\Phi))(j)$$
$$= \Phi\,|_{U_{ij}} - \Phi\,|_{U_{ij}} = 0.$$

Thus ψ can be regarded as a homomorphism

$$\psi\colon A(M) \to Z_D^{\cdot,0}$$

of graded algebras, and it is obvious that $\psi \circ \delta = \delta \circ \psi$.

Lemma XIV: ψ is an isomorphism of graded differential algebras.

Proof: Evidently ψ is injective. Moreover, if $f \in Z_D^{p,0}$, then the differential forms $f(i) \in A^p(U_i)$ satisfy

$$f(j)|_{U_{ij}} - f(i)|_{U_{ij}} = (-1)^p\, Df(i,j) = 0.$$

Hence a global p-form $\Phi \in A(M)$ is given by

$$\rho_{U_i}^{M}(\Phi) = f(i), \qquad i \in \mathcal{N}^0.$$

Clearly $\psi(\Phi) = f$, and so ψ is surjective.

Q.E.D.

Proof of the theorem: Lemmas XIII and XIV yield isomorphisms of graded algebras

$$\varphi_{\#}\colon H(\mathcal{N}) \xrightarrow{\cong} H(Z_\delta^{0,\cdot}, D)$$

and

$$\psi_{\#}\colon H(M) \xrightarrow{\cong} H(Z_D^{\cdot,0}, \delta).$$

On the other hand, Lemma XII provides an isomorphism

$$\tau_{\#}^{-1} \circ \sigma_{\#} : H(Z_\delta^{0,\cdot}, D) \xrightarrow{\cong} H(Z_D^{\cdot,0}, \delta)$$

of graded algebras. Combining these isomorphisms we obtain an isomorphism

$$H(\mathcal{N}) \xrightarrow{\cong} H(M)$$

of graded algebras, as was desired.

Q.E.D.

Problems

M and N are smooth manifolds.

1. Mayer–Vietoris sequence. Suppose $M = U \cup V$ (U, V open), and ∂ is the connecting homomorphism.

(i) Show that $\operatorname{Im} \partial$ is an ideal in $H(M)$. If $i_U: U \to M$, $i_V: V \to M$ are the inclusions and if $\alpha, \beta \in H(M)$ satisfy $i_U^{\#}\alpha = 0$, $i_V^{\#}\beta = 0$, show that $\alpha \cdot \beta = 0$. Conclude that, if $\gamma_1, \gamma_2 \in \operatorname{Im} \partial$, then $\gamma_1 \cdot \gamma_2 = 0$.

(ii) Suppose $M = U_1 \cup \cdots \cup U_p$, where each U_i is open and $H^+(U_i) = 0$. Show that, if $\alpha_i \in H^+(M)$ ($i = 1, ..., p$), then $\alpha_1 \cdot \cdots \cdot \alpha_p = 0$.

(iii) Suppose $M = U_1 \cup \cdots \cup U_p$ (U_i open). Assume that, for each sequence $1 \leqslant i_1 < \cdots < i_q \leqslant p$, the intersection $U_{i_1} \cap \cdots \cap U_{i_q}$ has finite dimensional cohomology and let $\chi(i_1, ..., i_q)$ denote its Euler characteristic. Show that M has finite dimensional cohomology and that

$$\chi_M = \sum_{q=1}^{p} (-1)^{q-1} \sum_{1 \leqslant i_1 < \cdots < i_q \leqslant p} \chi(i_1, ..., i_q).$$

2. Compute the cohomology of $M \# N$ in terms of $H(M)$ and $H(N)$. Thus obtain the cohomology of the compact surfaces $T^2 \# \cdots \# T^2$ and $T^2 \# \cdots \# T^2 \# \mathbb{R}P^2$ (T^2 is the 2-torus) (cf. problem 24, Chap. 3).

3. The Massey triple product. Let $\alpha, \beta, \gamma \in H(M)$ be of degrees p, q, r and represented by Φ, Ψ, X. Assume that $\Phi \wedge \Psi = \delta\Omega_1$, $\Psi \wedge X = \delta\Omega_2$.

(i) Show that $\Phi \wedge \Omega_2 - (-1)^p\Omega_1 \wedge X$ is closed and that the class it represents depends only α, β, and γ. It is called the *Massey triple product* and written $[\alpha, \beta, \gamma]$.

(ii) Define the Massey triple product in $H(\mathscr{N})$ ($\mathscr{N}$ is the nerve of a simple covering of M) and show that the de Rham isomorphism preserves the Massey triple product.

4. Open subsets of compact manifolds. (i) Let O be an open subset of a compact manifold. Show that the map $(\gamma_c)_{\#}: H_c(O) \to H(O)$ has finite dimensional image.

(ii) Find a manifold which is not an open subset of a compact manifold.

5. Wang sequence. Consider a smooth fibre bundle (E, π, S^n, F) which is trivial over $S^n - \{a\}$ $(a \in S^n)$.

Remark: It can be shown that every fibre bundle over a contractible base is trivial.

Obtain a Mayer–Vietoris triangle

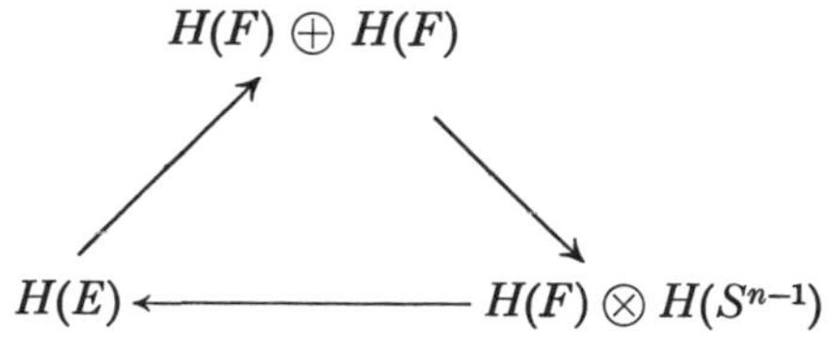

and derive the long exact *Wang sequence*

$$\cdots \to H^p(E) \to H^p(F) \to H^{p-n+1}(F) \to H^{p+1}(E) \to \cdots$$

6. (i) Suppose $\omega \in A^1(M)$ (M compact) satisfies $\omega(x) \neq 0$, $x \in M$. Show that ω is not exact.

(ii) Construct 1-forms ω_1, ω_2, ω_3 on S^3 such that for each x the $\omega_i(x)$ form a basis of $T_x(S^3)^*$ and such that $\omega_i \wedge \omega_j$ is exact.

(iii) Let Φ be a p-form on M such that for each $x \in M$, $\Phi(x)$ is the product of p independent covectors at x. Let $F_x = \{h \in T_x(M) \mid i(h)\Phi(x) = 0\}$. Show that the spaces F_x are the fibres of a subbundle of τ_M.

If Φ is exact and N is a compact p-submanifold of M, prove that for some $y \in N$, $T_y(N) \cap F_y \neq 0$. Does this hold if Φ is only assumed to be closed?

7. Hopf invariant. Assume that M and N are compact and oriented, $\dim M = m$, $\dim N = n$ with $m > n$. Let $\varphi: M \to N$ be smooth. Write $\ker \varphi^\# = K = \sum_p K^p$. Suppose $\alpha \in K^{p+1}$, $\beta \in K^{m-p}$ are represented by Φ, Ψ and write $\varphi^*\Phi = \delta\Omega$.

(i) Show that $\int_M \Omega \wedge \varphi^*\Psi$ depends only on α, β, and φ. Hence obtain a bilinear map

$$\langle\ ,\ \rangle_\varphi : K^{p+1} \times K^{m-p} \to \mathbb{R}.$$

(ii) Show that

$$\langle \alpha, \beta \rangle_\varphi = (-1)^{m(p+1)} \langle \beta, \alpha \rangle_\varphi .$$

(iii) Prove that $\langle\ ,\ \rangle_\varphi$ depends only on the homotopy class of φ.

(iv) Suppose $\varphi: S^{2n-1} \to S^n$ $(n \geqslant 2)$. Show that $\varphi^{\#}\omega = 0$, where ω is the orientation class of S^n. The number $h(\varphi) = \langle \omega, \omega \rangle_\varphi$ is called the *Hopf invariant of* φ.

(v) Compute the Hopf invariant of the Hopf fiberings $S^3 \to S^2$ and $S^7 \to S^4$ (cf. problems 10, 11, Chap. I).

8. The *n*-torus. Recall from sec. 1.4 the projection $\pi: \mathbb{R}^n \to T^n$. Let $\sigma_j: S^1 \to T^n$ be defined by

$$\sigma_j(\exp 2\pi it) = \pi(0, ..., t, ..., 0).$$

(i) Construct classes $\alpha_j \in H^1(T^n)$ such that $\int_{S^1}^{\#} \sigma_i^{\#}\alpha_j = \delta_{ij}$. Show that they are uniquely determined by these conditions.

(ii) Show that $\alpha_1 \cdot \cdots \cdot \alpha_n$ is an orientation class for T^n.

(iii) Interpret the α_i via the Künneth isomorphism ($T^n \cong S^1 \times \cdots \times S^1$). Show that $H(T^n)$ is isomorphic to the exterior algebra of an n-dimensional vector space.

(iv) Consider T^p as a submanifold of T^n and compute the cohomology algebra of the manifold $T^n - T^p$. Does it contain a compact manifold as retract? as deformation retract?

9. Let N be a closed submanifold of M. Let

$$\mathscr{I} = \{\Phi \in A(M) \mid \text{carr}\, \Phi \cap N = \varnothing\}.$$

(i) Show that $\mathscr{I}$ is an ideal in $A(M)$ and that the factor algebra, $A(N; M)$, is a graded differential algebra.

(ii) Show that the inclusion map $N \to M$ induces a homomorphism $A(N; M) \to A(N)$.

(iii) Show that the induced map $H(A(N; M)) \to H(N)$ is an isomorphism. *Hint*: Consider open subsets $U \subset N$ such that $U \cong \mathbb{R}^r$ and U has trivial normal bundle in M — cf. problem 20, Chap. III.

(iv) If M is compact, establish an exact triangle

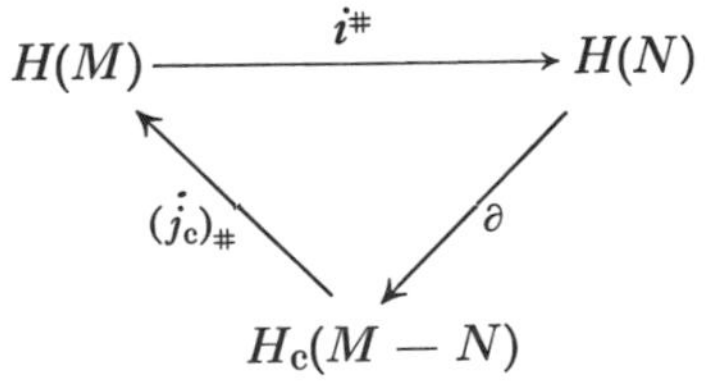

where ∂ is homogeneous of degree $+1$.

10. Manifolds-with-boundary. Let M be a compact n-manifold-with-boundary ∂M and interior $\mathring{M}$ (cf. problem 24, Chap. III). Let $i\colon \partial M \to M$ denote the inclusion map.

(i) Show that the inclusion map $\mathring{M} \to M$ induces an isomorphism $H(\mathring{M}) \xleftarrow{\cong} H(M)$. Is this true for cohomology with compact supports?

(ii) Show that the sequence

$$A_c(\mathring{M}) \longrightarrow A(M) \xrightarrow{i^*} A(\partial M)$$

leads to an exact triangle in cohomology (cf. problem 9, above).

(iii) If M is oriented interpret Stokes' theorem via the connecting homomorphism of (ii) (cf. problem 5, Chap. IV).

11. Harmonic forms. Let M be compact, oriented, and have a Riemannian metric. The space $\mathscr{H}^p(M)$ of *harmonic* forms consists of those forms Φ satisfying $\Delta\Phi = 0$ (cf. problem 9, Chap. IV). Show that $\dim \mathscr{H}^p(M) \leqslant b_p(M)$. (In fact, the *Hodge theorem* asserts that $\dim \mathscr{H}^p(M) = b_p(M)$.)

12. Smooth homology. Let E be a vector space with a fixed countable basis $a_0, a_1, \ldots$. The *p-dimensional standard simplex* is the set defined by

$$\Delta_p = \left\{ x \in E \;\middle|\; x = \sum_{i=0}^{p} \lambda^i a_i, \quad \lambda^i \geqslant 0, \quad \sum \lambda^i = 1 \right\}.$$

A *smooth p-simplex* on a manifold M is a smooth map $\sigma\colon \Delta_p \to M$ (a smooth map from a closed subset A of $\mathbb{R}^p$ into M is a map $A \to M$ which extends to a smooth map from a neighbourhood of A into M). The ith face of σ is the smooth $(p-1)$-simplex given by

$$(\partial_i\sigma)\left(\sum_0^{p-1} \lambda^j a_j\right) = \sigma\left(\sum_0^{i-1} \lambda^j a_j + \sum_i^{p-1} \lambda^j a_{j+1}\right).$$

Let R be any subring of $\mathbb{R}$ (eg. $R = \mathbb{Z}$ or $\mathbb{Q}$ or $\mathbb{R}$). Denote by $C_p(M; R)$ the free R-module with basis the smooth p-simplices on M. Define an R-linear map $\partial\colon C_p(M; R) \to C_{p-1}(M; R)$ $(p \geqslant 1)$ by

$$\partial\sigma = \sum_{i=0}^{p} (-1)^i\, \partial_i\sigma.$$

(i) Verify that $\partial^2 = 0$. The graded module $H_*(M; R) = \ker \partial/\operatorname{Im} \partial$ is called the *smooth homology of M with coefficients in R*. An element in

$C_p(M; R)$ (resp. $Z_p(M; R) = (\ker \partial)_p$, $B_p(M; R) = (\operatorname{Im} \partial)_p$, $H_p(M; R)$) is called a *p-chain* (resp. *p-cycle, p-boundary, p-dimensional homology class*).

(ii) Show that a smooth map $\varphi: M \to N$ induces a module homomorphism $\varphi_*: C_p(M; R) \to C_p(N; R)$ such that $\partial \circ \varphi_* = \varphi_* \circ \partial$. Obtain induced homomorphisms between the cycle, boundary, and homology modules. The last is written $\varphi_\#: H_*(M; R) \to H_*(N; R)$.

(iii) Let U be an open subset of a vector space, star-shaped with respect to some x_0. Define an R-linear map $k: C_p(U; R) \to C_{p+1}(U; R)$ by

$$(k\sigma)\left(\lambda^0 a_0 + \sum_1^{p+1} \lambda^i a_i\right) = f(\lambda^0)\, x_0 + (1 - f(\lambda^0))\sigma\left(\sum_1^{p+1} \frac{\lambda^i}{1 - \lambda^0} a_{i-1}\right),$$

where $f: \mathbb{R} \to \mathbb{R}$ satisfies $0 \leqslant f(t) \leqslant 1, f(0) = 0, f(t) = 1\ (t > 1 - \epsilon)$. Show that $k \circ \partial + \partial \circ k = \iota$ in $C_p(U; R)$ and conclude that $H_+(U; R) = 0$.

(iv) Establish a homotopy axiom, a disjoint union axiom, and a Mayer–Vietoris axiom for $H_*(M; R)$. *Hint*: If $M = U \cup V$ and σ is a smooth simplex in M, find smooth chains a in U, b in V and c in M such that $\sigma = a + b + \partial c$.

(v) If $H_p(M; \mathbb{Z})$ is finitely generated, show that $H_p(M; \mathbb{Z}) = F_p \oplus T_p$, where F_p is a free, finitely generated abelian group and T_p consists of the elements of finite order. F_p is called the ***pth Betti group*** and T_p is called the ***pth torsion group*** of M. A basis of F_p is called a ***homology basis*** of $H_p(M; \mathbb{Z})$.

(vi) If M is compact show that each $H_p(M; \mathbb{Z})$ is finitely generated and hence has a homology basis.

(vii) Construct canonical isomorphisms

$$H_p(M; \mathbb{Z}) \otimes_{\mathbb{Z}} \mathbb{Q} \xrightarrow{\cong} H_p(M; \mathbb{Q}) \qquad \text{and} \qquad H_p(M; \mathbb{Z}) \otimes_{\mathbb{Z}} \mathbb{R} \xrightarrow{\cong} H_p(M; \mathbb{R})$$

(universal coefficient theorem).

(viii) Suppose M has a finite simple cover $\{U_i\}$ such that each $U_{i_0} \cap \cdots \cap U_{i_r} = \varnothing$ (r is fixed). Prove by induction that $H_p(M; R) = 0$ for $p > r$.

(ix) Show that

$$H_p(S^n; \mathbb{Z}) = 0 \qquad (1 \leqslant p \leqslant n - 1)$$

$$H_0(S^n; \mathbb{Z}) \cong \mathbb{Z}$$

$$H_n(S^n; \mathbb{Z}) \cong \mathbb{Z}.$$

13. Integration and homology. The vertices $(a_0, \ldots, a_p)$ of the standard p-simplex Δ_p span an affine p-plane in E. This is oriented so that $(a_1 - a_0, \ldots, a_p - a_0)$ is a positive basis. The interior of Δ_p, $\mathring{\Delta}_p$, is an open subset of this plane. Hence, if σ is a smooth p-simplex in M and

$\Phi \in A^p(M)$, we can integrate the form $\sigma^*\Phi$ over $\mathring{\Delta}_p$. This integral is denoted by $\int_\sigma \Phi$ and is called the *integral of Φ over the smooth p-simplex σ*. If $c = \sum \lambda_i \sigma_i$, $\lambda_i \in R$, we set

$$\int_c \Phi = \sum \lambda_i \int_{\sigma_i} \Phi$$

and call $\int_c \Phi$ the integral of Φ over c.

(i) Establish Stokes' theorem for chains;

$$\int_c \delta\Psi = \int_{\partial c} \Psi \qquad \Psi \in A^p(M), \quad c \in C_{p+1}(M; R).$$

Conclude that $(\Phi, c) \mapsto \int_c \Phi$ defines an R-bilinear map

$$\langle\ ,\ \rangle : A^p(M) \times C_p(M; R) \to \mathbb{R}$$

and that $\langle \delta\Psi, c\rangle = \langle \Psi, \partial c\rangle$.

(ii) Show that $\int_c \Phi = 0$ if c is a cycle and Φ is exact, or if c is a boundary and Φ is closed. Thus obtain an R-bilinear map $H^p(M) \times H_p(M; R) \to \mathbb{R}$.

(iii) A cohomology class $\alpha \in H^p(M)$ is called *integral*, if $\langle \alpha, \beta\rangle \in \mathbb{Z}$ for every $\beta \in H_p(M; \mathbb{Z})$. Show that a closed p-form represents an integral class if and only if its integral over every integral cycle is an integer.

(iv) Define a map λ: $H^p(M) \to \mathrm{Hom}_{\mathbb{Z}}(H_p(M; \mathbb{Z}); \mathbb{R})$ by

$$(\lambda(\beta))(\alpha) = \langle \beta, \alpha\rangle, \qquad \beta \in H^p(M), \quad \alpha \in H_p(M; \mathbb{Z}).$$

Use problem 12, iv, to conclude that λ is an isomorphism of real vector spaces.

(v) If M is compact and $\alpha_1, \ldots, \alpha_k$ is a homology basis for $H_p(M; \mathbb{Z})$ (cf. problem 7, vi) show that there are unique classes $\beta_i \in H^p(M)$ such that $\langle \beta_i, \alpha_j\rangle = \delta_{ij}$. Conclude that the β_i are a basis of $H^p(M)$ consisting of integral classes.

(vi) *De Rham existence theorem*: Let z_i represent α_i (α_i as in (v)). If Φ is a closed p-form, the numbers $\int_{z_i} \Phi$ are called the *periods* of Φ with respect to the homology basis $\alpha_1, \ldots, \alpha_k$. Given real numbers $\lambda_1, \ldots, \lambda_k$, show that there exists a closed p-form Φ on M with the λ_i as periods. Show that Φ is uniquely determined up to an exact form.

14. Homology and densities. Assume M connected and oriented. Let $D_p^c(M)$ denote the space of p-densities on M with compact carrier (cf. problem 8, Chap. IV). Let $\mathscr{U}$ be a simple covering of M with nerve $\mathscr{N}$.

Define a graded differential space $(C_*(\mathcal{N}), \partial)$ as follows: the ordered p-simplices of $\mathcal{N}$ are a basis for $C_p(\mathcal{N})$ and ∂ is given by

$$\partial(i_0, ..., i_p) = \sum_{\nu=0}^{n} (-1)^\nu (i_0, ..., \hat{i}_\nu, ..., i_p).$$

$H(\mathcal{N}, \partial)$ is called the *simplicial homology of* $\mathcal{N}$. Let $C_{p,q} \subset D_p^c(M) \otimes C_q(\mathcal{N})$ be the subspace generated by the elements of the form $\Phi \otimes \sigma$ with carr $\Phi \subset \mathcal{U}_\sigma$. (*Note*: All vector spaces have coefficient field $\mathbb{R}$).

(i) Use the divergence operator and boundary operator in C_* to obtain a differential operator in $\sum_{p,q} C_{p,q}$ (cf. problem 8, Chap. IV).

(ii) Use integration to define a linear map $C_{0,q} \to C_q(\mathcal{N})$. Use the linear map $C_0(\mathcal{N}) \to R$ which sends each simplex to 1 to define a linear map $C_{p,0} \to D_p^c(M)$.

(iii) Mimic the proof of the de Rham theorem to construct an isomorphism $H_p(D^c(M)) \xrightarrow{\cong} H_p(\mathcal{N})$. Derive from this another proof of the Poincaré duality theorem.

(iv) Replace $D^c(M)$ by the group of integral smooth chains and replace $C_*(\mathcal{N})$ by the free abelian group generated by the simplices of $\mathcal{N}$. Repeat the argument and show that (even if M is not orientable) the integral smooth homology of M is isomorphic to the integral simplicial homology of $\mathcal{N}$.

15. Line integrals. M is a connected n-manifold, and $a\colon [0, 1] \to M$ is a smooth path. The *line integral* along a of $\omega \in A^1(M)$ is the number

$$\int_a \omega = \int_0^1 \omega(a(t); \dot{a}(t))\, dt.$$

(i) Suppose that ω is closed and $H_1(M, \mathbb{Z})$ consists only of torsion elements. Show that $\int_a \omega$ depends only on the endpoints of a. Fix $x_0 \in M$ and set $f(x) = \int_{a_x} \omega$, where a_x is any smooth path joining x_0 to x. Show that f is smooth, and that $\delta f = \omega$. Conclude that $H_1(\mathbb{R}^2 - \{0\}; \mathbb{Z})$ is an infinite group.

(ii) Let $h \in \mathcal{S}(M; \mathbb{C})$ satisfy $h(x) \neq 0$, $x \in M$. Show that $\omega = (1/2\pi i)(1/h)\, \delta h$ is a closed $\mathbb{C}$-valued 1-form, and that ω differs from a real form by a coboundary. Prove that $\int_z \omega \in \mathbb{Z}$ for each integral cycle z.

(iii) Let $\omega \in A^1(M)$ be closed and integral. Construct $h \in \mathcal{S}(M; \mathbb{C})$ such that $|h(x)| = 1$, $x \in M$, and $(1/2\pi i)(1/h)\, \delta h = \omega$.

(iv) Assume that $H_1(M; \mathbb{Z}) = 0$. Let f and g be functions on M satisfying $f(x)^2 + g(x)^2 = 1$, $x \in M$. Construct a function θ on M such that

$$\cos \circ \theta = f, \qquad \sin \circ \theta = g.$$

Show that if θ_1 is another such function, then $\theta_1 - \theta = 2\pi k$, where k is a fixed integer.

16. Assume that $H(M)$ (M an n-manifold) is isomorphic to the exterior algebra over an n-dimensional subspace of $H(M)$. Find a smooth map $\varphi: M \to T^n$ such that $\varphi^\#$ is an isomorphism.

17. Assume M compact and connected. Let ω be a closed 1-form such that $\omega(x) \neq 0$, $x \in M$. Construct a submersion $\pi: M \to S^1$. Show that π is the projection of a smooth fibre bundle over S^1 with M as total space.

18. Cauchy's integral theorem. Let O be an open subset of $\mathbb{C}$ and let $\delta z \in A^1(O; \mathbb{C})$ denote the gradient of the identity function.

(i) If $f \in \mathscr{S}(O; \mathbb{C})$, show that $(f\,\delta z)(x; h) = f(x) \cdot h$, $x \in O$, $h \in \mathbb{C}$.
(ii) Show that $f \cdot \delta z$ is closed if and only if f is complex differentiable.
(iii) Let a: $[0, 1] \to O$ be a smooth path in O. Show that $\int_a f \cdot \delta z = \int_a f(z)\, dz$.
(iv) Prove *Cauchy's integral theorem*: If f is complex differentiable in O, then $\int_{\partial c} f(z)\, dz = 0$, where c is a 2-chain in O.
(v) Let f be a complex differentiable function in $|z| < 1$ which extends to a continuous function in $|z| \leqslant 1$. Show that $\int_{S^1} f(z)\, dz = 0$.

19. Simply-connected manifolds. Let M be connected. Two smooth paths a: $[0, 1] \to M$ and b: $[0, 1] \to M$ having the same initial point x_0 and the same endpoint x_1 are called *homotopic*, if there is a homotopy connecting a and b and leaving x_0 and x_1 fixed. A manifold is called *simply-connected*, if every closed path is homotopic to the constant path.

(i) Show that if M is simply-connected, then $H_1(M; \mathbb{Z}) = 0$.
(ii) Let M be simply-connected. Let $\Phi \in A^1(M; L_F)$ (F a vector space). Assume that Φ satisfies $\delta\Phi + \Phi \circ \Phi = 0$ (cf. sec. 4.7). Given points $a \in M$ and $b \in F$, show that there exists precisely one smooth map $\varphi: M \to F$ such that $d\varphi = \Phi(\varphi)$ and $\varphi(a) = b$.
(iii) Assume that P is a parallelism on M (cf. problem 14, Chap. IV). The torsion S of P is called *parallel*, if

$$P(x, y)\, S(x; \xi, \eta) = S(y; P(x, y)\xi, P(x, y)\eta), \qquad x, y \in M, \quad \xi, \eta \in T_x(M).$$

Show that if P admits a conjugate parallelism, then the torsion is parallel. If M is simply-connected show the converse.

20. Simplicial complexes in $\mathbb{R}^n$. Let $\Delta = (a_0, a_1, ..., a_p)$, where $a_i \in \mathbb{R}^n$ and the vectors $a_i - a_0$ $(i = 1, ..., p)$ are linearly independent. Then the set

$$|\Delta| = \left\{\sum_{i=0}^{p} \lambda^i a_i \mid \lambda^i \geqslant 0, \sum_{i=0}^{p} \lambda^i = 1\right\},$$

is called an *affine simplex* with vertices a_i. If in addition the a_i are ordered, then $|\Delta|$ is called an *ordered affine simplex*. The interior of Δ is given by

$$\mathring{\Delta} = \left\{\sum_{i=0}^{p} \lambda^i a_i \mid \lambda^i > 0, \sum_{i=0}^{p} \lambda^i = 1\right\}$$

and its boundary is $\Delta - \mathring{\Delta}$. Let $A \subset \mathbb{R}^n$ be a closed set which is the disjoint union of finitely many $\mathring{\sigma}_i$ (σ_i is an affine p_i-simplex). If the σ_i form an abstract simplicial complex, K, then A is called a *finite affine simplicial complex* and is written $|K|$. The maximum of the p_i is called the *dimension of* $|K|$. The union of the affine simplices of dimension $\leqslant k$ is called the *k-skeleton* of $|K|$.

Let $|K| \subset \mathbb{R}^n$ be an affine simplicial complex of dimension r. Find an open subset $O \subset \mathbb{R}^n$ such that $|K| \subset O$ and such that O admits a simple covering with nerve K. Conclude that $H^p(O) = 0$, if $p > r$.

21. Degenerate and invisible chains. A smooth p-simplex σ on M is called *degenerate* if, for each $a \in \Delta_p$, $\operatorname{rank}(d\sigma)_a < p$. A p-chain $c = \sum_i \lambda^i \sigma_i$ $(\lambda^i \neq 0)$ is called *degenerate*, if all p-simplices σ_i are. A p-chain c is called *invisible* if, for every p-form Φ, $\int_c \Phi = 0$.

(i) Show that a degenerate chain is invisible and that the boundary of an invisible chain is invisible.

(ii) Let $\varphi\colon \Delta_p \to F$ be smooth, where F denotes the plane through $a_0, ..., a_p$. Assume that $\operatorname{Im} \varphi \subset \Delta_p$ and that φ restricts to the identity near the boundary of Δ_p. For each smooth p-simplex σ, find a degenerate integral $(p + 1)$-chain c such that $\sigma - \sigma \circ \varphi = \partial c$. Conclude that $\sigma - \sigma \circ \varphi$ is invisible. Extend the result to smooth chains.

(iii) Let $c = \sum \lambda^i \sigma_i$ $(\lambda_i \neq 0)$ be a p-chain on M. A point $x \in M$ is called a regular value for c, if, for each i, $\sigma_i^{-1}(x) \subset \mathring{\Delta}_p$ and x is a regular value for σ_i. Otherwise x will be called a *critical value* for c. Denote by $\operatorname{Crit}(c)$ the set of critical values. Let $N \subset M$ be a p-dimensional submanifold. Show that $N \cap \operatorname{Crit}(c)$ has measure zero in N. *Hint*: Compare problem 12, Chap. III.

(iv) Let R be a subring of $\mathbb{R}$. Given $c = \sum \lambda^i \sigma_i \in C_p(M; R)$ $(\lambda_i \neq 0)$ define $\operatorname{Im} c = \bigcup_i \operatorname{Im} \sigma_i$ and call it the *image* of c. Let $x \in \operatorname{Im} c$ be a regular value for c. Assume that there is a neighbourhood U of x

and a p-submanifold $N \subset M$ such that $U \cap \operatorname{Im} c \subset N$. Show that $c = a + \lambda\sigma + \partial b$, where

(1) $a \in C_p(M; R)$, $b \in C_{p+1}(M; R)$, $\lambda \in R$.
(2) $\sigma: \Delta_p \to M$ is a smooth embedding and $x \in \sigma(\mathring{\Delta}_p)$.
(3) $x \notin \operatorname{Im} a$.
(4) b is degenerate.

Conclude that c is invisible if and only if $\lambda = 0$ and a is invisible.

22. **The invisibility theorem.** Show that every invisible cycle, z, in $C_p(M; R)$ is the boundary of some $c \in C_{p+1}(M; R)$ (invisibility theorem).

Hint: One method is as follows:

(i) Use a tubular neighbourhood of an embedding of M in $\mathbb{R}^N$ to reduce to the case when M is an open subset of $\mathbb{R}^N$ (cf. problem 20, Chap. III).

(ii) Construct an affine simplicial complex, $|K| \subset \mathbb{R}^N$, such that $\operatorname{Im} z \subset |K|$ (cf. problem 20).

(iii) Suppose $\operatorname{Im} z$ is contained in the r-skeleton of $|K|$. If $r > p$, no r-simplex of $|K|$ is contained in $\operatorname{Im} z$. Hence find a smooth map of $M \to M$, homotopic to ι_M, which carries z into the $(r-1)$-skeleton $|K|$ (cf. problem 13, Chap. III).

(iv) Suppose $\operatorname{Im} z$ is contained in the p-skeleton of $|K|$. Modify problem 21, iv, to show that there is a $b \in C_{p+1}(M; R)$ such that

(1) $\operatorname{Im} \partial b$ is contained in the p-skeleton of $|K|$ and
(2) No p-simplex of $|K|$ is contained in $\operatorname{Im}(z + \partial b)$. Thus find a smooth map of M, homotopic to ι_M, which carries $z + \partial b$ into the $(p-1)$-skeleton of $|K|$.

(v) Use problem 20 and problem 12, viii, to complete the proof.

23. **The fundamental cycle.** Let M be connected. Use the invisibility theorem, and problem 21, to establish the following results:

(i) $H_p(M; \mathbb{Z}) = 0$, $p > n$.
(ii) $H_n(M; \mathbb{Z}) \cong \mathbb{Z}$, if M is compact and oriented and $H_n(M; \mathbb{Z}) = 0$ otherwise.

(iii) Let M be compact and oriented. Show that there is a unique generator $\omega_M^* \in H_n(M; \mathbb{Z})$ such that

$$\int_M \Phi = \int_z \Phi, \qquad \Phi \in A^n(M),$$

where z is a representing cycle. ω_M^* is called the *fundamental class* of the oriented manifold M. Conclude that $\langle \omega_M, \omega_M^* \rangle = 1$, where ω_M is the orientation class. In particular, show that the orientation class is integral.

(iv) Let $(M, \partial M)$ be a compact orientable manifold-with-boundary. Let $(\partial M)_1, \ldots, (\partial M)_r$ be the components of ∂M with inclusion maps $i_\lambda: (\partial M)_\lambda \to M$ and induced fundamental classes α_λ. Show that $\sum_\lambda (i_\lambda)_\# \alpha_\lambda = 0$.

(v) Let $\varphi: N \to M$ be a smooth map (N a compact oriented r-manifold). Suppose $\Phi \in A^r(M)$ is an integral closed form. Show that $\int_N \varphi^* \Phi \in \mathbb{Z}$.

24. Direct limits. Let $\mathscr{I}$ be a partially ordered set such that, for all $\alpha, \beta \in \mathscr{I}$, there is some $\gamma \in \mathscr{I}$ with $\alpha < \gamma$ and $\beta < \gamma$. Let $\{A_\alpha\}$ be a family of vector spaces, indexed by $\mathscr{I}$. Let $\{\rho_\beta^\alpha: A_\alpha \to A_\beta\}_{\alpha<\beta}$ be a family of linear maps such that if $\alpha < \beta < \gamma$, $\rho_\gamma^\alpha = \rho_\gamma^\beta \circ \rho_\beta^\alpha$. Then $\{A_\alpha, \rho_\beta^\alpha\}$ is called a *directed system of vector spaces*. Its direct limit, written $\varinjlim A_\alpha$, is the space A/B, where $A = \bigoplus_\alpha A_\alpha$ and $B \subset A$ is the subspace generated by the vectors of the form $x_\beta - \rho_\beta^\alpha(x_\alpha)$.

(i) Define canonical linear maps $i_\alpha: A_\alpha \to \varinjlim A_\alpha$. If $\varphi_\alpha: A_\alpha \to C$ are linear maps satisfying $\varphi_\beta \circ \rho_\beta^\alpha = \varphi_\alpha$, show that they induce a unique linear map $\varphi: \varinjlim A_\alpha \to C$ such that $\varphi \circ i_\alpha = \varphi_\alpha$ for each $\alpha \in \mathscr{I}$.

(ii) Suppose the A_α are algebras (resp. graded algebras, differential algebras) and assume that the ρ_β^α are homomorphisms. Make $\varinjlim A_\alpha$ into an algebra (resp. a graded algebra, differential algebra) so that each i_α is a homomorphism. In the third case show that

$$H(\varinjlim A_\alpha) = \varinjlim H(A_\alpha).$$

25. Čech cohomology. Let $\mathscr{U} = \{U_\alpha \mid \alpha \in \mathscr{I}\}$ be a star-finite open covering of M and let $\mathscr{V} = \{V_j \mid j \in \mathscr{J}\}$ be a refinement of $\mathscr{U}$. Let $\mathscr{N}_\mathscr{U}$ and $\mathscr{N}_\mathscr{V}$ denote the corresponding nerves. Choose a map $\alpha: \mathscr{J} \to \mathscr{I}$ such that $V_j \subset U_{\alpha(j)}$.

(i) Show that a map $\lambda_\mathscr{V}^\mathscr{U}: C(\mathscr{N}_\mathscr{U}) \to C(\mathscr{N}_\mathscr{V})$ is defined by

$$(\lambda_\mathscr{V}^\mathscr{U} f)(i_0, \ldots, i_p) = f(\alpha(i_0), \ldots, \alpha(i_p))$$

and that $\lambda_\mathscr{V}^\mathscr{U}$ is a homomorphism of graded differential algebras.

(ii) Show that $(\lambda_\mathscr{V}^\mathscr{U})^\#$ is independent of the choice of α. Show that the algebras $H(\mathscr{N}_\mathscr{U})$ form a direct system of graded algebras. The direct limit is called the *Čech cohomology* of M and is written

$$\check{H}(M) = \varinjlim H(\mathscr{N}_\mathscr{U}).$$

(iii) Given a star-finite open covering $\mathscr{U}$ of M imitate the proof of the de Rham theorem to obtain a homomorphism $H(\mathscr{N}_{\mathscr{U}}) \to H(M)$. Hence obtain a homomorphism $\check{H}(M) \to H(M)$. Show that this is an isomorphism, $\check{H}(M) \cong H(M)$.

26. Dimension theory. Show that the Lebesgue dimension (cf. sec. 1.2) of an n-manifold is n. *Hint*: Proceed as follows:

(i) Show that, if $O \subset M$ is open, then $\dim O \leqslant \dim M$.

(ii) Show that $\dim M = \dim \mathbb{R}^n = \dim S^n$.

(iii) By brutal force show that $\dim \mathbb{R}^n \leqslant n$.

(iv) Use problem 25 to show that, if $\dim M \leqslant p$, then $H^i(M) = 0$, $i > p$. Conclude that $\dim S^n \geqslant n$.

Chapter VI

Mapping Degree

§1. Global degree

All manifolds in this article are connected, compact, and oriented, unless otherwise stated.

6.1. Definition. Let $\varphi: M \to N$ be a smooth map between n-manifolds. The linear isomorphisms

$$\int_M^{\#}: H^n(M) \xrightarrow{\cong} \mathbb{R}, \qquad \int_N^{\#}: H^n(N) \xrightarrow{\cong} \mathbb{R}$$

(cf. sec. 5.13) determine a unique linear map

$$f_\varphi: \mathbb{R} \to \mathbb{R}$$

which makes the diagram

$$\begin{array}{ccc} H^n(M) & \xleftarrow{\varphi^{\#}} & H^n(N) \\ {\scriptstyle \int_M^{\#}}\downarrow & & \downarrow{\scriptstyle \int_N^{\#}} \\ \mathbb{R} & \xleftarrow[f_\varphi]{} & \mathbb{R} \end{array}$$

commute.

The (mapping) *degree* of φ is defined by

$$\deg \varphi = f_\varphi(1).$$

It follows from the definition that

$$\deg \varphi = \int_M \varphi^* \Phi,$$

where $\Phi \in A^n(N)$ satisfies $\int_N \Phi = 1$. More generally, if Φ is any n-form on N we have

$$\int_M \varphi^* \Phi = \deg \varphi \int_N \Phi.$$

Note that the sign of the mapping degree changes if either the orientation of M, or that of N, is reversed. In particular, the degree of a map $\varphi: M \to M$ is independent of the orientation of M.

Proposition I: The mapping degree has the following properties:

(1) Let $\varphi: M \to N$ and $\psi: N \to Q$ be smooth maps between oriented n-manifolds. Then

$$\deg(\psi \circ \varphi) = \deg \varphi \cdot \deg \psi.$$

(2) The degree of the identity map is one,

$$\deg \iota_M = 1.$$

(3) If the maps $\varphi, \psi: M \to N$ are homotopic, then

$$\deg \varphi = \deg \psi.$$

(4) If $\deg \varphi \neq 0$, then φ is surjective.

(5) Let $\varphi_1: M_1 \to N_1$ and $\varphi_2: M_2 \to N_2$ be smooth maps, where $\dim M_1 = \dim N_1$ and $\dim M_2 = \dim N_2$. Then the degree of the map $\varphi_1 \times \varphi_2: M_1 \times M_2 \to N_1 \times N_2$ is given by

$$\deg(\varphi_1 \times \varphi_2) = \deg \varphi_1 \cdot \deg \varphi_2 .$$

(6) If $\varphi: M \to N$ is a diffeomorphism, then $\deg \varphi = +1$ when φ preserves orientations, and $\deg \varphi = -1$ when φ reverses orientations.

Proof: Properties (1)–(3) are immediate consequences of the definition. Property (6) follows from Proposition XII, sec. 4.13.

(4) Assume that φ is not surjective. Then $\varphi(M)$ is a proper compact subset of N (since M is compact). Let Φ be a nonnegative nonzero n-form on N with

$$\operatorname{carr} \Phi \subset N - \varphi(M).$$

Then $\varphi^*\Phi = 0$. Hence

$$0 = \int_M \varphi^*\Phi = \deg \varphi \cdot \int_N \Phi.$$

Since $\int_N \Phi > 0$, $\deg \varphi = 0$.

(5) Choose differential forms $\Psi_1 \in A(N_1)$ and $\Psi_2 \in A(N_2)$ such that

$$\int_{N_1} \Psi_1 = 1 \qquad \text{and} \qquad \int_{N_2} \Psi_2 = 1.$$

Then according to Proposition XIII, sec. 4.13,

$$\int_{N_1 \times N_2} \Psi_1 \times \Psi_2 = 1.$$

Hence

$$\deg(\varphi_1 \times \varphi_2) = \int_{M_1 \times M_2} \varphi_1{}^*\Psi_1 \times \varphi_2{}^*\Psi_2$$

$$= \int_{M_1} \varphi_1{}^*\Psi_1 \cdot \int_{M_2} \varphi_2{}^*\Psi_2 = \deg \varphi_1 \cdot \deg \varphi_2 .$$

Q.E.D.

Corollary I: If $\varphi: M \to N$ and $\psi: M \to N$ are inverse diffeomorphisms, then $\deg \varphi \cdot \deg \psi = 1$.

Corollary II: If a map $\varphi: M \to N$ is homotopic to a constant map, then $\deg \varphi = 0$. If a map is homotopic to the identity, then $\deg \varphi = 1$.

Remark: In sec. 6.3 it will be shown that the mapping degree is always an integer.

6.2. Examples: 1. Let $M = N = S^n$ be the unit sphere in a Euclidean space E of dimension $n + 1$ and let $\tau: E \to E$ be a Euclidean rotation. Then $\deg \tau = \det \tau$. (Recall that $\det \tau = \pm 1$.)

In fact, let Δ_E be a determinant function in E. Then the n-form

$$\Omega(x; \xi_1, \ldots, \xi_n) = \Delta_E(x, \xi_1, \ldots, \xi_n), \qquad x \in S^n, \quad \xi_i \in T_x(S^n),$$

orients S^n (cf. Example 2, sec. 3.21). Since τ is linear, $(d\tau)\xi = \tau(\xi)$, whence

$$\tau^*\Omega = \det \tau \cdot \Omega.$$

It follows that $\int_{S^n} \tau^*\Omega = \det \tau \int_{S^n} \Omega$; i.e. $\deg \tau = \det \tau$.

In particular, let τ be the rotation given by $\tau(x) = -x$. Then $\deg \tau = (-1)^{n+1}$ and so τ is not homotopic to the identity if n is even.

2. Let $\varphi, \psi: S^n \to S^n$ be smooth maps such that

$$\varphi(x) \neq -\psi(x), \qquad x \in S^n.$$

Then φ and ψ are homotopic (cf. Example 7, sec. 5.5) and hence Proposition I, part 3, sec. 6.1 implies that

$$\deg \varphi = \deg \psi.$$

In particular, if $\varphi(x) \neq -x$ ($x \in S^n$) we have $\deg \varphi = 1$, while if $\varphi(x) \neq x$ ($x \in S^n$), it follows that $\deg \varphi = (-1)^{n+1}$ (cf. Example 1). Hence if n is even, no smooth map $\varphi: S^n \to S^n$ satisfies $\langle \varphi(x), x \rangle = 0$ ($x \in S^n$) (for then both the conditions $\varphi(x) \neq x$, $\varphi(x) \neq -x$ would be satisfied).

3. Let $\mathbb{C}$ be the complex plane and consider the unit circle $S^1 = \{z \in \mathbb{C} \mid |z| = 1\}$. Define $f: S^1 \to S^1$ by $f(z) = z^p$. Then $\deg f = p$.

In fact, a determinant function Δ is given in $\mathbb{C} = \mathbb{R}^2$ by

$$\Delta(z_1, z_2) = \operatorname{Im}(\bar{z}_1 z_2).$$

Hence an orienting one-form Ω on S^1 is defined by

$$\Omega(z; \zeta) = \operatorname{Im}(\bar{z}\zeta), \qquad z \in S^1, \quad \zeta \in T_z(S^1),$$

and

$$(f^*\Omega)(z; \zeta) = \Omega(z^p; pz^{p-1}\zeta) = p\Omega(z; \zeta).$$

It follows that $\deg f = p$.

A similar argument shows that the map $\varphi: S^1 \to S^1$ obtained by restricting the map $z \mapsto \bar{z}^p$ has degree $-p$.

4. Let $\varphi, \psi: S^1 \to S^1$ be smooth. If $w \in \mathbb{C} - \{0\}$, w^{-1} denotes its complex inverse. Define $\chi: S^1 \to S^1$ by

$$\chi(z) = \varphi(z)\, \psi(z)^{-1}.$$

Then

$$\deg \chi = \deg \varphi - \deg \psi.$$

In fact, for $\zeta \in T_z(S^1)$,

$$\chi(z)^{-1}\,(d\chi)\zeta = \varphi(z)^{-1}\,(d\varphi)\zeta - \psi(z)^{-1}\,(d\psi)\zeta.$$

Now (in the notation of Example 3) a simple calculation shows that $\chi^*\Omega = \varphi^*\Omega - \psi^*\Omega$, whence

$$\deg \chi \cdot \int_{S^1} \Omega = \int_{S^1} \varphi^*\Omega - \int_{S^1} \psi^*\Omega = (\deg \varphi - \deg \psi) \cdot \int_{S^1} \Omega.$$

Proposition II: If two smooth maps $\varphi, \psi: S^1 \to S^1$ have the same degree, they are homotopic.

Lemma I: Assume $\varphi: S^1 \to S^1$ has zero degree. Let $\alpha: \mathbb{R} \to S^1$ be the map given by $\alpha(t) = \exp(2\pi i t)$ ($t \in \mathbb{R}$). Then there exists a smooth map $g: S^1 \to \mathbb{R}$ such that $\varphi = \alpha \circ g$.

Proof: Since

$$\int_{S^1} \varphi^*\Omega = \deg \varphi \cdot \int_{S^1} \Omega = 0,$$

Theorem II, sec. 5.13, implies that $\varphi^*\Omega = 2\pi\delta g$ for some smooth $g: S^1 \to \mathbb{R}$. Choose g so that

$$\alpha(g(1)) = \varphi(1).$$

Now let $\delta t \in A^1(\mathbb{R})$ be the exterior derivative of the identity function—cf. Example 1, sec. 3.17. Then $g^*\delta t = \delta g$. Moreover, $\alpha^*\Omega = 2\pi\delta t$. These formulae yield

$$(\alpha \circ g)^*\Omega = g^*(2\pi\delta t) = 2\pi\delta g = \varphi^*\Omega.$$

Define $\chi: S^1 \to S^1$ by $\chi(z) = \varphi(z)[\alpha(g(z))]^{-1}$. Then, as in Example 4 above,

$$\chi^*\Omega = \varphi^*\Omega - (\alpha \circ g)^*\Omega = 0.$$

Hence $d\chi = 0$ and so χ is constant;

$$\chi(z) = \chi(1) = 1.$$

It follows that $\varphi = \alpha \circ g$.

Q.E.D.

Proof of the proposition: Define $\chi: S^1 \to S^1$ by $\chi(z) = \varphi(z)\,\psi(z)^{-1}$. Then (Example 4)

$$\deg \chi = \deg \varphi - \deg \psi = 0$$

Hence for some smooth map $g: S^1 \to \mathbb{R}$, $\chi = \alpha \circ g$ (Lemma I). It follows that

$$\varphi(z) = \psi(z) \cdot \alpha(g(z)), \qquad z \in S^1.$$

Since $\mathbb{R}$ is contractible, $\alpha \circ g$ is homotopic to the constant map $S^1 \to 1$ via $H: \mathbb{R} \times S^1 \to S^1$. Now

$$K(t, z) = \psi(z) \cdot H(t, z)$$

is a homotopy connecting φ and ψ.

Q.E.D.

6.3. Regular values. Let $\varphi: M \to N$ be a smooth map between (not necessarily compact or oriented) n-manifolds. A point $b \in N$ is called a *regular value*, if either $b \notin \operatorname{Im} \varphi$ or $(d\varphi)_x$ is a linear isomorphism for every $x \in \varphi^{-1}(b)$. Otherwise b is called a *critical value* of φ. Sard's theorem (cf. [10, Lemma 3.2, p. 47]) asserts that every smooth map between n-manifolds has infinitely many regular values (if $n \geqslant 1$).

Suppose $b \in \operatorname{Im} \varphi$ is a regular value for φ. Let $x \in \varphi^{-1}(b)$. Since $(d\varphi)_x \colon T_x(M) \to T_b(N)$ is a linear isomorphism, there exist neighbourhoods U_x of x and U_b of b such that φ maps U_x diffeomorphically onto U_b. In particular

$$U_x \cap \varphi^{-1}(b) = x.$$

It follows that $\varphi^{-1}(b)$ is a closed, *discrete* subset of M. Thus, if M is compact, $\varphi^{-1}(b)$ must be finite.

Now assume that M and N are oriented. Then φ determines an integer-valued function $x \mapsto \epsilon(x)$ on M, given by

$$\epsilon(x) = \begin{cases} 0 & \text{if } (d\varphi)_x \text{ is not a linear isomorphism} \\ +1 & \text{if } (d\varphi)_x \text{ preserves the orientation} \\ -1 & \text{if } (d\varphi)_x \text{ reverses the orientation.} \end{cases}$$

Theorem I: Let M and N be compact connected oriented n-manifolds and let $\varphi \colon M \to N$ be a smooth map. Assume that $b \in N$ is a regular value for φ and let $\{x_1, x_2, \ldots, x_p\}$ be the preimage of b (if $b \in \operatorname{Im} \varphi$). Then

$$\deg \varphi = 0 \qquad \text{if} \quad b \notin \operatorname{Im} \varphi$$

and

$$\deg \varphi = \sum_{i=1}^{p} \epsilon(x_i) \qquad \text{if} \quad b \in \operatorname{Im} \varphi.$$

In particular the mapping degree is an integer.

Proof: If $b \notin \operatorname{Im} \varphi$, φ is not surjective and $\deg \varphi = 0$ (Proposition I, sec. 6.1). Assume $b \in \operatorname{Im} \varphi$. Choose neighbourhoods U_i of x_i $(i = 1, \ldots, p)$ so that $U_i \cap U_j = \varnothing$ $(i \neq j)$ and so that the restriction of φ to U_i is a diffeomorphism onto $\varphi(U_i)$. Then $A = M - \bigcup_i U_i$ is a closed, and hence compact, subset of M. Thus $\varphi(A)$ is compact.

Now since $b \notin A$, there is a neighbourhood V of b in N such that $\varphi(A) \cap V = \varnothing$; i.e. $\varphi^{-1}(V) \subset \bigcup_i U_i$. Since each $\varphi(U_i)$ is a neighbourhood of b, we can choose V so that

$$V \subset \bigcap_{i=1}^{p} \varphi(U_i).$$

Let $W_i = \varphi^{-1}(V) \cap U_i$. Then $\varphi^{-1}(V)$ is the disjoint union of the W_i, and the restriction of φ to each W_i is a diffeomorphism of W_i onto V.

Now let $\Delta \in A^n(N)$ satisfy $\operatorname{carr} \Delta \subset V$ and $\int_N \Delta = 1$. Since $\operatorname{carr} \varphi^*\Delta \subset \varphi^{-1}(V) = \bigcup_i W_i$, it follows that

$$\deg \varphi = \int_M \varphi^*\Delta = \sum_{i=1}^{p} \int_{W_i} \varphi^*\Delta.$$

But $\varphi\colon W_i \to V$ is a diffeomorphism. Thus by Proposition XII, sec. 4.13,

$$\int_{W_i} \varphi^*\Delta = \epsilon(x_i) \int_V \Delta = \epsilon(x_i) \int_N \Delta = \epsilon(x_i)$$

and so

$$\deg \varphi = \sum_{i=1}^{p} \epsilon(x_i).$$

Q.E.D.

Corollary: Let $\varphi\colon M \to N$ be a local diffeomorphism. Then $\deg \varphi = 1$ if and only if φ is an orientation preserving diffeomorphism.

Proof: If φ is an orientation preserving diffeomorphism, then $\deg \varphi = 1$ follows immediately from Proposition I,6, sec. 6.1. Conversely, assume φ is a local diffeomorphism with $\deg \varphi = 1$. Define $\epsilon(x)$ $(x \in M)$ as above and set

$$U_+ = \{x \in M \mid \epsilon(x) = 1\} \qquad \text{and} \qquad U_- = \{x \in M \mid \epsilon(x) = -1\}.$$

Then U_+ and U_- are open. Since φ is a local diffeomorphism,

$$M = U_+ \cup U_- .$$

Since M is connected, it follows that $M = U_+$ or $M = U_-$. Set

$$\epsilon(\varphi) = \begin{cases} +1 & \text{if } M = U_+ \\ -1 & \text{if } M = U_- . \end{cases}$$

Now let $b \in N$ be arbitrary. Since φ is a local diffeomorphism and M is compact, the set $\varphi^{-1}(b)$ is finite,

$$\varphi^{-1}(b) = \{a_1, ..., a_m\}.$$

Moreover, in view of the theorem,

$$\deg \varphi = \sum_{i=1}^{m} \epsilon(a_i) = \epsilon(\varphi) \cdot m,$$

whence

$$\epsilon(\varphi) = m = 1.$$

Thus φ is injective and preserves orientations at each point. On the other hand, since $\deg \varphi \neq 0$, φ is surjective. Thus φ is a diffeomorphism.

Q.E.D.

6.4. Examples: 1. Consider S^2 as the one point compactification of the complex plane $\mathbb{C}$ (cf. Example 10, sec. 1.5). Every polynomial with complex coefficients and leading coefficient 1 determines a smooth map $\varphi: S^2 \to S^2$ given by

$$\varphi(z) = z^n + \sum_{j=0}^{n-1} a_j z^j, \qquad z \in \mathbb{C}$$

and

$$\varphi(z_\infty) = z_\infty .$$

We show that φ is homotopic to the map $\psi: S^2 \to S^2$ given by

$$\psi(z) = z^n, \qquad z \in \mathbb{C}$$

$$\psi(z_\infty) = z_\infty .$$

Define a smooth map $H: \mathbb{R} \times S^2 \to S^2$ by

$$H(t, z) = z^n + \sum_{j=0}^{n-1} a_j t\, z^j, \qquad z \in \mathbb{C}$$

$$H(t, z_\infty) = z_\infty .$$

Then H is a homotopy connecting φ and ψ. In particular, $\deg \varphi = \deg \psi$.

To compute $\deg \psi$ observe that 1 is a regular value for ψ and that $\psi^{-1}(1) = \{a_1, ..., a_n\}$, where $a_k = \exp(2k\pi i/n)$ $(k = 0, 1, ..., n-1)$. It is easy to show that ψ is orientation preserving at each of these points and so we have $\deg \varphi = \deg \psi = n$.

In particular it follows that φ must be surjective if $n \geqslant 1$ and so there exists at least one zero of φ ("fundamental theorem of algebra").

2. Consider S^3 as the unit sphere in the space of quaternions (cf. sec. 0.2). Denote the unit quaternion by e. Let $\varphi: S^3 \to S^3$ be the map given by $\varphi(x) = x^3$. To determine the degree of φ, let $a \in S^3$ be a fixed vector such that $\langle a, e \rangle = 0$. We shall construct the solutions of the equation $x^3 = a$.

Write

$$x = \lambda e + y, \qquad \langle e, y \rangle = 0, \qquad \lambda \in \mathbb{R}.$$

Then $x^3 = \lambda^3 e + 3\lambda^2 y + 3\lambda y^2 + y^3$. But

$$y^2 = -\langle y, y\rangle e = -(1 - \lambda^2)e$$

and so

$$x^3 = [\lambda^3 - 3\lambda(1 - \lambda^2)]e + [3\lambda^2 - (1 - \lambda^2)]y.$$

Thus $x^3 = a$ is equivalent to

$$\lambda(4\lambda^2 - 3) = 0, \qquad (4\lambda^2 - 1)y = a.$$

It follows that the solutions of $x^3 = a$ are given by

$$x_1 = \tfrac{1}{2}(\sqrt{3}\, e + a), \qquad x_2 = \tfrac{1}{2}(-\sqrt{3}\, e + a), \qquad x_3 = -a.$$

Moreover, it is easy to see that a is a regular value for φ, and that $\epsilon(x_i) = 1$ $(i = 1, 2, 3)$. Hence

$$\deg \varphi = 3.$$

3. Let T^n be the n-dimensional torus (cf. Example 3, sec. 1.4). Recall that the points $x \in T^n$ are n-tuples of residue classes of $\mathbb{R}$ (mod 1). Let p be a positive integer and define $\varphi: T^n \to T^n$ by

$$\varphi([\xi^1], ..., [\xi^n]) = ([p\xi^1], ..., [p\xi^n]).$$

Then all points of T^n are regular values and φ preserves orientations. Since the equation $\varphi(x) = 0$ has precisely the solutions

$$[\xi^i] = [\nu/p], \qquad \nu = 0, ..., p - 1, \quad i = 1, ..., n,$$

it follows that $\deg \varphi = p^n$.

4. Let M be any compact connected oriented n-manifold. We shall construct a smooth map $\varphi: M \to S^n$ which has degree 1. Let (V, ψ, E) be a chart on M (E a Euclidean n-space). Denote by A the closed subset of M which corresponds to $\{z \in E \mid |z| \leqslant 2\}$ under this map.

Consider S^n as the one-point compactification of E:

$$S^n = E \cup \{z_\infty\}$$

(Example 10, sec. 1.5) and construct a smooth map $\alpha: E \to S^n$ so that

(i) $\alpha(z) = z, \ |z| < 1$
(ii) $\alpha^{-1}(0) = 0$

and

(iii) $\alpha(z) = z_\infty, \ |z| > 2.$

Define $\varphi\colon M \to S^n$ by

$$\varphi(x) = \begin{cases} \alpha(\psi(x)), & x \in V \\ z_\infty , & x \in M - A. \end{cases}$$

φ is well defined and smooth, as follows from (iii). Since ψ is a diffeomorphism, conditions (i) and (ii) imply that $0 \in E \subset S^n$ is a regular value for φ, and that $\varphi^{-1}(0) = \psi^{-1}(0)$. Hence $\deg \varphi = \pm 1$. In view of Example 1, sec. 6.2, there is a smooth map $\beta\colon S^n \to S^n$ with $\deg \beta = -1$. Hence, if $\deg \varphi = -1$, set $\varphi_1 = \beta \circ \varphi$. Then Proposition I yields $\deg \varphi_1 = 1$.

6.5. Poincaré duality. Let $\varphi\colon M \to N$ be a smooth map between compact oriented n-manifolds. Dualizing $\varphi^{\#}\colon H(M) \leftarrow H(N)$ we obtain a linear map $(\varphi^{\#})^*\colon H(M)^* \to H(N)^*$.

Proposition III: The diagram

$$\begin{array}{ccc} H(M) & \xleftarrow{\ \varphi^{\#}\ } & H(N) \\ {\scriptstyle D_M}\big\downarrow{\scriptstyle \cong} & & \big\downarrow{\scriptstyle \deg\varphi \cdot D_N} \\ H(M)^* & \xrightarrow[(\varphi^{\#})^*]{} & H(N)^* \end{array}$$

commutes (cf. sec. 5.11 for D_M, D_N).

Proof: Let $\alpha \in H(N)$ be arbitrary. Then we have, for $\beta \in H(N)$,

$$\begin{aligned} \langle (\varphi^{\#})^* D_M \varphi^{\#}\alpha, \beta \rangle &= \langle D_M \varphi^{\#}\alpha, \varphi^{\#}\beta \rangle \\ &= \int_M^{\#} \varphi^{\#}\alpha \cdot \varphi^{\#}\beta = \int_M^{\#} \varphi^{\#}(\alpha \cdot \beta) \\ &= \deg \varphi \cdot \int_N^{\#} \alpha \cdot \beta = \deg \varphi \cdot \langle D_N \alpha, \beta \rangle. \end{aligned}$$

Q.E.D.

Corollary I: Let $\varphi\colon M \to N$ be a smooth map between compact oriented connected n-manifolds, with $\deg \varphi \neq 0$. Then $\varphi^{\#}\colon H(M) \leftarrow H(N)$ is injective.

Proof: Since $\deg \varphi \neq 0$, Proposition III implies that $(\varphi^{\#})^* \circ D_M \circ \varphi^{\#}$ is a linear isomorphism. Hence $\varphi^{\#}$ is injective.

Q.E.D.

Corollary II: Let $\varphi: S^n \to M$ be a smooth map (M, a compact connected oriented n-manifold) with $\deg \varphi \neq 0$. Then $H^p(M) = 0$ $(1 \leqslant p \leqslant n-1)$.

6.6. Mappings through products of spheres. Let S_1, S_2 denote oriented n-spheres with base points $a \in S_1$, $b \in S_2$. Denote the projections $S_1 \times S_2 \to S_i$ by ρ_i and let the inclusions $S_i \to S_1 \times S_2$ opposite a, b be denoted by j_a, j_b.

Consider smooth maps

$$M \xrightarrow{\varphi} S_1 \times S_2 \xrightarrow{\psi} N$$

(M, N compact oriented n-manifolds) and write

$$\varphi_1 = \rho_1 \circ \varphi, \qquad \psi_a = \psi \circ j_a$$

$$\varphi_2 = \rho_2 \circ \varphi, \qquad \psi_b = \psi \circ j_b .$$

Proposition IV: With the hypotheses and notation above,

$$\deg(\psi \circ \varphi) = \deg \psi_b \cdot \deg \varphi_1 + \deg \psi_a \cdot \deg \varphi_2 .$$

Lemma II: Let $\alpha \in H^n(S_1 \times S_2)$. Then

$$\alpha = \rho_1{}^\# j_b{}^\# \alpha + \rho_2{}^\# j_a{}^\# \alpha.$$

Proof: By the Künneth theorem (cf. sec. 5.20), an isomorphism

$$\kappa_\#: H(S_1) \otimes H(S_2) \xrightarrow{\cong} H(S_1 \times S_2)$$

is given by

$$\kappa_\#(\alpha_1 \otimes \alpha_2) = \rho_1{}^\# \alpha_1 \cdot \rho_2{}^\# \alpha_2 .$$

Thus $\kappa_\#$ restricts to the isomorphism

$$(H^n(S_1) \otimes 1) \oplus (1 \otimes H^n(S_2)) \xrightarrow{\cong} H^n(S_1 \times S_2)$$

given by

$$\kappa_\#(\alpha_1 \otimes 1 + 1 \otimes \alpha_2) = \rho_1{}^\# \alpha_1 + \rho_2{}^\# \alpha_2 .$$

Now let $\alpha \in H^n(S_1 \times S_2)$ be arbitrary. Write $\alpha = \rho_1{}^\# \alpha_1 + \rho_2{}^\# \alpha_2$ and apply the equations

$$j_a{}^\# \circ \rho_2{}^\# = \iota, \qquad j_a{}^\# \circ \rho_1{}^\# = 0$$

$$j_b{}^\# \circ \rho_2{}^\# = 0, \qquad j_b{}^\# \circ \rho_1{}^\# = \iota.$$

Q.E.D.

Proof of the proposition: Let ω_N be the orientation class, $\int_N^{\#} \omega_N = 1$. Then

$$\deg(\psi \circ \varphi) = \int_M^{\#} \varphi^{\#}\psi^{\#}\omega_N .$$

Apply Lemma II with $\alpha = \psi^{\#}\omega_N$: this gives

$$\varphi^{\#}\psi^{\#}\omega_N = \varphi_1{}^{\#}\psi_b{}^{\#}\omega_N + \varphi_2{}^{\#}\psi_a{}^{\#}\omega_N .$$

Hence

$$\begin{aligned}\deg(\psi \circ \varphi) &= \deg(\psi_b \circ \varphi_1) + \deg(\psi_a \circ \varphi_2)\\ &= \deg \psi_b \cdot \deg \varphi_1 + \deg \psi_a \cdot \deg \varphi_2 .\end{aligned}$$

Q.E.D.

§2. The canonical map α_M

6.7. Definition: Let M be an oriented n-manifold and let $a \in M$ be a point. We shall denote $M - \{a\}$ by $\dot{M}$. In this section we shall construct a linear map

$$\alpha_M: H^{n-1}(\dot{M}) \to \mathbb{R}.$$

First, choose a smooth function f on M so that $f = 0$ in a neighbourhood of a, and $f = 1$ outside a compact set. Then δf is closed and has compact carrier contained in $\dot{M}$; thus δf represents an element $\alpha_a \in H_c^1(\dot{M})$. If another such function g is chosen, then $f - g \in \mathscr{S}_c(\dot{M})$ and so δg also represents α_a. It follows that α_a is independent of the choice of f: α_a is called the *localizing class at a*.

Now define α_M by setting (cf. sec. 5.9)

$$\alpha_M(\beta) = \int_{\dot{M}}^{\#} \alpha_a * \beta, \qquad \beta \in H^{n-1}(\dot{M}).$$

If $\Phi \in A^{n-1}(\dot{M})$ represents β and δf represents α_a, then

$$\alpha_M(\beta) = \int_{\dot{M}} \delta f \wedge \Phi = \int_M \delta f \wedge \Phi.$$

Next consider a smooth map between oriented n-manifolds, $\varphi: M \to N$. Let $\dot{M} = M - \{a\}$ and $\dot{N} = N - \{\varphi(a)\}$ and assume that φ restricts to a map $\dot{\varphi}: \dot{M} \to \dot{N}$.

Proposition V: Assume that $(d\varphi)_a$ is a linear isomorphism. Then the diagram

$$\begin{array}{ccc} H^{n-1}(\dot{M}) & \xleftarrow{\dot{\varphi}^{\#}} & H^{n-1}(\dot{N}) \\ {\scriptstyle \alpha_M}\downarrow & & \downarrow{\scriptstyle \alpha_N} \\ \mathbb{R} & \xleftarrow[\epsilon]{} & \mathbb{R} \end{array}$$

commutes, where $\epsilon(t) = t$ $(t \in \mathbb{R})$ if $(d\varphi)_a$ preserves the orientations and $\epsilon(t) = -t$ if $(d\varphi)_a$ reverses the orientations.

Proof: Since $(d\varphi)_a$ is a linear isomorphism there are connected neighbourhoods U of a and V of $\varphi(a)$ such that φ restricts to a diffeomorphism of U onto V (cf. Theorem I, sec. 3.8). Choose $g \in \mathscr{S}(N)$ so that g is zero in a neighbourhood of $\varphi(a)$ and so that carr$(g - 1)$ is compact and contained in V. Define $f \in \mathscr{S}(M)$ by

$$f(x) = \begin{cases} g(\varphi(x)), & x \in U \\ 1, & x \notin U. \end{cases}$$

Then δf and δg represent α_a and $\alpha_{\varphi(a)}$.

Now consider a closed form $\Psi \in A^{n-1}(\dot{N})$. Then

$$\delta g \wedge \Psi \in A_c^n(V)$$

and thus, by Proposition XII, sec. 4.13,

$$\epsilon \cdot \int_N \delta g \wedge \Psi = \epsilon \cdot \int_V \delta g \wedge \Psi = \int_U \varphi^*(\delta g \wedge \Psi) = \int_U \delta\varphi^* g \wedge \dot{\varphi}^* \Psi.$$

On the other hand, since $\delta f(x) = (\delta\varphi^* g)(x)$ $(x \in U)$ and carr $\delta f \subset U$, we have

$$\int_U \delta\varphi^* g \wedge \dot{\varphi}^* \Psi = \int_U \delta f \wedge \dot{\varphi}^* \Psi = \int_M \delta f \wedge \dot{\varphi}^* \Psi.$$

It follows that

$$\int_M \delta f \wedge \dot{\varphi}^* \Psi = \epsilon \left(\int_N \delta g \wedge \Psi \right),$$

whence

$$\int_{\dot{M}}^{\#} \alpha_a * \dot{\varphi}^{\#} \beta = \epsilon \int_{\dot{N}}^{\#} \alpha_{\varphi(a)} * \beta, \qquad \beta \in H^{n-1}(\dot{N}).$$

Q.E.D.

6.8. Euclidean spaces. Let S denote the unit sphere of an oriented n-dimensional Euclidean space E $(n \geqslant 2)$ and let $\dot{E} = E - \{0\}$. The inclusion map $i: S \to E$ induces an isomorphism

$$i^{\#}: H^{n-1}(S) \xleftarrow{\cong} H^{n-1}(\dot{E})$$

(cf. Example 5, sec. 5.5). On the other hand, consider the isomorphism

$$\int_S^{\#}: H^{n-1}(S) \xrightarrow{\cong} \mathbb{R},$$

and the canonical linear map $\alpha_E\colon H^{n-1}(\dot{E}) \to \mathbb{R}$, where S is given the orientation induced by the orientation of E.

Proposition VI:

$$\alpha_E = \int_S^{\#} \circ\, i^{\#} : H^{n-1}(\dot{E}) \to \mathbb{R}.$$

In particular, α_E is an isomorphism.

Proof: Let $\Phi \in A^{n-1}(\dot{E})$ be a closed form and let $f \in \mathscr{S}(E)$ satisfy

$$f(x) = \begin{cases} 0, & |x| < \tfrac{1}{4} \\ 1, & |x| > \tfrac{1}{2}. \end{cases}$$

We must show that

$$\int_E \delta f \wedge \Phi = \int_S i^*\Phi.$$

Let B be the open unit ball in E. Then $\delta f \wedge \Phi$ has carrier in B. Moreover, since Φ is closed, $\delta f \wedge \Phi = \delta(f \cdot \Phi)$. Now Stokes' theorem (Theorem II, sec. 4.17) yields

$$\int_E \delta f \wedge \Phi = \int_B \delta(f \cdot \Phi) = \int_S i^*(f \cdot \Phi) = \int_S i^*\Phi.$$

Q.E.D.

In view of Proposition V, sec. 6.7, Proposition VI has the obvious

Corollary: Let M be an oriented n-manifold ($n \geqslant 2$) diffeomorphic to $\mathbb{R}^n$. Let $a \in M$, $\dot{M} = M - \{a\}$. Then the map

$$\alpha_M\colon H^{n-1}(\dot{M}) \to \mathbb{R}$$

is a linear isomorphism.

6.9. Mayer–Vietoris sequences. Let M be a compact oriented n-manifold. Suppose U_i ($i = 1, ..., r$) are disjoint open subsets of M and that $a_i \in U_i$ ($i = 1, ..., r$). Set

$$U = \bigcup_{i=1}^{r} U_i, \qquad V = M - \{a_1, ..., a_r\}$$

and

$$\dot{U}_i = U_i - \{a_i\}, \qquad i = 1, ..., r.$$

Then

$$U \cup V = M \qquad \text{and} \qquad U \cap V = \bigcup_{i=1}^{r} \dot{U}_i .$$

Thus according to sec. 5.4 we have the exact triangle

$$\begin{array}{ccc} H(M) & \longrightarrow & H(U) \oplus H(V) \\ & {\scriptstyle\partial}\nwarrow \quad \swarrow & \\ & \bigoplus_{i=1}^{r} H(\dot{U}_i). & \end{array}$$

On the other hand, consider the linear maps

$$\alpha_{U_i}: H^{n-1}(\dot{U}_i) \to \mathbb{R}, \qquad i = 1, \ldots, r,$$

as defined above. We denote them simply by α_i . These maps determine the linear map

$$\alpha : \bigoplus_{i=1}^{r} H^{n-1}(\dot{U}_i) \to \mathbb{R}$$

given by

$$\alpha(\beta_1 \oplus \cdots \oplus \beta_r) = \sum_{1}^{r} \alpha_i(\beta_i), \qquad \beta_i \in H^{n-1}(\dot{U}_i).$$

Proposition VII: With the hypotheses and notation defined above, the diagram

$$\begin{array}{ccc} & \bigoplus_{i=1}^{r} H^{n-1}(\dot{U}_i) & \\ & {\scriptstyle\partial}\swarrow \qquad \searrow{\scriptstyle\alpha} & \\ H^n(M) & \xrightarrow[\int_M^{\#}]{} & \mathbb{R} \end{array}$$

commutes.

Proof: It is sufficient to show that

$$\int_M^{\#} \partial\beta = \alpha_1(\beta), \qquad \beta \in H^{n-1}(\dot{U}_1).$$

Let $\Phi \in A^{n-1}(\dot{U}_1)$ be a closed form representing β. Extend Φ to $U \cap V$ by setting

$$\Phi(x) = 0, \qquad x \in \bigcup_{i=2}^{r} \dot{U}_i .$$

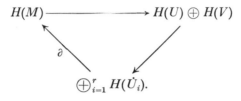

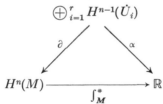

Choose a partition of unity $\{f, g\}$ for M subordinate to the open covering by U and V. Then

$$f \cdot \Phi \in A^{n-1}(V), \qquad g \cdot \Phi \in A^{n-1}(U),$$

and the element $\partial\beta$ is represented by the form $\Omega \in A^n(M)$ given by

$$\Omega(x) = \begin{cases} \delta(g\Phi)(x), & x \in U \\ -\delta(f\Phi)(x), & x \in V. \end{cases}$$

Hence

$$\int_M^{\#} \partial\beta = \int_M \Omega.$$

On the other hand, carr $\Omega \subset U_1$. Thus

$$\int_M \Omega = \int_{U_1} \Omega = \int_{U_1} \delta g \wedge \Phi.$$

Moreover, f and g can be chosen so that $g = 0$ in a neighbourhood of a_1, and $g = 1$ in $U_1 - K$ for some compact set $K \subset U_1$. With this choice

$$\int_M^{\#} \partial\beta = \int_M \Omega = \int_{U_1} \delta g \wedge \Phi = \alpha_1(\beta).$$

Q.E.D.

6.10. Products. Let M and N be oriented manifolds of dimensions m and n respectively, and give $M \times N$ the product orientation. Choose points $a \in M$, $b \in N$ and set

$$\dot{M} = M - \{a\}, \qquad \dot{N} = N - \{b\}, \qquad M \dot{\times} N = M \times N - \{(a, b)\}.$$

Then we have linear maps

$$\alpha_M: H^{m-1}(\dot{M}) \to \mathbb{R}, \qquad \alpha_N: H^{n-1}(\dot{N}) \to \mathbb{R}$$

and

$$\alpha_{M\times N}: H^{m+n-1}(M \dot{\times} N) \to \mathbb{R}$$

Now consider the open covering U, V of $M \dot{\times} N$ given by

$$U = M \times \dot{N}, \qquad V = \dot{M} \times N.$$

Then

$$U \cap V = \dot{M} \times \dot{N}.$$

Hence we obtain the exact triangle

$$\begin{array}{ccc} H(M \dot{\times} N) & \longrightarrow & H(U) \oplus H(V) \\ & {}_{\partial}\nwarrow \quad \swarrow & \\ & H(\dot{M} \times \dot{N}) & \end{array}$$

(cf. sec. 5.4) and the exact triangle

$$\begin{array}{ccc} H_c(\dot{M} \times \dot{N}) & \longrightarrow & H_c(U) \oplus H_c(V) \\ & {}_{\partial_c}\nwarrow \quad \swarrow & \\ & H_c(M \dot{\times} N) & \end{array}$$

(cf. sec. 5.10). Moreover, the Künneth homomorphisms

$$\kappa_{\#}: H(\dot{M}) \otimes H(\dot{N}) \to H(\dot{M} \times \dot{N}), \qquad (\kappa_c)_{\#}: H_c(\dot{M}) \otimes H_c(\dot{N}) \to H_c(\dot{M} \times \dot{N})$$

are defined (cf. sec. 5.17 and sec. 5.18).

Proposition VIII: With the hypotheses and notation above the diagram

$$\begin{array}{ccc} H^{m-1}(\dot{M}) \otimes H^{n-1}(\dot{N}) & \xrightarrow{\kappa_{\#}} & H^{m+n-2}(\dot{M} \times \dot{N}) \\ {\scriptstyle (-1)^m \alpha_M \otimes \alpha_N} \downarrow & & \downarrow {\scriptstyle \partial} \\ \mathbb{R} & \xleftarrow[\alpha_{M \times N}]{} & H^{m+n-1}(M \dot{\times} N) \end{array}$$

commutes.

Lemma III: The localizing classes $\alpha_a \in H_c^1(\dot{M})$, $\alpha_b \in H_c^1(\dot{N})$, and $\alpha_{(a,b)} \in H_c^1(M \dot{\times} N)$ are related by

$$\partial_c(\alpha_{(a,b)}) = -(\kappa_c)_{\#}(\alpha_a \otimes \alpha_b).$$

Proof: Choose $f \in \mathscr{S}(M)$ so that f is zero near a and $f - 1$ has compact carrier. Choose $g \in \mathscr{S}(N)$ to be zero near b and so that $g - 1$ has compact carrier. Then $\delta f \times \delta g$ represents $(\kappa_c)_{\#}(\alpha_a \otimes \alpha_b)$.

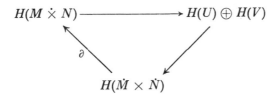

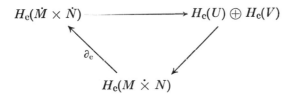

Next observe that a closed form $\omega \in A^1_c(M \dot{\times} N)$ is defined by

$$\omega = (f - 1) \times \delta g + \delta f \times (g - 1).$$

Moreover,

$$(f - 1) \times \delta g \in A^1_c(M \times \dot{N}) \qquad \text{and} \qquad \delta f \times (g - 1) \in A^1_c(\dot{M} \times N).$$

Thus, if ω represents $\gamma \in H^1_c(M \dot{\times} N)$, then $\partial_c \gamma$ is represented by $\delta f \times \delta g$; i.e.

$$\partial_c \gamma = (\kappa_c)_\#(\alpha_a \otimes \alpha_b).$$

Finally, define $h \in \mathscr{S}(M \times N)$ by

$$h = f \times 1 + 1 \times g - f \times g$$

and note that

$$\omega = -\delta h.$$

When $f(x) = 0$, $g(y) = 0$, then $h(x, y) = 0$; thus h is zero in a neighbourhood of (a, b). $1 - h$ can be written

$$1 - h = (1 - f) \times (1 - g)$$

and so $\operatorname{carr}(1 - h) \subset \operatorname{carr}(1 - f) \times \operatorname{carr}(1 - g)$ is compact. It follows that δh represents $\alpha_{(a,b)}$; i.e.

$$\partial_c \alpha_{(a,b)} = \partial_c(-\gamma) = -(\kappa_c)_\#(\alpha_a \otimes \alpha_b).$$

Q.E.D.

Proof of the proposition: Let $\sigma \in H^{m-1}(\dot{M})$, $\tau \in H^{n-1}(\dot{N})$. Then it follows from Proposition VII, sec. 5.11, that

$$\begin{aligned}(\alpha_{M \times N} \partial \kappa_\#)(\sigma \otimes \tau) &= \int^{\#}_{M \dot{\times} N} \alpha_{(a,b)} * \partial \kappa_\#(\sigma \otimes \tau) \\ &= \int^{\#}_{\dot{M} \times \dot{N}} \partial_c \alpha_{(a,b)} * \kappa_\#(\sigma \otimes \tau).\end{aligned}$$

Applying the lemma, and using Proposition XIII, sec. 4.13, yields

$$\begin{aligned}(\alpha_{M \times N} \partial \kappa_\#)(\sigma \otimes \tau) &= - \int^{\#}_{\dot{M} \times \dot{N}} (\kappa_c)_\#(\alpha_a \otimes \alpha_b) * \kappa_\#(\sigma \otimes \tau) \\ &= (-1)^m \int^{\#}_{\dot{M}} \alpha_a * \sigma \int^{\#}_{\dot{N}} \alpha_b * \tau \\ &= (-1)^m \alpha_M(\sigma)\, \alpha_N(\tau). \\ &= (-1)^m (\alpha_M \otimes \alpha_N)(\sigma \otimes \tau).\end{aligned}$$

Q.E.D.

§3. Local degree

6.11. Definition. Let $\varphi\colon M \to N$ be a smooth map between oriented n-manifolds ($n \geqslant 2$). A point $a \in M$ will be called *isolated for* φ, if there exists a neighbourhood U of a such that

$$\varphi(x) \neq \varphi(a), \qquad x \in U - \{a\}.$$

We shall define the *local degree* of φ at an isolated point a. Choose charts $(U, u, \mathbb{R}^n)$ on M and $(V, v, \mathbb{R}^n)$ on N so that $a \in U$, $\varphi(a) \in V$ and φ restricts to a smooth map

$$\dot{\varphi}\colon U - \{a\} \to V - \{\varphi(a)\}.$$

Write $U - \{a\} = \dot{U}$, $V - \{\varphi(a)\} = \dot{V}$.

The corollary to Proposition VI, sec. 6.8, gives linear isomorphisms

$$\alpha_U\colon H^{n-1}(\dot{U}) \xrightarrow{\cong} \mathbb{R}, \qquad \alpha_V\colon H^{n-1}(\dot{V}) \xrightarrow{\cong} \mathbb{R}.$$

Thus a linear map $f_\varphi\colon \mathbb{R} \to \mathbb{R}$ is determined by the commutative diagram

$$\begin{array}{ccc} H^{n-1}(\dot{U}) & \xleftarrow{\dot{\varphi}^{\#}} & H^{n-1}(\dot{V}) \\ {\scriptstyle \alpha_U}\downarrow{\scriptstyle \cong} & & {\scriptstyle \cong}\downarrow{\scriptstyle \alpha_V} \\ \mathbb{R} & \xleftarrow[f_\varphi]{} & \mathbb{R} \end{array}$$

Lemma IV: The map f_φ is independent of the choice of U and V.

Proof: Let $(U', u', \mathbb{R}^n)$ and $(V', v', \mathbb{R}^n)$ be a second pair of charts satisfying the conditions above. It is easy to reduce to the case $U' \subset U$, $V' \subset V$. Then Proposition V, sec. 6.7, implies that the diagram

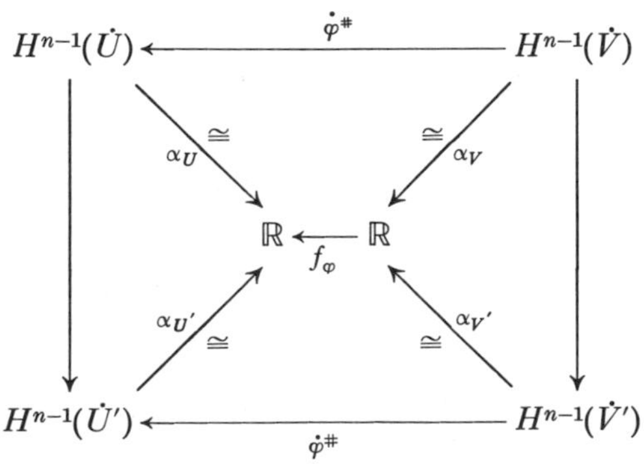

commutes, and the lemma follows. Q.E.D.

Definition: The number $f_\varphi(1)$ is called the *local degree of* φ *at* a and will be denoted by $\deg_a \varphi$.

The defining diagram shows that

$$\deg_a \varphi = \alpha_U \dot{\varphi}^\#(\gamma),$$

where $\gamma \in H^{n-1}(\dot{V})$ is the unique element satisfying $\alpha_V(\gamma) = 1$.

Example: *Maps between Euclidean spaces:* Suppose $\varphi: E \to F$ is a smooth map between oriented Euclidean spaces of dimension n ($n \geqslant 2$) such that $\varphi^{-1}(0) = 0$. Set $\dot{E} = E - \{0\}$, $\dot{F} = F - \{0\}$ and denote the unit spheres of E and F by S_E and S_F, respectively, with inclusion maps

$$i_E: S_E \to \dot{E} \qquad \text{and} \qquad i_F: S_F \to \dot{F}.$$

Let $\rho_F: \dot{F} \to S_F$ be the projection $y \mapsto y/|y|$. Then

$$\psi = \rho_F \circ \varphi \circ i_E : S_E \to S_F$$

is a smooth map.

Proposition IX: With the hypotheses and notation above

$$\deg \psi = \deg_0 \varphi.$$

Proof: Let $\dot{\varphi}: \dot{E} \to \dot{F}$ denote the restriction of φ. Let

$$\beta = \alpha_F^{-1}(1) \in H^{n-1}(\dot{F}).$$

Then, using Proposition VI, sec. 6.8, we find

$$\deg_0 \varphi = \alpha_E \dot{\varphi}^\#(\beta) = \int_{S_E}^\# i_E^\# \dot{\varphi}^\# \beta.$$

On the other hand, recall from Example 5, sec. 5.5, that $\rho_F^\# = (i_F^\#)^{-1}$. Thus

$$\deg_0 \varphi = \int_{S_E}^\# (i_E^\# \dot{\varphi}^\# \rho_F^\# i_F^\#)(\beta) = \int_{S_E}^\# (\psi^\# i_F^\#)(\beta)$$

$$= \deg \psi \cdot \int_{S_F}^\# i_F^\#(\beta) = \deg \psi \cdot \alpha_F(\beta) = \deg \psi.$$

Q.E.D.

Proposition X: Let $\varphi: M \to N$ be a smooth map between oriented n-manifolds ($n \geqslant 2$). Suppose $a \in M$ is isolated for φ.

(1) If $(d\varphi)_a$ is a linear isomorphism, then

$$\deg_a \varphi = \epsilon,$$

where $\epsilon = +1$ if $(d\varphi)_a$ preserves the orientations, and $\epsilon = -1$ if $(d\varphi)_a$ reverses the orientations.

(2) If $\psi: N \to Q$ is a second smooth map between oriented n-manifolds and $\varphi(a)$ is isolated for ψ, then a is an isolated point for $\psi \circ \varphi$ and

$$\deg_a(\psi \circ \varphi) = \deg_{\varphi(a)} \psi \cdot \deg_a \varphi.$$

(3) $\deg_a \varphi$ is an integer.

(4) Let $\psi: P \to Q$ be a smooth map between oriented r-manifolds. Suppose $b \in P$ is an isolated point for ψ. Then (a, b) is an isolated point for $\varphi \times \psi$ and

$$\deg_{(a,b)}(\varphi \times \psi) = \deg_a \varphi \cdot \deg_b \psi.$$

Proof: (1) follows from Proposition V, sec. 6.7. (2) is obvious. (3) follows from (1) and (2), together with Proposition IX and Theorem I, sec. 6.3. (4) is a consequence of Proposition VIII, sec. 6.10, formula 5.3, sec. 5.4, and some elementary diagram chasing.

Q.E.D.

6.12. Examples: **1.** Let E and F be oriented n-dimensional vector spaces ($n \geqslant 2$) and let $\varphi: E \to F$ be a linear isomorphism. Then $\deg_0 \varphi = 1$ if φ preserves the orientations, and $\deg_0 \varphi = -1$ if φ reverses the orientations, as follows from Proposition X, (1), sec. 6.11.

In particular, if $\sigma: E \to E$ denotes the map $x \mapsto -x$, then

$$\deg_0 \sigma = (-1)^n.$$

2. Let $\varphi: E \to F$ be a smooth map between oriented n-dimensional vector spaces ($n \geqslant 2$) such that $\varphi^{-1}(0) = 0$. Define a map $-\varphi: E \to F$ by

$$(-\varphi)(x) = -\varphi(x), \qquad x \in E.$$

Then

$$\deg_0(-\varphi) = (-1)^n \deg_0 \varphi.$$

In fact, write $-\varphi = \sigma \circ \varphi$ where $\sigma: F \to F$ is the linear isomorphism given by $y \mapsto -y$. Then, Proposition X, (2), together with Example 1, yields

$$\deg_0(-\varphi) = \deg_0 \sigma \cdot \deg_0 \varphi = (-1)^n \deg_0 \varphi.$$

3. Let $\varphi: \mathbb{C} \to \mathbb{C}$ be defined by

$$\varphi(z) = z^p, \qquad p = 1, 2, \ldots, .$$

Then Proposition IX, sec. 6.11, shows that

$$\deg_0 \varphi = p,$$

since the restriction of φ to the unit circle S^1 has degree p (cf. Example 3, sec. 6.2).

4. *Maps $S^n \to S^n$ with given degree:* Let E be an oriented $(n+1)$-dimensional vector space, $n \geqslant 1$. Write $E = \mathbb{C} \oplus F$, where F is an $(n-1)$-dimensional vector space. Introduce an orientation in F so that the product orientation coincides with the given orientation of E.

Let $\varphi: \mathbb{C} \to \mathbb{C}$ be the map given by

$$\varphi(z) = z^p, \qquad p = 1, 2, \ldots,$$

and consider the map $\varphi \times \iota: E \to E$. Combining Example 3 with Proposition X, (4), we obtain

$$\deg_0(\varphi \times \iota) = \deg_0 \varphi = p.$$

Next, endow E with a Euclidean metric, and let S^n be the unit sphere. Then a smooth map $\psi: S^n \to S^n$ is given by

$$\psi(x) = \frac{(\varphi \times \iota)(x)}{|(\varphi \times \iota)(x)|}, \qquad x \in S^n.$$

It follows from Proposition IX, sec. 6.11, that

$$\deg \psi = \deg_0(\varphi \times \iota) = p.$$

In a similar way a map $\psi: S^n \to S^n$ of degree $-p$ can be constructed.

Remark: Let M be any compact connected oriented n-manifold. According to Example 4, sec. 6.4, there exists a smooth map $\varphi: M \to S^n$ with $\deg \psi = 1$. Composing φ with the map ψ yields a map $M \to S^n$ of degree p.

Example 5: Let E be an oriented Euclidean n-space. Define $\varphi: E \to E$ by

$$\varphi(x) = -\langle x, x\rangle a + 2\langle x, a\rangle x, \qquad x \in E,$$

where a is a fixed unit vector. Then φ restricts to a map $\dot{\varphi}: \dot{E} \to \dot{E}$. We shall show that

$$\deg_0 \varphi = 1 + (-1)^n.$$

Let S^{n-1} denote the unit sphere of E and define $\psi: S^{n-1} \to S^{n-1}$ by

$$\psi(x) = \frac{\varphi(x)}{|\varphi(x)|}, \qquad x \in S^{n-1}.$$

Proposition IX, sec. 6.11, shows that

$$\deg \psi = \deg_0 \varphi.$$

To compute $\deg \psi$, we use Theorem I of sec. 6.3.

First it will be shown that $\psi^{-1}(a) = \{a, -a\}$. Suppose $\psi(x) = a$. Then

$$-\langle x, x\rangle a + 2\langle x, a\rangle x = \lambda a, \qquad \lambda \in \mathbb{R}, \quad \lambda > 0;$$

i.e.,

$$(\lambda + \langle x, x\rangle)a = 2\langle x, a\rangle x.$$

Since $\lambda > 0$, it follows that $\lambda + |x|^2 > 0$, whence $\langle x, a\rangle \neq 0$. Thus we obtain

$$x = \frac{\lambda + \langle x, x\rangle}{2\langle x, a\rangle}\, a.$$

Since $|a| = 1$, it follows that $x = \pm a$. On the other hand, clearly, $\psi(a) = \psi(-a) = a$ and so $\psi^{-1}(a) = \{-a, a\}$.

Next we compute $(d\psi)_a$ and $(d\psi)_{-a}$. Observe that the linear maps $\varphi'(a)$ and $\varphi'(-a)$ are given by

$$\varphi'(a) = 2\iota \qquad \text{and} \qquad \varphi'(-a) = -2\iota.$$

Now let $h \in T_a(S^{n-1})$. Evidently

$$(d\psi)_a h = 2h.$$

Similarly,

$$(d\psi)_{-a}\, h = -2h, \qquad h \in T_{-a}(S^{n-1}).$$

These relations show that $(d\psi)_a$ is orientation preserving, and that $(d\psi)_{-a}$ is orientation preserving if and only if n is even. Thus a is a regular value for ψ, and Theorem I yields

$$\deg \psi = 1 + (-1)^n.$$

6.13. Local degree and global degree. In this section we shall prove:

Theorem II: Let $\varphi: M \to N$ be a smooth map between compact, connected, oriented n-manifolds ($n \geqslant 2$). Let $b \in N$ be a point for which the set $\{\varphi^{-1}(b)\}$ is finite,

$$\varphi^{-1}(b) = \{a_1, ..., a_r\}.$$

Then

$$\deg \varphi = \sum_{i=1}^{r} \deg_{a_i} \varphi.$$

Remark: This generalizes Theorem I of sec. 6.3.

Proof: Choose charts $(U_i, u_i, \mathbb{R}^n)$ on M and $(W, w, \mathbb{R}^n)$ on N so that the U_i are disjoint and $a_i \in U_i$ ($i = 1, ..., r$) and $b \in W$. Set

$$\dot{U}_i = U_i - \{a_i\}, \qquad V = M - \{a_1, ..., a_r\}, \qquad U = \bigcup_{i=1}^{r} U_i .$$

Then

$$U \cup V = M, \qquad U \cap V = \bigcup_{i=1}^{r} \dot{U}_i .$$

Assume the choices of U_i, W have been made so that φ restricts to smooth maps $\varphi_i: U_i \to W$. Then φ_i restricts to $\dot{\varphi}_i: \dot{U}_i \to \dot{W}$ ($\dot{W} = W - \{b\}$). By definition

$$\deg_{a_i} \varphi = \alpha_{U_i} \dot{\varphi}_i{}^{\#} \alpha_{\dot{W}}^{-1}(1).$$

The $\dot{\varphi}_i$ define a map $\varphi_{U \cap V}: U \cap V \to \dot{W}$ (simply the restriction of φ). On the other hand set $\alpha_i = \alpha_{U_i}$ and write

$$\alpha = \bigoplus_{i=1}^{r} \alpha_i : H^{n-1}(U \cap V) \to \mathbb{R}.$$

We have

$$\sum_{i=1}^{r} \deg_{a_i} \varphi = (\alpha \circ \varphi_{U \cap V}^{\#} \circ \alpha_{\dot{W}}^{-1})(1).$$

Finally, the triples (M, U, V) and $(N, W, N - \{b\})$ lead to exact Mayer–Vietoris triangles. Denote the connecting homomorphisms by $\partial_M: H(U \cap V) \to H(M)$, $\partial_N: H(\dot{W}) \to H(N)$. Then (cf. sec. 5.4)

$$\partial_M \varphi^{\#}_{U \cap V} = \varphi^{\#} \partial_N .$$

Thus we can apply Proposition VII, sec. 6.9, twice to obtain

$$\sum_{i=1}^{r} \deg_{a_i} \varphi = \int_M^{\#} \partial_M \varphi^{\#}_{U \cap V} \alpha_{\dot{W}}^{-1}(1)$$

$$= \deg \varphi \int_N^{\#} \partial_N \alpha_{\dot{W}}^{-1}(1) = \deg \varphi.$$

Q.E.D.

§4. The Hopf theorem

Let M be a compact connected oriented n-manifold. Recall, from Proposition I, sec. 6.1, that any two homotopic smooth maps $\varphi, \psi: M \to S^n$ have the same degree. The converse is true as well: if $\varphi, \psi: M \to S^n$ have the same degree then φ and ψ are homotopic. This is a theorem of H. Hopf, cf. [8, p. 149]. It is the purpose of this article to establish this result in the case $M = S^n$. (The reader is invited to attempt the proof of the general theorem.)

6.14. Suspension. Consider the unit sphere S^n ($n \geqslant 1$) of an $(n + 1)$-dimensional Euclidean space E. Choose two fixed points $x_N \in S^n$ and $x_S = -x_N \in S^n$, called the *north pole* and the *south pole.* The $(n - 1)$-sphere given by

$$S^{n-1} = \{x \in S^n \mid \langle x, x_N \rangle = 0\}$$

will be called the *equator sphere.* The closed subsets of S^n given by

$$H_N = \{x \in S^n \mid \langle x, x_N \rangle \geqslant 0\}$$

and

$$H_S = \{x \in S^n \mid \langle x, x_N \rangle \leqslant 0\}$$

will be called the *north* and the *south hemisphere.*

Now fix a smooth function $\omega: \mathbb{R} \to \mathbb{R}$ which satisfies the conditions:

(1) $\omega(-t) = -\omega(t)$
(2) $|\omega(t)| \leqslant \pi/2$, $t \in \mathbb{R}$, and $\omega(t) = \pi/2$, $t > 1 - \epsilon > 0$,
(3) $\omega^{-1}(0) = 0$.

Set $\Phi(t) = \sin \omega(t)$ and $\Psi(t) = \cos \omega(t)$, $t \in \mathbb{R}$, and, for each smooth map $f: S^{n-1} \to S^{n-1}$, define a smooth map $\sigma_f: S^n \to S^n$ as follows: let $\gamma = \langle x, x_N \rangle$ and put

$$\sigma_f(x) = \begin{cases} x_N, & x = x_N \\ \Phi(\gamma)\, x_N + \Psi(\gamma)\, f\left(\dfrac{x - \gamma x_N}{|x - \gamma x_N|}\right), & x \neq x_N, x_S \\ x_S, & x = x_S. \end{cases}$$

σ_f will be called the *suspension of f*. It extends f and preserves the north and south hemispheres of S^n. Moreover $\sigma_f^{-1}(S^{n-1}) = S^{n-1}$.

Lemma V: If $f: S^{n-1} \to S^{n-1}$ and $g: S^{n-1} \to S^{n-1}$ are homotopic maps, then so are the suspensions σ_f and σ_g.

Proof: Let $h: \mathbb{R} \times S^{n-1} \to S^{n-1}$ be a homotopy connecting f and g. Define $H: \mathbb{R} \times S^n \to S^n$ by

$$H(t, x) = \sigma_{h_t}(x),$$

where $h_t(y) = h(t, y)$. Then H is a homotopy connecting σ_f and σ_g.

Q.E.D.

Proposition XI: Let $f: S^{n-1} \to S^{n-1}$ $(n \geqslant 2)$ be a smooth map and let σ_f be the suspension of f. Then

$$\deg \sigma_f = \deg f.$$

Proof: Define open sets U, V on S^n by

$$U = S^n - \{x_S\}, \qquad V = S^n - \{x_N\}.$$

Similarly, set (for $0 < a < 1$)

$$U_a = \{x \in S^n \mid \langle x, x_N \rangle > -a\}$$

and

$$V_a = \{x \in S^n \mid \langle x, x_N \rangle < a\}.$$

Since σ_f preserves north and south hemispheres, for some $a \in (0, 1)$,

$$\sigma_f(U_a) \subset U \qquad \text{and} \qquad \sigma_f(V_a) \subset V.$$

The triples (S^n, U, V) and (S^n, U_a, V_a) induce exact cohomology triangles (cf. sec. 5.4). In view of formula 5.3, sec. 5.4, we obtain a commutative diagram:

$$\begin{array}{ccc} H^{n-1}(U_a \cap V_a) & \xrightarrow{\partial_a} & H^n(S^n) \\ {\scriptstyle \tilde{\sigma}_f^{\#}}\uparrow & & \uparrow{\scriptstyle \sigma_f^{\#}} \\ H^{n-1}(U \cap V) & \xrightarrow[\partial]{} & H^n(S^n) \end{array} \qquad (6.1)$$

,

where $\tilde{\sigma}_f\colon U_a \cap V_a \to U \cap V$ is the restriction of σ_f. Since U_a, V_a, U, and V are all contractible and $n \geqslant 2$, it follows from the exactness of the cohomology triangles that the linear maps ∂ and ∂_a are isomorphisms.

Since σ_f extends f, thc inclusions

$$i_a\colon S^{n-1} \to U_a \cap V_a, \qquad i\colon S^{n-1} \to U \cap V,$$

induce a commutative diagram

$$\begin{array}{ccc} H^{n-1}(S^{n-1}) & \xleftarrow{\ i_a^{\#}\ } & H^{n-1}(U_a \cap V_a) \\ {\scriptstyle f^{\#}}\uparrow & & \uparrow{\scriptstyle \tilde{\sigma}_f^{\#}} \\ H^{n-1}(S^{n-1}) & \xleftarrow[\ i^{\#}\ ]{} & H^{n-1}(U \cap V). \end{array} \tag{6.2}$$

Moreover, according to Example 6, sec. 5.5, the horizontal arrows are isomorphisms. Combining (6.1) and (6.2) gives the commutative diagram

$$\begin{array}{ccc} H^{n-1}(S^{n-1}) & \xrightarrow[\cong]{\ \partial_a \circ (i_a^{\#})^{-1}\ } & H^n(S^n) \\ {\scriptstyle f^{\#}}\uparrow & & \uparrow{\scriptstyle \sigma_f^{\#}} \\ H^{n-1}(S^{n-1}) & \xrightarrow[\partial \circ (i^{\#})^{-1}]{\cong} & H^n(S^n). \end{array} \tag{6.3}$$

Finally, the inclusion $(S^n, U_a, V_a) \to (S^n, U, V)$ induces an isomorphism of Mayer–Vietoris sequences: in particular we have the commutative diagram

$$\begin{array}{ccccc} & & H^{n-1}(U_a \cap V_a) & & \\ & {\scriptstyle i_a^{\#}}\swarrow & \uparrow & \searrow{\scriptstyle \partial_a} & \\ H^{n-1}(S^{n-1}) & & \cong & & H^n(S^n) \\ & {\scriptstyle i^{\#}}\nwarrow & | & \nearrow{\scriptstyle \partial} & \\ & & H^{n-1}(U \cap V) & & \end{array}$$

It follows that $\partial_a \circ (i_a^{\#})^{-1} = \partial \circ (i^{\#})^{-1}$. The proposition is now an obvious consequence of (6.3).

Q.E.D.

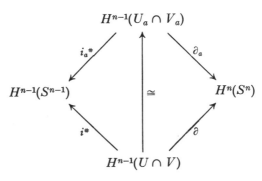

6.15. Proof of the Hopf theorem. **Lemma VI:** Let $\varphi: S^n \to S^n$ be a smooth map such that $\varphi(H_N) \subset H_N$ and $\varphi(H_S) \subset H_S$ and let $f: S^{n-1} \to S^{n-1}$ be the restriction of φ. Then φ is homotopic to σ_f.

Proof: Since

$$\varphi(H_N) \subset H_N, \qquad \varphi(H_S) \subset H_S$$

and

$$\sigma_f(H_N) \subset H_N, \qquad \sigma_f(H_S) \subset H_S, \qquad \sigma_f^{-1}(S^{n-1}) = S^{n-1},$$

it follows that

$$\varphi(x) \neq -\sigma_f(x), \qquad x \in S^n.$$

Hence, according to Example 7, sec. 5.5, φ is homotopic to σ_f.

Q.E.D.

Lemma VII: Every map $\varphi: S^n \to S^n$ is homotopic to a map $\psi: S^n \to S^n$ which satisfies

$$\psi: H_N \to S^n - \{x_S\}, \qquad \psi: H_S \to S^n - \{x_N\}.$$

Proof: Choose regular values a, b for φ. Then the sets $\varphi^{-1}(a)$ and $\varphi^{-1}(b)$ are finite (possibly empty):

$$\varphi^{-1}(a) = \{a_1, \ldots, a_p\}, \qquad \varphi^{-1}(b) = \{b_1, \ldots, b_q\}.$$

In view of the corollary to Theorem III, sec. 1.12, there exists a diffeomorphism $\alpha: S^n \to S^n$ homotopic to the identity and satisfying

$$\alpha(a_i) \in S^n - H_S, \quad \alpha(b_j) \in S^n - H_N, \qquad i = 1, \ldots, p; \quad j = 1, \ldots, q.$$

Similarly, there is a diffeomorphism $\beta: S^n \to S^n$, homotopic to the identity, such that

$$\beta(a) = x_N \qquad \text{and} \qquad \beta(b) = x_S.$$

The map $\psi = \beta \circ \varphi \circ \alpha^{-1}$ satisfies the required conditions (cf. sec. 1.10).

Q.E.D.

Lemma VIII: Every smooth map $\varphi: S^n \to S^n$ is homotopic to a smooth map $\psi: S^n \to S^n$ which satisfies $\psi(H_N) \subset H_N$ and $\psi(H_S) \subset H_S$.

Proof: In view of Lemma VII we may assume that

$$\varphi(H_N) \subset S^n - \{x_S\}, \qquad \varphi(H_S) \subset S^n - \{x_N\}.$$

Since $\varphi(H_N)$ and $\varphi(H_S)$ are compact, there is an $a \in (0, 1)$ such that

$$\varphi(H_N) \subset U_a \qquad \text{and} \qquad \varphi(H_S) \subset V_a .$$

Here U_a, V_a are the sets in the proof of Proposition XI.

Let $\lambda: \mathbb{R} \to \mathbb{R}$ be a smooth function satisfying the conditions

(1) $0 \leqslant \lambda(t) \leqslant 1$
(2) $\lambda(t) = 0, \ |t| > 1 - \epsilon \quad (0 < \epsilon < 1 - a)$
(3) $\lambda(t) = 1, \ |t| \leqslant a.$

Define a map $\chi: S^n \to S^n$ by

$$\chi(x) = \frac{x - \lambda(\gamma(x))\, \gamma(x)\, x_N}{|\, x - \lambda(\gamma(x))\, \gamma(x)\, x_N \,|},$$

where $\gamma(x) = \langle x, x_N \rangle$. Then χ is homotopic to the identity and

$$\chi(U_a) \subset H_N, \qquad \chi(V_a) \subset H_S .$$

Set $\psi = \chi \circ \varphi$. Then $\psi \sim \varphi$ and

$$\psi(H_N) \subset H_N, \qquad \psi(H_S) \subset H_S .$$

Q.E.D.

Proposition XII: Let $\varphi: S^n \to S^n$ be smooth ($n \geqslant 2$). Then there is a smooth map $f: S^{n-1} \to S^{n-1}$ such that φ is homotopic to σ_f.

Proof: Choose $\psi: S^n \to S^n$ to satisfy the conditions of Lemma VIII. Let f be the restriction of ψ to S^{n-1}. Then Lemma VI gives

$$\varphi \sim \psi \sim \sigma_f .$$

Q.E.D.

Theorem III (Hopf): Let $\varphi: S^n \to S^n$ and $\psi: S^n \to S^n$ ($n \geqslant 1$) be smooth maps such that

$$\deg \varphi = \deg \psi.$$

Then φ and ψ are homotopic.

Proof: The case $n = 1$ has been settled in Proposition II, sec. 6.2. Now we proceed by induction on n. Assume the theorem holds for some $n - 1 \geqslant 1$ and let $\varphi: S^n \to S^n$, $\psi: S^n \to S^n$ be maps such that

$$\deg \varphi = \deg \psi.$$

According to Proposition XII, there are smooth maps $f: S^{n-1} \to S^{n-1}$, $g: S^{n-1} \to S^{n-1}$ such that

$$\varphi \sim \sigma_f \qquad \text{and} \qquad \psi \sim \sigma_g .$$

Now Proposition XI implies that

$$\deg f = \deg \varphi = \deg \psi = \deg g.$$

Hence, by induction, the maps f and g are homotopic. Applying Lemma V, sec. 6.14, we find

$$\varphi \sim \sigma_f \sim \sigma_g \sim \psi,$$

whence $\varphi \sim \psi$.

Q.E.D.

Corollary I: A map $\varphi: S^n \to S^n$ of degree 1 is homotopic to the identity map.

Corollary II: Let $\varphi: S^n \to S^n$ be a smooth map of degree zero. Then φ can be extended to a smooth map

$$\psi: \mathbb{R}^{n+1} \to S^n$$

so that

$$\psi(x) = \varphi\left(\frac{x}{|x|}\right), \qquad |x| \geqslant 1.$$

Proof: Fix $e \in S^n$. In view of Theorem III there is a smooth map $h: \mathbb{R} \times S^n \to S^n$ such that

$$h(t, x) = e \quad (t \leqslant 0) \qquad \text{and} \qquad h(t, x) = \varphi(x) \quad (t \geqslant 1), \qquad x \in S^n.$$

Choose a smooth function $\lambda: \mathbb{R} \to \mathbb{R}$ such that

$$\lambda(t) = 1, \qquad |t| \geqslant 1,$$

and

$$\lambda(t) = 0, \qquad |t| < \epsilon.$$

(some $\epsilon \in (0, 1)$). Define ψ by

$$\psi(x) = \begin{cases} h(\lambda(|x|), |x|^{-1}x), & x \in E - \{0\} \\ e, & |x| < \epsilon. \end{cases}$$

Q.E.D.

Next, let E, F be oriented Euclidean spaces of dimension $n + 1$ ($n \geqslant 1$). Let S_E and S_F denote the unit spheres of E and F and let $\dot{E} = E - \{0\}$, $\dot{F} = F - \{0\}$. Assume $\varphi\colon \dot{E} \to \dot{F}$ is a smooth map and define $\psi\colon S_E \to S_F$ by

$$\psi(x) = \frac{\varphi(x)}{|\varphi(x)|}, \qquad x \in S_E .$$

Corollary III: With the notation defined above, assume

$$\deg \psi = 0.$$

Then there is a smooth map $\tilde{\varphi}\colon E \to \dot{F}$ so that

$$\tilde{\varphi}(x) = \varphi(x), \qquad |x| \geqslant 1.$$

Problems

1. Let $\hat{B}$, B be closed oriented unit balls in Euclidean n-spaces. Let $\varphi: \hat{B} \to B$ be a smooth map which restricts to a map $\psi: \hat{S}^{n-1} \to S^{n-1}$.

(i) Show that

$$\int_{\hat{B}} \varphi^* \Psi = \deg \psi \int_B \Psi, \qquad \Psi \in A^n(B).$$

(ii) Assume that $b \in B$ is a point such that $\varphi^{-1}(b)$ consists of finitely many points $\{a_i\}$, all in the interior of $\hat{B}$. Show that

$$\deg \psi = \sum_i \deg_{a_i} \varphi.$$

(iii) Generalize these results to arbitrary oriented connected compact manifolds-with-boundary.

2. Let $(M, \partial M)$ be a compact connected oriented $(n+1)$-manifold-with-boundary. Let N be a compact connected oriented n-manifold and assume that $\psi: \partial M \to N$ extends to a smooth map $\varphi: M \to N$. Show that $\deg \psi = 0$.

3. **Proper maps.** Let $\varphi: M \to N$ be a proper smooth map between connected, oriented n-manifolds. Define $\deg \varphi$ by the relation

$$(\varphi_c)^\# \omega_N = \deg \varphi \cdot \omega_M ,$$

where ω_M and ω_N are the orientation classes.

(i) Show that the properties of the mapping degree generalize appropriately. In particular, show that $\deg \varphi$ is an integer, invariant under proper homotopies.

(ii) Find examples of proper maps which are homotopic but not properly homotopic.

(iii) If $\varphi^{-1}(b)$ consists of finitely many points, $\{a_i\}$, show that

$$\deg \varphi = \sum_i \deg_{a_i} \varphi.$$

4. Suppose $\varphi: M \to N$ is a smooth injective map between compact oriented n-manifolds. Prove that $\deg \varphi = \pm 1$ and conclude that φ is

bijective. Show that $\deg_a \varphi = \deg \varphi$, $a \in M$. Is φ necessarily a diffeomorphism?

5. Suppose $a \in M$ is an isolated point for a smooth map $\varphi: M \to N$ of n-manifolds. If $\deg_a \varphi \neq 0$, show that, if $U \subset M$ is any open set containing a, there is an open set V in N such that $\varphi(a) \in V \subset \varphi(U)$.

6. Let (E, π, B, F) be a smooth bundle. Assume that E and B are compact, oriented and connected, and that F consists of p points. Compute the degree of π.

7. **Degree mod 2.** Let $\varphi: M \to N$ be a smooth map between compact connected n-manifolds (not necessarily orientable).

(i) Let $a \in M$ be an isolated point for φ. Choose local orientations in charts about a and $\varphi(a)$ and define $\deg_a \varphi$ with respect to these. Let $\operatorname{Deg}_a \varphi$ be the mod 2-reduction of the integer $\deg_a \varphi$. Show that it is independent of the choice of local orientations.

(ii) Let $b \in N$ be such that $\varphi^{-1}(b) = \{a_i\}$ is finite. Show that the sum $\sum_i \operatorname{Deg}_{a_i} \varphi$ (addition in $\mathbb{Z}_2$) is independent of the choice of b. It is called the *mod 2-degree of* φ.

(iii) Show that, if the mod 2-degree of a map is nonzero, then the map is surjective.

(iv) Let $U \subset N$ be chosen so that U and $\varphi^{-1}U$ are orientable and let ψ be the restriction of φ to $\varphi^{-1}U$. Show that ψ is proper. Show that reduction mod 2 of $\deg \psi$ is independent of the choice of orientations in U and $\varphi^{-1}U$ and coincides with the mod 2-degree of φ.

8. **Complex functions.** In this problem S^2 is the Riemann sphere.

(i) Let f and g be polynomials with complex coefficients. Interpret f/g as a smooth map $S^2 \to S^2$ and compute its degree.

(ii) Let f be a complex differentiable function with an isolated zero or pole (of order > 0) at a and regard f as a map into S^2. Show that a is an isolated point for f and that

$$\deg_a f = \left| \frac{1}{2\pi i} \int_c \frac{f'(z)}{f(z)}\, dz \right|,$$

where c is a sufficiently small positively oriented circle about a. How is $\deg_a f$ related to the order of the zero or pole a?

(iii) Let f be a complex differentiable function defined for $|z| < r$,

$r > 1$, with no zeros or poles on S^1. Define $\varphi: S^1 \to S^1$ by $\varphi(z) = f(z)/|f(z)|$. Show that

$$\deg \varphi = \frac{1}{2\pi i} \int_{S^1} \frac{f'(z)}{f(z)}\, dz.$$

Interpret this in terms of the zeros and poles of f inside S^1 (cf. problem 1, ii).

9. **Rouché's theorem.** Let $\varphi, \psi: \dot{E} \to \dot{F}$ (E, F oriented Euclidean n-spaces) satisfy (for some $a > 0$)

$$|\varphi(x) - \psi(x)| < |\varphi(x)|, \qquad |x| = a.$$

Show that

$$\deg_0 \varphi = \deg_0 \psi$$

10. **Quaternions.** Consider S^3 as the unit sphere in the space of quaternions and consider S^4 as the 1-point compactification of $\mathbb{H}$.

(i) Find the degree of the map $S^3 \to S^3$ given by $x \to x^p$ ($p \in \mathbb{Z}$).
(ii) Let f and g be polynomials with quaternionic coefficients. Interpret $g^{-1}f$ and fg^{-1} as smooth maps $S^4 \to S^4$ and compute the degrees.
(iii) Show that if $q \in \mathbb{H}$ and f is as in (ii) then the equation $f(x) = q$ has a solution.

11. **Smooth maps from S^n.** (i) Let $\varphi: S^n \to \mathbb{R}^{n+1}$ be a smooth map (n even). Show that $\varphi(a) = \lambda a$ for some $a \in S^n$, $\lambda \in \mathbb{R}$; conclude that every vector field on S^n has a zero. Is this true for n odd?
(ii) Let $\varphi: S^n \to S^n$ (any n) satisfy $\varphi(-x) \neq \varphi(x)$, for $x \in S^n$. Show that φ has odd degree and conclude that φ is surjective. (*Hint*: Reduce to the case $\varphi(-x) = -\varphi(x)$.)
(iii) (Borsuk–Ulam theorem) Let $\varphi: S^n \to \mathbb{R}^n$ (any n). Show that $\varphi(a) = \varphi(-a)$ for some $a \in S^n$. Conclude that if $f_i \in \mathscr{S}(S^n)$ ($i = 1, \ldots, n$) are odd functions, they have a common zero.

12. Consider a covering of S^n by $n + 2$ closed sets $A_1, \ldots, A_{n+2}$ such that none of the A_i contains a pair of antipodal points. Show that

$$A_1 \cap \cdots \cap A_{n+2} = \varnothing,$$

while for each j ($1 \leqslant j \leqslant n + 2$)

$$A_1 \cap \cdots \cap \hat{A}_j \cdots \cap A_{n+2} \neq \varnothing.$$

Express this property in terms of the nerve of the covering. Conclude that, for every covering of S^n by $n+1$ closed sets, at least one of these sets contains a pair of antipodal points (theorem of Lusternik–Schnirelmann–Borsuk). *Hint*: Use problem 11.

13. Linking number. Let M, N be compact, connected, oriented manifolds, respectively of dimensions p and q. Let E be an oriented Euclidean $(n+1)$-space, where $n = p + q$. If $\varphi\colon M \to E$, $\psi\colon N \to E$ are smooth maps such that $\varphi(x) \neq \psi(y)$ $(x \in M, y \in N)$, *their linking number*, $l(\varphi, \psi)$, is defined by $l(\varphi, \psi) = \deg \chi$, where $\chi\colon M \times N \to S^n$ is given by

$$\chi(x, y) = \frac{\psi(y) - \varphi(x)}{|\psi(y) - \varphi(x)|}.$$

(i) Show that $l(\varphi, \psi) = (-1)^{(p+1)(q+1)} l(\psi, \varphi)$.

(ii) If φ_t, ψ_t are homotopies such that $\varphi_t(x) \neq \psi_t(y)$ $(x \in M, y \in N, t \in \mathbb{R})$, show that $l(\varphi_0, \psi_0) = l(\varphi_1, \psi_1)$.

(iii) If $\varphi(M)$ and $\psi(N)$ can be separated by a hyperplane, show that $l(\varphi, \psi) = 0$.

(iv) Define $\chi_1\colon M \times N \to \dot{E}$ by $\chi_1(x, y) = \psi(y) - \varphi(x)$. Let Δ_E be the positive normed determinant function in E. Define $\Omega \in A^n(\dot{E})$ by

$$\Omega(x; h_1, \ldots, h_n) = \frac{1}{|x|^{n+1}} \Delta_E(x, h_1, \ldots, h_n).$$

Show that

$$l(\varphi, \psi) = \frac{1}{\kappa_n} \int_{M \times N} \chi_1^* \Omega,$$

where $\kappa_n = \operatorname{vol} S^n$ (cf. sec. 4.15, and problem 6, Chap. IV).

(v) If $M = N = S^1$, show that

$$l(\varphi, \psi) = \frac{1}{4\pi} \int_0^1 \int_0^1 \frac{\langle \psi(\tau) - \varphi(t), \dot{\psi}(\tau) \times \dot{\varphi}(t) \rangle}{|\psi(\tau) - \varphi(t)|^3} \, d\tau \, dt.$$

14. Winding number. Let M be a compact oriented connected n-manifold and let $\mathbb{R}^{n+1}$ be an oriented Euclidean $(n+1)$-space. Let $\varphi\colon M \to \mathbb{R}^{n+1}$ be smooth and assume that $a \notin \operatorname{Im} \varphi$. The integer

$$\omega_a(\varphi) = l(a, \varphi)$$

(where a is regarded as the constant map $\{a\} \to \mathbb{R}^{n+1}$) is called the *winding number of* φ *about* a.

(i) Show that $\omega_a(\varphi)$ is the degree of the map $M \to S^n$ given by $x \mapsto (\varphi(x) - a)/|\varphi(x) - a|$.

(ii) Show that if φ and ψ: $M \to \mathbb{R}^{n+1} - \{a\}$ are homotopic (as maps into $\mathbb{R}^{n+1} - \{a\}$), then $\omega_a(\varphi) = \omega_a(\psi)$.

(iii) Interpret $\omega_a(\varphi)$ via an integral (problem 13, iv), and via solid angles (problem 6, Chap. IV).

(iv) Show that the function $a \mapsto \omega_a(\varphi)$ is constant in each component of $\mathbb{R}^{n+1} - \operatorname{Im} \varphi$.

(v) If φ is an embedding, show that for some $a \in \mathbb{R}^{n+1} - \operatorname{Im} \varphi$, $\omega_a(\varphi) \neq 0$. Conclude that $\mathbb{R}^{n+1} - \operatorname{Im} \varphi$ has at least two components. Use a tubular neighbourhood of $\operatorname{Im} \varphi$ to show that $\mathbb{R}^{n+1} - \operatorname{Im} \varphi$ consists of exactly two components.

(vi) Show that every compact n-dimensional submanifold of $\mathbb{R}^{n+1}$ is orientable. (*Hint*: Use problem 7.)

15. Rotation number. Let $\mathbb{R}^2$ be an oriented Euclidean plane. Let Δ be the positive normed determinant function. Suppose φ: $S^1 \to \mathbb{R}^2$ is an immersion and define f: $\mathbb{R} \to \mathbb{R}^2$ by $f(t) = \varphi(\exp 2\pi i t)$. The *rotation number* of φ is defined by

$$\rho(\varphi) = \frac{1}{2\pi} \int_0^1 \frac{\Delta(\dot{f}(t), \ddot{f}(t))}{|\dot{f}(t)|^2}\, dt.$$

(i) Show that $\rho(\varphi)$ is the winding number of $\dot{\varphi}$: $S^1 \to \mathbb{R}^2$ about the origin. Conclude that $\rho(\varphi) \in \mathbb{Z}$.

(ii) (Whitney–Graustein theorem) Two immersions φ_0 and φ_1 of S^1 into $\mathbb{R}^2$ are called *i-homotopic*, if there exists a connecting homotopy φ_τ $(0 \leqslant \tau \leqslant 1)$ such that each map φ_τ: $S^1 \to \mathbb{R}^2$ is an immersion. Show that two immersions are i-homotopic if and only if they have the same rotation number. *Hint*: To show that the condition $\rho(\varphi_0) = \rho(\varphi_1)$ is sufficient consider first the case that the rotation number is different from zero. Establish the following lemma: Let z: $\mathbb{R} \to \mathbb{R}^2$ be a non-constant smooth map satisfying $z(t+1) = z(t)$ and $|z(t)| = 1$. Then, $\left| \int_0^1 z(t)\, dt \right| < 1$.

(iii) Consider the map f: $\mathbb{R} \to \mathbb{R}^2$ given by

$$f(t) = \left(-\frac{1}{\sqrt{2}} \sqrt{1 + \cos^2 2\pi t}\ \cos 2\pi t,\ \frac{1}{2\sqrt{2}} \sin 4\pi t \right).$$

Show that f determines an immersion $S^1 \to \mathbb{R}^2$. Show that this immersion is not i-homotopic to the standard immersion $S^1 \to \mathbb{R}^2$.

(iv) Show that the rotation number of an embedding is $\pm 2\pi$. *Hint*: Use parts v and vi.

(v) Let $f: \mathbb{R} \to \mathbb{R}^2$ be a smooth map satisfying $|f(t)| = 1$ and $f(t) = f(\tau)$ if and only if $t - \tau \in \mathbb{Z}$. Define $F: \mathbb{R} \to S^1$ by

$$F(t) = \begin{cases} \epsilon(t) \dfrac{f(t) - f(0)}{|f(t) - f(0)|}, & t \notin \mathbb{Z}, \\ \epsilon(t)\dot{f}(t), & t \in \mathbb{Z}, \end{cases}$$

where $\epsilon(t) = 1$, if $2k \leqslant t < 2k + 1$ and $\epsilon(t) = -1$, if $2k + 1 \leqslant t < 2k$, $k \in \mathbb{Z}$. Show that F is smooth. Find a function $\omega \in \mathscr{S}(\mathbb{R})$ such that $F(t) = (\cos \omega(t), \sin \omega(t))$ (cf. problem 15, Chap. V). Prove that $\omega(1) - \omega(0) = \pm\pi$.

(vi) Let f, ϵ be as in (v). Define $\Phi: \mathbb{R} \times \mathbb{R} \to \mathbb{R}^2$ by

$$\Phi(t, \tau) = \begin{cases} \epsilon(t - \tau) \dfrac{f(t) - f(\tau)}{|f(t) - f(\tau)|}, & t - \tau \notin \mathbb{Z}, \\ \epsilon(t - \tau)\dot{f}(t), & t - \tau \in \mathbb{Z}. \end{cases}$$

Show that Φ is smooth. Find $\Omega \in \mathscr{S}(\mathbb{R}^2)$ such that

$$\Phi(t) = (\cos \Omega(t), \sin \Omega(t)) \qquad \text{and} \qquad \Omega(0, 0) = \omega(0).$$

Conclude that $\Omega(1, 1) - \Omega(0, 0) = \pm 2\pi$.

16. Parallelisms. Let M be a compact connected oriented manifold with parallelism P (cf. problem 14, Chap. IV).

(i) Show that there is a unique parallel n-form, Δ, on M such that $\int_M \Delta = 1$. If $f \in \mathscr{S}(M)$, set $\int_M f(x)\,dx = \int_M f \cdot \Delta$.

(ii) Let $\varphi: M \to M$ be smooth and fix $a \in M$. Set $P(a, x) = P(x)$ and define $F: M \to L(T_a(M))$ by

$$F(x) = P(\varphi(x))^{-1} \circ (d\varphi)_x \circ P(x).$$

Show that for $f \in \mathscr{S}(M)$

$$\int_M (\varphi^* f)(x) \det F(x)\,dx = \deg \varphi \int_M f(x)\,dx.$$

17. Fundamental class. Let $z = \sum_i k^i \sigma_i$ be a smooth integral cycle on a compact oriented n-manifold representing the fundamental class. Suppose $b \in M$ satisfies the following condition: For each i, $\sigma_i^{-1}(b) = \{a_{ij}\}$ is finite and contained in $\mathring{\Delta}_n$ (cf. problem 12, Chap. V). Show that

$$\sum_{i,j} k^i \deg_{a_{ij}} \sigma_i = 1.$$

18. Show that two maps from a compact oriented connected n-manifold into S^n are homotopic if and only if they have the same degree.

19. Let $\varphi: M \to N$ be a smooth map between compact connected oriented n-manifolds. Show that φ is homotopic to a map $\psi: M \to N$ such that for some $b \in N$, $\psi^{-1}(b)$ contains $|\deg \varphi|$ points. In particular, show that if $\deg \varphi = 0$ then φ is homotopic to a map which is not surjective.

20. Denote by $\mathbb{C}_\infty$ the Riemann sphere.

(i) Show that the 2-form Φ in $\mathbb{C}$ given by

$$\Phi(z; \zeta_1, \zeta_2) = \frac{1}{\pi} \frac{1}{(1 + |z|^2)^2} \operatorname{Im}(\bar{\zeta}_1 \zeta_2)$$

extends to a 2-form, Ψ, in $\mathbb{C}_\infty$ such that $\Psi(z_\infty) \neq 0$.

(ii) Use Ψ to show that the map $\varphi: \mathbb{C}_\infty \to \mathbb{C}_\infty$ given by

$$\varphi(z) = z^p, \qquad p \in \mathbb{Z},$$

has degree $|p|$.

(iii) Let S^2 denote the sphere of radius $\frac{1}{2}$ in $\mathbb{R}^3$ with north pole N and south pole S and let σ be the stereographic projection of S^2 from N to $T_S(S^2)$. Identify $T_S(S^2)$ with $\mathbb{C}$ and show that $\sigma * \Psi$ is the restriction of the positive normed determinant function in $\mathbb{R}^3$ (with respect to an appropriate orientation) to S^2.

Chapter VII

Integration over the Fibre

§1. Tangent bundle of a fibre bundle

7.1. The vertical subbundle. Let (E, π, B, F) be a smooth fibre bundle with $\dim B = n$, $\dim F = r$. The derivative of π is a bundle map between the tangent bundles;

$$d\pi: \tau_E \to \tau_B .$$

Definition: The space

$$V_z(E) = \ker(d\pi)_z , \qquad z \in E,$$

is called the *vertical subspace* of $T_z(E)$. The vectors of $V_z(E)$ are called *vertical.*

The linear maps $(d\pi)_z$ are all surjective; hence

$$\dim V_z(E) = \dim E - \dim B = \dim F.$$

Recall from Example 4, sec. 3.10, that for each $a \in B$ the fibre $F_a = \pi^{-1}(a)$ is a submanifold of E. Denote the inclusion by

$$j_a: F_a \to E.$$

Lemma I: For $a \in B$, $z \in F_a$

$$V_z(E) = \operatorname{Im}(dj_a)_z .$$

Proof: Since $\pi \circ j_a$ is the constant map $F_a \to a$,

$$d\pi \circ dj_a = 0.$$

Hence

$$V_z(E) \supset \operatorname{Im}(dj_a)_z .$$

On the other hand, since $(dj_a)_z$ is injective

$$\dim \operatorname{Im}(dj_a)_z = \dim F = \dim V_z(E).$$

The lemma follows.

Q.E.D.

Now consider the subset $V_E \subset T_E$ given by

$$V_E = \bigcup_{z \in E} V_z(E).$$

We shall make V_E into a subbundle of the tangent bundle, τ_E; it will be called the *vertical subbundle*. Let $\{(U_\alpha, \psi_\alpha)\}$ be a coordinate representation for (E, π, B, F). Then the commutative diagram

$$\begin{array}{ccc} T_{U_\alpha} \times T_F & \xrightarrow[\cong]{d\psi_\alpha} & T_{\pi^{-1}(U_\alpha)} \\ \downarrow & & \downarrow \\ U_\alpha \times F & \xrightarrow[\psi_\alpha]{\cong} & \pi^{-1}(U_\alpha) \end{array}$$

restricts to a commutative diagram

$$\begin{array}{ccc} U_\alpha \times T_F & \xrightarrow[\cong]{} & V_E\,|_{\pi^{-1}(U_\alpha)} \\ \downarrow & & \downarrow \\ U_\alpha \times F & \xrightarrow[\psi_\alpha]{\cong} & \pi^{-1}(U_\alpha) \end{array}$$

from which the subbundle structure of V_E is obvious.

V_E is a submanifold of T_E and

$$\dim V_E = n + 2r.$$

Lemma I states that the maps $dj_a: T_{F_a} \to T_E$ can be considered as bundle maps $dj_a: T_{F_a} \to V_E$ inducing linear isomorphisms on the fibres. For this reason V_E is often called the *bundle along the fibres*.

If $(\hat{E}, \hat{\pi}, \hat{B}, \hat{F})$ is a second fibre bundle and $\varphi: E \to \hat{E}$ is a fibre preserving map, then $d\varphi$ restricts to a bundle map

$$(d\varphi)_V: V_E \to V_{\hat{E}}.$$

A vector field Z on E is called *vertical*, if for every $z \in E$ the vector $Z(z)$ is vertical, or equivalently, if

$$Z \underset{\pi}{\sim} 0$$

(cf. sec. 3.13). The Lie product of two vertical vector fields Z_1 and Z_2 is again vertical. In fact, if $Z_1 \underset{\pi}{\sim} 0$ and $Z_2 \underset{\pi}{\sim} 0$ then, by Proposition VIII, sec. 3.13,

$$[Z_1, Z_2] \underset{\pi}{\sim} 0.$$

Thus the vertical vector fields form a subalgebra, $\mathscr{X}_V(E)$, of the Lie algebra $\mathscr{X}(E)$.

On the other hand, since the vertical vector fields are the cross-sections in V_E, $\mathscr{X}_V(E)$ is a finitely generated module over the ring $\mathscr{S}(E)$ (cf. the corollary to Theorem I, sec. 2.23).

7.2. Horizontal subbundles. If (E, π, B, F) is a smooth fibre bundle, a subbundle H_E of τ_E will be called *horizontal* if

$$\tau_E = H_E \oplus V_E.$$

Proposition VII, sec. 2.18, guarantees the existence of horizontal subbundles. The fibres $H_z(E)$ $(z \in E)$ of a horizontal subbundle will be called the *horizontal subspaces* (with respect to the choice of H_E).

Suppose now that a horizontal subbundle, H_E, has been fixed. Then the derivative $d\pi\colon \tau_E \to \tau_B$ restricts to a bundle map $H_E \to \tau_B$; this map induces linear isomorphisms in each fibre. Hence H_E is strongly isomorphic to the pull-back (via π) of τ_B. The manifold H_E has dimension $2n + r$ ($n = \dim B$, $r = \dim F$). A vector field Z on E is called *horizontal* if

$$Z(z) \in H_z(E), \qquad z \in E.$$

The horizontal vector fields on E form a finitely generated projective module $\mathscr{X}_H(E)$ over $\mathscr{S}(E)$. However, they do not, in general, form a subalgebra of the Lie algebra $\mathscr{X}(E)$.

Every vector field Z on E can be uniquely decomposed in the form

$$Z = Z_V + Z_H, \qquad Z_V \in \mathscr{X}_V(E), \quad Z_H \in \mathscr{X}_H(E).$$

The vector fields Z_V and Z_H are called the *vertical* and *horizontal components* of Z.

Examples: 1. Consider the product bundle $E = B \times F$. Then the vertical subbundle is given by $V_{B\times F} = B \times T_F$, and $H_{B\times F} = T_B \times F$ is a horizontal subbundle of $\tau_{B\times F}$.

2. Let (E, π, B, F) be any fibre bundle and choose a Riemannian metric on E. Let $H_z(E)$ denote the orthogonal complement of $V_z(E)$ in $T_z(E)$ with respect to the inner product. Then

$$H_E = \bigcup_{z \in E} H_z(E)$$

is a horizontal subbundle.

7.3. Differential forms. Let (E, π, B, F) be a smooth fibre bundle. A differential form $\Phi \in A(E)$ is called *horizontal* if

$$i(X)\Phi = 0, \qquad X \in \mathscr{X}_V(E).$$

Since each $i(X)$ is a homogeneous antiderivation, the horizontal forms are a graded subalgebra of $A(E)$. This algebra is called the *horizontal subalgebra*, and is denoted by $A_H(E)$.

Now assume that a horizontal subbundle H_E of τ_E has been chosen, and let $\mathscr{X}_H(E)$ be the $\mathscr{S}(E)$-module of horizontal vector fields on E. Define a graded subalgebra $A_V(E) \subset A(E)$ by

$$A_V(E) = \{\Phi \in A(E) \mid i(X)\Phi = 0, \quad X \in \mathscr{X}_H(E)\}.$$

$A_V(E)$ is called the *vertical subalgebra* of $A(E)$, and depends on the choice of H_E.

Now form the graded anticommutative algebra $A_H(E) \otimes_E A_V(E)$ (anticommutative tensor product of algebras).

Proposition I: The multiplication map $\Phi \otimes \Psi \mapsto \Phi \wedge \Psi$ defines an isomorphism

$$\mu: A_H(E) \otimes_E A_V(E) \xrightarrow{\cong} A(E)$$

of graded algebras.

Proof: μ is clearly a homomorphism of graded algebras. To show that μ is bijective, let

$$H_z: T_z(E) \to H_z(E) \qquad \text{and} \qquad V_z: T_z(E) \to V_z(E)$$

be the projections induced by the decomposition $\tau_E = H_E \oplus V_E$. Then isomorphisms

$$f_H: \operatorname{Sec} \wedge H_E^* \xrightarrow{\cong} A_H(E) \qquad \text{and} \qquad f_V: \operatorname{Sec} \wedge V_E^* \xrightarrow{\cong} A_V(E)$$

of $\mathscr{S}(E)$-modules are given by

$$f_H\Phi(z; \zeta_1, \ldots, \zeta_p) = \Phi(z; H_z\zeta_1, \ldots, H_z\zeta_p), \qquad \zeta_\nu \in T_z(E),$$

and

$$f_V \Psi(z; \zeta_1, \dots, \zeta_p) = \Psi(z; V_z\zeta_1, \dots, V_z\zeta_p), \qquad \zeta_i \in T_z(E).$$

Moreover, the diagram

$$\begin{array}{ccc} A_H(E) \otimes_E A_V(E) & \xrightarrow{\mu} & A(E) \\ {\scriptstyle \cong}\nwarrow {\scriptstyle f_H \otimes f_V} & & {\scriptstyle f}\nearrow{\scriptstyle \cong} \\ & \operatorname{Sec} \wedge H_E^* \otimes_E \operatorname{Sec} \wedge V_E^* & \end{array}$$

commutes, where f denotes the isomorphism induced by the bundle isomorphism

$$\wedge H_E^* \otimes \wedge V_E^* \xrightarrow{\cong} \wedge \tau_E^*$$

(cf. Proposition XIV, sec. 2.24). Hence μ is an isomorphism.

Q.E.D.

Next, observe that the inclusion $\wedge V_E \to \wedge \tau_E$ induces a homomorphism of $\mathscr{S}(E)$-algebras

$$\rho_V : \operatorname{Sec} \wedge V_E^* \leftarrow A(E);$$

$$\rho_V \Phi(z; \zeta_1, \dots, \zeta_p) = \Phi(z; \zeta_1, \dots, \zeta_p), \qquad z \in E, \quad \zeta_i \in V_z(E)$$

(independent of the choice of a horizontal subbundle). The map ρ_V is surjective: in fact, let H_E be a horizontal subbundle and let $A_V(E)$ be the corresponding vertical subalgebra. Then the restriction of ρ_V to $A_V(E)$ is inverse to the isomorphism f_V defined in the proof of Proposition I above.

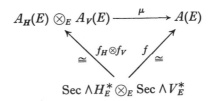

§2. Orientation in fibre bundles

7.4. Orientable fibre bundles. Let $\mathscr{B} = (E, \pi, B, F)$ be a smooth fibre bundle with $\dim B = n$, $\dim F = r$. Recall that the fibre F_x at $x \in B$ is a submanifold of E (Example 4, sec. 3.10) and denote the inclusion by $j_x: F_x \to E$.

Consider those differential forms $\Psi \in A^r(E)$ such that for each $x \in B$ the differential form $j_x^*\Psi \in A^r(F_x)$ orients F_x (there may be none). Two such forms Ψ_1, Ψ_2 are called equivalent, if $j_x^*\Psi_1$ and $j_x^*\Psi_2$ induce the same orientation on F_x for every $x \in B$.

Definition: The bundle $\mathscr{B}$ is called *orientable* if there exists an r-form Ψ on E such that $j_x^*\Psi$ orients F_x for every $x \in B$. An equivalence class of such r-forms is called an *orientation* for the bundle and a member of the equivalence class is said to *represent the orientation.*

Remark: It will be shown in sec. 7.8 that this definition coincides with the definition of sec. 2.16 if $\mathscr{B}$ is a vector bundle.

An orientation in the bundle specifies an orientation in each fibre F_x. In particular, the typical fibre of an orientable bundle is orientable.

If $\Psi \in A^r(E)$ orients the bundle (E, π, B, F) and U is an open subset of B, then the restriction of Ψ to $\pi^{-1}U$ orients the bundle $(\pi^{-1}U, \pi, U, F)$.

Example: The trivial bundle $(B \times F, \pi, B, F)$ is orientable if and only if F is orientable.

In fact, we have seen above that if the bundle is orientable then so is F. Conversely, assume that F is orientable and let $\Delta_F \in A^r(F)$ be an orienting r-form. Then

$$j_x^*(1 \times \Delta_F) = \Delta_F, \qquad x \in B.$$

It follows that $1 \times \Delta_F$ orients the bundle.

Recall the map $\rho_V: A(E) \to \operatorname{Sec} \wedge V_E^*$ defined in sec. 7.3. For $z \in E$, we can regard $\wedge(dj_x)_z$ ($x = \pi z$) as an isomorphism

$$\sigma_z: \wedge T_z(F_x) \xrightarrow{\cong} \wedge(V_z(E)).$$

Then for $\Psi \in A(E)$,

$$\sigma_z^*(\rho_V \Psi(z)) = (j_x^* \Psi)(z), \qquad z \in E, \quad x = \pi z. \tag{7.1}$$

Proposition II: (1) If $\Psi \in A^r(E)$ orients the fibre bundle (E, π, B, F) then $\rho_V \Psi \in \operatorname{Sec} \wedge^r V_E^*$ orients the vector bundle V_E in the sense of sec. 2.16.

(2) Φ and Ψ represent the same orientations of the bundle if and only if $\rho_V \Phi$ and $\rho_V \Psi$ represent the same orientation of V_E.

(3) The map so obtained, from orientations of the fibre bundle to orientations of V_E, is a bijection.

Proof: (1) Since Ψ orients the bundle we have

$$(j_x^* \Psi)(z) \neq 0, \qquad z \in F_x, \quad x \in B.$$

It follows from this and (7.1) that $(\rho_V \Psi)(z) \neq 0$, $z \in E$. Thus $\rho_V \Psi$ orients V_E.

(2) If Φ and Ψ orient the bundle, then there are unique nonzero scalars λ_z $(z \in E)$ such that

$$(j_x^* \Psi)(z) = \lambda_z \cdot (j_x^* \Phi)(z), \qquad z \in F_x, \quad x \in B.$$

It follows from (7.1) that

$$(\rho_V \Psi)(z) = \lambda_z \cdot (\rho_V \Psi)(z), \qquad z \in E.$$

Hence both conditions of (2) are equivalent to

$$\lambda_z > 0, \qquad z \in E.$$

(3) We have already shown in (1) and (2) that $\Psi \mapsto \rho_V \Psi$ defines an injection from orientations of the bundle to orientations of V_E. Let $\Omega \in \operatorname{Sec} \wedge^r V_E^*$ orient the vertical bundle. Choose a horizontal subbundle, and let $A_V(E) \subset A(E)$ be the corresponding vertical subalgebra (cf. sec. 7.3). Then ρ_V maps $A_V(E)$ isomorphically onto $\operatorname{Sec} \wedge V_E^*$, and so, for a unique $\Psi \in A_V^r(E)$,

$$\rho_V \Psi = \Omega.$$

With the aid of formula (7.1), it is simple to verify that Ψ orients the bundle.

Q.E.D.

Proposition III: Let (E, π, B, F) be a smooth bundle over a connected base B. Let Φ, $\Psi \in A^r(E)$ orient the bundle, and assume that for some $a \in B$, $j_a^*\Phi$ and $j_a^*\Psi$ represent the same orientation in F_a. Then Φ and Ψ represent the same orientation in the bundle.

Proof: For any component O of E the restriction of π to O defines a smooth bundle (O, π_O, B, F_O) where F_O is the union of components of F. Moreover, since $j_a^*\Phi$ and $j_a^*\Psi$ represent the same orientation in F_a, they represent the same orientation in $(F_O)_a$. Thus we may reduce to the case that E is connected.

Now assume that E is connected. Since $\rho_V\Phi$, $\rho_V\Psi$ orient the vector bundle V_E in the sense of sec. 2.16 (cf. Proposition II), it follows that

$$\rho_V\Phi = f \cdot \rho_V\Psi,$$

where $f \in \mathscr{S}(E)$ has no zeros. Thus, because E is connected, either $f > 0$ or $f < 0$. By hypothesis there are positive numbers λ_z such that

$$(j_a{}^*\Phi)(z) = \lambda_z \cdot (j_a{}^*\Psi)(z), \qquad z \in F_a.$$

In view of formula (7.1),

$$f(z) = \lambda_z > 0$$

and hence $f > 0$.

Thus $\rho_V\Phi$ and $\rho_V\Psi$ represent the same orientation in V_E; the proposition follows now from Proposition II, (2).

Q.E.D.

Corollary: Let B be connected and assume $\Psi \in A^r(B \times F)$ orients the trivial bundle $\mathscr{B} = (B \times F, \pi, B, F)$. Fix $a \in B$ and let $\Delta_F = j_a^*\Psi \in A^r(F)$. Then $1 \times \Delta_F$ represents the same orientation in $\mathscr{B}$ as Ψ.

7.5. Orientation preserving maps. Let $\mathscr{B} = (E, \pi, B, F)$ and $\hat{\mathscr{B}} = (\hat{E}, \hat{\pi}, \hat{B}, \hat{F})$ be smooth bundles. Assume $\varphi: E \to \hat{E}$ is a smooth fibre-preserving map (cf. sec. 1.13) which induces $\psi: B \to \hat{B}$. Suppose further that φ restricts to local diffeomorphisms

$$\varphi_x : F_x \to \hat{F}_{\psi(x)}, \qquad x \in B.$$

If $\mathscr{B}$ and $\hat{\mathscr{B}}$ are oriented, φ will be said to *preserve* (resp. *reverse*) *the bundle orientations* if each φ_x is orientation preserving (resp. reversing).

Suppose $\Psi \in A^r(\hat{E})$ represents the orientation of $\hat{\mathscr{B}}$. Since for each $x \in B$

$$j_x^* \varphi^* \Psi = \varphi_x^* j_{\psi(x)}^* \Psi,$$

it follows that $\varphi^*\Psi$ orients $\mathscr{B}$ and that φ preserves (resp. reverses) the bundle orientations if and only if $\varphi^*\Psi$ (resp. $-\varphi^*\Psi$) represents the orientation of $\mathscr{B}$.

Proposition IV: Let $\mathscr{B} = (E, \pi, B, F)$ and $\hat{\mathscr{B}} = (\hat{E}, \hat{\pi}, \hat{B}, \hat{F})$ be oriented bundles, with B connected. Assume that $\varphi: E \to \hat{E}$ is a fibre preserving map which restricts to local diffeomorphisms

$$\varphi_x: F_x \longrightarrow \hat{F}_{\psi(x)}, \qquad x \in B.$$

If φ_a is orientation preserving (resp. orientation reversing) for some $a \in B$, then φ preserves (resp. reverses) the bundle orientations.

Proof: Assume first that φ_a is orientation preserving. Let Ψ represent the orientation of $\hat{\mathscr{B}}$ and let Φ represent the orientation of $\mathscr{B}$. By hypothesis

$$j_a^* \Phi \qquad \text{and} \qquad j_a^* \varphi^* \Psi = \varphi_a^* j_{\psi(a)}^* \Psi$$

represent the same orientation of F_a. Hence, by Proposition III, sec. 7.4, $\varphi^*\Psi$ represents the orientation of $\mathscr{B}$, i.e., φ is orientation preserving.

The case that φ_a reverses orientations is treated in the same way.

Q.E.D.

7.6. Local product orientation. Let (E, π, B, F) be a fibre bundle, oriented by an r-form Ψ. Assume further that $\Delta_B \in A^n(B)$ orients B.

Lemma II: The $(n + r)$-form

$$\Delta_E = \pi^*\Delta_B \wedge \Psi$$

orients the manifold E. The orientation of E represented by Δ_E depends only on the orientation of the bundle and the orientation of B.

Proof: It is clearly sufficient to consider the case $E = B \times F$ where B is connected. Fix $a \in B$ and set $\Delta_F = j_a^*\Psi$. Then Δ_F orients F. According to the corollary to Proposition III, sec. 7.4, $1 \times \Delta_F$ represents the same orientation as Ψ. Hence by Proposition II, sec. 7.4, $\rho_V \Psi$ and $\rho_V(1 \times \Delta_F)$ represent the same orientation in V_E:

$$\rho_V \Psi = f \cdot \rho_V(1 \times \Delta_F),$$

where $f \in \mathscr{S}(B \times F)$ satisfies $f > 0$.

Using formula (7.1), sec. 7.4, we find

$$(j_x^* \Psi)(y) = f(x, y) \cdot \Delta_F(y), \qquad x \in B, \quad y \in F,$$

and it follows easily that

$$\pi^* \Delta_B \wedge \Psi = f \cdot \pi^* \Delta_B \wedge \pi_F^* \Delta_F .$$

On the other hand, according to Example 8, sec. 3.21, the form $\pi^* \Delta_B \wedge \pi_F^* \Delta_F$ orients $B \times F$. Moreover, the orientation so obtained depends only on the orientations of B and F represented by Δ_B and Δ_F. Since f is strictly positive, the lemma follows.

Q.E.D.

Definition: Let (E, π, B, F) be an oriented bundle over an oriented base B. Let $\Psi \in A^r(E)$ represent the orientation of the bundle and let $\Delta_B \in A^n(B)$ represent the orientation of B. Then the orientation of E represented by $\pi^* \Delta_B \wedge \Psi$ is called the *local product orientation.*

Next, consider oriented bundles (E, π, B, F) and $(\hat{E}, \hat{\pi}, \hat{B}, \hat{F})$ over oriented bases B and $\hat{B}$. Let $\varphi: E \to \hat{E}$ be a fibre-preserving map which restricts to local diffeomorphisms

$$\varphi_x: F_x \to \hat{F}_{\psi(x)} , \qquad x \in B.$$

Assume further that the induced map $\psi: B \to \hat{B}$ is a local diffeomorphism. Then φ is a local diffeomorphism.

Proposition V: With the hypotheses and notation above, assume further that φ preserves the bundle orientations. Then $\varphi: E \to \hat{E}$ preserves (resp. reverses) the local product orientations if $\psi: B \to \hat{B}$ preserves (resp. reverses) the orientations.

Proof: Let $\Psi \in A^r(\hat{E})$ represent the orientation of the bundle $(\hat{E}, \hat{\pi}, \hat{B}, \hat{F})$. Then (cf. sec. 7.5) $\varphi^* \Psi$ represents the orientation of the bundle (E, π, B, F). Next, let $\Delta \in A^n(\hat{B})$ represent the orientation of $\hat{B}$.

Then $\epsilon \cdot \psi^*\Delta$ represents the orientation of B, where $\epsilon = +1$ (resp. $\epsilon = -1$) if ψ is orientation preserving (resp. orientation reversing).

Thus the local product orientations of E and $\hat{E}$ are represented by

$$\Delta_{\hat{E}} = \hat{\pi}^*\Delta \wedge \Psi$$

and

$$\Delta_E = \epsilon \cdot \pi^*\psi^*\Delta \wedge \varphi^*\Psi = \epsilon \cdot \varphi^*\Delta_{\hat{E}}.$$

Thus φ preserves (resp. reverses) orientations if ψ does.

Q.E.D.

Example: Let K be the Klein bottle (cf. Example 4, sec. 3.21). Define a smooth map $K \to S^1$ by the commutative diagram

$$\begin{array}{ccc} \mathbb{R}^2 & \xrightarrow{\varphi} & K \\ {\scriptstyle \pi_1}\downarrow & \swarrow{\scriptstyle \pi} & \\ S^1 & & \end{array},$$

where φ is the projection defined in sec. 3.21 and

$$\pi_1(x, y) = \exp(2\pi i x), \qquad (x, y) \in \mathbb{R}^2.$$

It is easy to see that (K, π, S^1, S^1) is a fibre bundle. Since K is non-orientable and S^1 is orientable, it follows that the bundle is non-orientable.

§3. Vector bundles and sphere bundles

7.7. The bundle maps α and β. Let $\xi = (E, \pi, B, F)$ be a vector bundle of rank r over an n-manifold and consider the tangent bundle $\tau_E = (T_E, \pi_E, E, \mathbb{R}^{n+r})$. For $x \in B$, $z \in F_x$, we may identify the vector spaces F_x and $T_z(F_x)$, and regard $(dj_x)_z$ as a linear isomorphism

$$\omega_z: F_{\pi(z)} \xrightarrow{\cong} V_z(E).$$

Denote ω_z^{-1} by α_z. The isomorphisms α_z define a bundle map $\alpha: V_E \to E$ inducing π as map of base manifolds:

$$\begin{array}{ccc} V_E & \xrightarrow{\alpha} & E \\ {\scriptstyle \pi_V}\downarrow & & \downarrow{\scriptstyle \pi} \\ E & \xrightarrow[\pi]{} & B \end{array}$$

(the smoothness of α is easily shown).

On the other hand, if σ is a cross-section in ξ, a bundle map

$$\begin{array}{ccc} E & \xrightarrow{\beta_\sigma} & V_E \\ {\scriptstyle \pi}\downarrow & & \downarrow{\scriptstyle \pi_V} \\ B & \xrightarrow[\sigma]{} & E \end{array}$$

is defined by

$$(\beta_\sigma)_x = \omega_{\sigma(x)}, \qquad x \in B.$$

The bundle maps α and β_σ restrict to isomorphisms in the fibres and satisfy

$$\alpha \circ \beta_\sigma = \iota.$$

The bundle map induced from the zero cross-section will be denoted simply by β.

Next, define a vertical vector field Z on E by setting

$$Z(z) = \omega_z(z), \qquad z \in E$$

(use a coordinate representation for ξ to check that Z is smooth). Z is called the *radial vertical vector field.* Evidently

$$\alpha(Z(z)) = z, \qquad z \in E,$$

and Z is the unique vertical vector field which satisfies this relation.

Now fix $t \in \mathbb{R}$ and define a strong bundle map μ_t, in ξ by setting

$$\mu_t(z) = tz, \qquad z \in E.$$

Then

$$Z \underset{\mu_t}{\sim} Z$$

or equivalently

$$d\mu_t(Z(z)) = Z(tz), \qquad z \in E.$$

In fact, this relation follows, after a simple computation, from

$$Z(z) = \omega_z(z) = (dj_x)_z(z), \qquad z \in F_x, \quad x \in B.$$

7.8. Orientations in vector bundles. Let $\xi = (E, \pi, B, F)$ be a vector bundle of rank r. In sec. 2.16 and in sec. 7.4 we gave different definitions for orientations in ξ. Now it will be shown that these definitions coincide.

In fact, in Proposition II, sec. 7.4, there was established a canonical bijection between orientations of the fibre bundle ξ (in the sense of sec. 7.4), and orientations of the vector bundle V_E (in the sense of sec. 2.16).

It remains to construct a bijection between the orientations of the vector bundles ξ and V_E. Without loss of generality we may assume B (and hence E) is connected. Thus the set $\mathcal{O}(\xi)$ of orientations in ξ contains two elements, or is void. Similarly $\mathcal{O}(V_E)$ contains two elements, or is void. Consider the bundle maps

$$\alpha: V_E \to E \qquad \text{and} \qquad \beta: E \to V_E.$$

Since α and β restrict to isomorphisms in each fibre they induce maps

$$\hat{\alpha}: \mathcal{O}(V_E) \leftarrow \mathcal{O}(\xi) \qquad \text{and} \qquad \hat{\beta}: \mathcal{O}(\xi) \leftarrow \mathcal{O}(V_E).$$

(cf. sec. 2.16). Since $\alpha \circ \beta = \iota$, it follows that

$$\hat{\beta} \circ \hat{\alpha} = \iota.$$

Thus either $\mathcal{O}(V_E) = \mathcal{O}(\xi) = \varnothing$, or else both sets have two elements and $\hat{\alpha}$, $\hat{\beta}$ are inverse bijections.

7.9. Associated sphere bundle. Let $\xi = (E, \pi, B, F)$ be an oriented Riemannian vector bundle of rank r ($r \geqslant 1$), with $\dim B = n$. Let $\xi_S = (E_S, \pi_S, B, S)$ be the associated sphere bundle with inclusion map $i: E_S \to E$ (cf. Example 6, sec. 3.10).

Lemma III: Suppose $\Omega \in A^r(E)$ orients ξ. Then

$$\Omega_S = i^*(i(Z)\Omega) \in A^{r-1}(E_S)$$

orients ξ_S (Z is the radial vertical vector field of sec. 7.7).

Proof: Z restricts to the vector field Z_x on the vector space F_x given by $Z_x(z) = z$. Since Ω orients ξ, Ω restricts to an orienting r-form $\Omega_x \in A^r(F_x)$. Thus

$$\Omega_x = f \cdot \Delta_x$$

where $f > 0$ and Δ_x is a positive determinant function in F_x.

Hence the restriction of Ω_S to a fibre S_x is the $(r-1)$-form $(\Omega_S)_x$ given by

$$\begin{aligned}(\Omega_S)_x(y; \eta_1, \ldots, \eta_{r-1}) &= f(y) \cdot \Delta_x(Z_x(y), \eta_1, \ldots, \eta_{r-1}) \\ &= f(y) \cdot \Delta_x(y, \eta_1, \ldots, \eta_{r-1}), \quad y \in S_x, \quad \eta_i \in T_y(S_x).\end{aligned}$$

Thus according to Example 2, sec. 3.21, $(\Omega_S)_x$ orients S_x. Hence Ω_S orients ξ_S.

Q.E.D.

Definition: The orientation of the associated sphere bundle defined by the $(r-1)$-form Ω_S is called the *induced orientation.*

Remark: If B consists of a point, then the definition of the induced orientation coincides with that of sec. 3.21.

Next, assume that an orientation is defined in the *bundle manifold E* of ξ. Let $\Delta_E \in A^{n+r}(E)$ represent the orientation and again let Z be the radial vertical vector field on E.

Lemma IV: Let

$$\Delta_S = (-1)^n i^*(i(Z)\Delta_E).$$

Then Δ_S orients E_S. Moreover, if B and ξ are oriented and if Δ_E represents the local product orientation, then Δ_S represents the local product orientation in E_S determined by B and the induced orientation of ξ_S.

Proof: It is sufficient to consider the case that ξ is trivial and B connected. In this case we may without loss of generality suppose that Δ_E represents the local product orientation with respect to orientations in ξ and B. Thus

$$\Delta_E = f \cdot \pi^* \Delta_B \wedge \Psi,$$

where $\Psi \in A^r(E)$ orients ξ, Δ_B orients B, and $f \in \mathscr{S}(E)$ is strictly positive.

It follows that

$$\begin{aligned}\Delta_S &= i^*f \cdot i^*(\pi^* \Delta_B) \wedge i^*(i(Z)\Psi) \\ &= i^*f \cdot \pi_S^* \Delta_B \wedge i^*(i(Z)\Psi).\end{aligned}$$

According to Lemma III, $i^*(i(Z)\Psi)$ orients ξ_S. Hence Lemma II, sec. 7.6, implies that Δ_S orients E_S. It obviously represents the local product orientation.

Q.E.D.

Definition: The orientation of E_S represented by the differential form Δ_S is called the *induced orientation of* E_S.

§4. Fibre-compact carrier

7.10. Let (E, π, B, F) be a smooth fibre bundle. A differential form $\Omega \in A(E)$ will be said to have *fibre-compact carrier* (*or support*) if, for every compact subset $K \subset B$, the intersection $\pi^{-1}(K) \cap \operatorname{carr} \Omega$ is compact. The set of forms with fibre-compact support is a graded ideal in $A(E)$, which will be denoted by $A_F(E)$. It is stable under the operators $i(Z)$, $\theta(Z)$ $(Z \in \mathscr{X}(E))$ and δ. The cohomology algebra $H(A_F(E), \delta)$ will be denoted by $H_F(E)$.

Evidently $A_c(E) \subset A_F(E) \subset A(E)$. If B is compact, then $A_F(E) = A_c(E)$; if F is compact then $A_F(E) = A(E)$.

Lemma V: Let $\{(U_\alpha, \psi_\alpha)\}$ be a coordinate representation of E. Then $\Omega \in A_F(E)$ if and only if, for each α,

$$\psi_\alpha^* \Omega \in A_F(U_\alpha \times F).$$

Proof: Obvious.

Now let $(\hat{E}, \hat{\pi}, \hat{B}, \hat{F})$ be a second smooth bundle. Assume $\varphi: E \to \hat{E}$ is a smooth fibre preserving map inducing $\psi: B \to \hat{B}$. Suppose further that each φ_x maps F_x diffeomorphically onto an open subset of $\hat{F}_{\psi(x)}$.

Proposition VI: With the notation and hypotheses above assume $\Psi \in A_{\hat{F}}(\hat{E})$ satisfies

$$\hat{F}_{\psi(x)} \cap \operatorname{carr} \Psi \subset \operatorname{Im} \varphi_x, \qquad x \in B.$$

Then $\varphi^* \Psi \in A_F(E)$.

Lemma VI: Suppose $E = B \times F$ and $\hat{E} = \hat{B} \times \hat{F}$. Define $\chi: B \times F \to B \times \hat{F}$ by

$$\chi(x, y) = (x, \varphi_x(y)).$$

Then χ is a fibre preserving diffeomorphism of $B \times F$ onto an open subset of $B \times \hat{F}$. Moreover

$$\varphi = (\psi \times \iota) \circ \chi.$$

Proof: Let $\rho: \hat{B} \times \hat{F} \to \hat{F}$ be the projection. Then

$$\chi(x, y) = (x, \rho\varphi(x, y))$$

and so χ is smooth. Since each φ_x is injective it follows that χ is injective.

It remains to check that each linear map,

$$d\chi: T_{(x,y)}(B \times F) \to T_{\chi(x,y)}(B \times \hat{F}),$$

is an isomorphism. But, for $\xi \in T_x(B)$, $\eta \in T_y(F)$,

$$d\chi(\xi, \eta) = (\xi, \gamma(\xi) + d\varphi_x(\eta))$$

where $\gamma: T_x(B) \to T_{\varphi_x(y)}(\hat{F})$ is a linear map. By hypothesis each $(d\varphi_x)_y$ is an isomorphism. Hence so is $(d\chi)_{(x,y)}$.

Q.E.D.

Corollary: If each φ_x is a diffeomorphism, then χ is a diffeomorphism.

Proof of the proposition: In view of Lemmas V and VI it is sufficient to consider the case $E = B \times F$, $\hat{E} = \hat{B} \times \hat{F}$, and $\varphi = (\psi \times \iota) \circ \chi$; here $\chi(x, y) = (x, \varphi_x(y))$ and χ is a diffeomorphism of E onto an open subset of $B \times \hat{F}$.

Let $K \subset B$ be compact. Choose a compact subset $L \subset \hat{F}$ so that $\text{carr}\,\Psi \cap (\psi(K) \times \hat{F}) \subset \psi(K) \times L$. Then

$$\text{carr}((\psi \times \iota)^*\Psi) \cap (K \times \hat{F}) \subset K \times L,$$

as follows from a straightforward computation. Thus

$$(\psi \times \iota)^*\Psi \in A_{\hat{F}}(B \times \hat{F}).$$

Set $(\psi \times \iota)^*\Psi = \Phi$. Apply $(\psi \times \iota)^{-1}$ to the relation of the proposition, to obtain

$$\hat{F}_x \cap \text{carr}\,\Phi \subset \text{Im}\,\chi_x, \qquad x \in B.$$

It follows that $\text{carr}\,\Phi \subset \text{Im}\,\chi$. Since χ is a diffeomorphism (onto an open subset of $B \times \hat{F}$), $\chi^{-1}(C)$ is compact whenever C is a compact subset of $\text{carr}\,\Phi$. Thus, since Φ has fibre-compact support, so does $\chi^*\Phi = \varphi^*\Psi$.

Q.E.D.

Corollary: If $F = \hat{F}$, and each φ_x is a diffeomorphism, then φ^* restricts to a homomorphism

$$\varphi_F^*: A_F(E) \leftarrow A_F(\hat{E}).$$

In particular it induces a homomorphism

$$\varphi_F^{\#}: H_F(E) \leftarrow H_F(\hat{E}).$$

Finally, assume that (E, π, B, F) and $(\hat{E}, \hat{\pi}, \hat{B}, \hat{F})$ are smooth bundles, and that $\varphi: E \to \hat{E}$ is a smooth, fibre preserving map which maps E diffeomorphically onto an open subset, U, of $\hat{E}$. (Thus φ satisfies the hypotheses of Proposition VI.) If $\Phi \in A_F(E)$, we can form $(\varphi^{-1})^*\Phi \in A(U)$ and then

$$\operatorname{carr}(\varphi^{-1})^*\Phi = \varphi(\operatorname{carr} \Phi)$$

is closed in $\hat{E}$. Extend $(\varphi^{-1})^*\Phi$ to $\hat{E}$ by making it zero outside U.

The resulting form has fibre-compact support; thus in this way we obtain a homomorphism

$$(\varphi_F)_*: A_F(E) \to A_{\hat{F}}(\hat{E}).$$

It induces

$$(\varphi_F)_{\#}: H_F(E) \to H_{\hat{F}}(\hat{E}).$$

§5. Intcgration over the fibre

7.11. The general fibre integral. Let $\mathscr{B} = (E, \pi, B, F)$ be a smooth fibre bundle with $\dim B = n$ and $\dim F = r$; V_E is the vertical subbundle of the tangent bundle τ_E of E. Let $\xi = (M, \pi_M, B, H)$ be a vector bundle over the same base.

Consider a bundle map $\Phi\colon \wedge^r V_E \to \xi$ inducing $\pi\colon E \to B$ in the base manifolds:

$$\begin{array}{ccc} \wedge^r V_E & \xrightarrow{\Phi} & M \\ \downarrow & & \downarrow{\scriptstyle \pi_M} \\ E & \xrightarrow[\pi]{} & B \end{array}$$

The *carrier* of Φ is the closure in E of the set

$$\{z \in E \mid \Phi_z \neq 0\}.$$

We say Φ has *fibre-compact carrier* if, for all compact subsets $A \subset B$,

$$\pi^{-1}A \cap \operatorname{carr} \Phi$$

is compact.

Now assume that the bundle $\mathscr{B}$ is oriented and Φ has fibre-compact carrier. We shall define a cross-section $\sigma \in \operatorname{Sec} \xi$, which will be called the *integral over the fibre* of Φ.

For each $x \in B$, Φ determines an H_x-valued r-form on F_x, $\Phi_x \in A^r(F_x; H_x)$, given by

$$\Phi_x(z; \eta_1, \ldots, \eta_r) = \Phi(\eta_1 \wedge \cdots \wedge \eta_r), \qquad z \in F_x, \quad \eta_i \in T_z(F_x) = V_z(E),$$

(cf. Lemma I, sec. 7.1).

If Φ has fibre-compact carrier then each Φ_x has compact carrier. Since $\mathscr{B}$ is oriented, an orientation is induced in each F_x. Thus we can define a map $\sigma\colon B \to M$ by

$$\sigma(x) = \int_{F_x} \Phi_x, \qquad x \in B.$$

In particular $\sigma(x) \in H_x$ and so $\pi_M \circ \sigma = \iota$. We write $\sigma = {\int\!\!\!\!\!-}_F \Phi$.

If $\Psi\colon \wedge^r V_E \to M$ is a second bundle map inducing $\pi\colon E \to B$, and with fibre-compact carrier, then $\Phi + \Psi$ has fibre-compact carrier, and

$$\oint_F (\Phi + \Psi) = \oint_F \Phi + \oint_F \Psi. \tag{7.2}$$

Proposition VII. With the notation and hypotheses above, the map

$$\sigma\colon x \mapsto \int_{F_x} \Phi_x$$

is a cross-section in ξ.

Proof: We need only show that σ is smooth. Now smoothness is a local property. Since $\mathscr{B}$, ξ are locally trivial, and Φ has fibre-compact carrier, it is sufficient to consider the case that

(i) $B = \mathbb{R}^n$
(ii) $E = B \times F$ and $M = B \times H$

and

(iii) $\operatorname{carr} \Phi \subset B \times K$ where $K \subset F$ is compact.

Cover K by finitely many chart neighborhoods U_i $(i = 1, \ldots, p)$ and write $\Phi = \Phi_1 + \cdots + \Phi_p$, where

$$\operatorname{carr} \Phi_i \subset B \times K_i \subset B \times U_i \qquad (K_i \text{ compact}).$$

In view of condition (7.2), it is now sufficient to consider the case that, in addition to (i), (ii), and (iii)

(iv) F is an oriented vector space.

Using (i), (ii), and (iv), observe that

$$\wedge^r V_E = B \times \wedge^r T_F = \mathbb{R}^n \times F \times \wedge^r F.$$

Let Δ_F be a positive determinant function in F and define a smooth map

$$f\colon \mathbb{R}^n \times F \to H$$

by

$$f(x, y) \cdot \Delta_F(v_1, \ldots, v_r) = \Phi(x, y, v_1 \wedge \cdots \wedge v_r), \qquad x \in \mathbb{R}^n, \quad y \in F, \quad v_i \in F.$$

Then if $y_1, \ldots, y_r$ are coordinate functions on F corresponding to a suitable basis, we have

$$\sigma(x) = \int_F \Phi_x = \int_F f(x, y)\, dy_1 \cdots dy_r .$$

In view of condition (iii) it follows from standard calculus that σ is smooth.

Q.E.D.

Definition: The cross-section $\sigma = \oint_F \Phi$ is called the *integral over the fibre* of Φ.

7.12. Differential forms. Let $\mathscr{B} = (E, \pi, B, F)$ be an oriented smooth fibre bundle with $\dim F = r$, $\dim B = n$. We shall define a linear map

$$\oint_F : A_F(E) \to A(B),$$

homogeneous of degree $-r$, to be called *integration over the fibre.*

Let $\Omega \in A_F^{r+p}(E)$ $(p \geqslant 0)$. For each $x \in B$, Ω determines a compactly supported r-form, Ω_x, on F_x with values in $\wedge^p T_x(B)^*$. Ω_x is defined as follows: Fix $z \in F_x$ and fix tangent vectors

$$\eta_1, \ldots, \eta_r \in V_z(E) \qquad \text{and} \qquad \xi_1, \ldots, \xi_p \in T_x(B).$$

Let $\zeta_i \in T_z(E)$ satisfy $d\pi\zeta_i = \xi_i$. Since $V_z(E)$ $(= \ker(d\pi)_z)$ has dimension r, the number $\Omega(z; \zeta_1, \ldots, \zeta_p, \eta_1, \ldots, \eta_r)$ is independent of the choice of the ζ_i. Define Ω_x by setting

$$\langle \Omega_x(z; \eta_1, \ldots, \eta_r), \xi_1 \wedge \cdots \wedge \xi_p \rangle = \Omega(z; \zeta_1, \ldots, \zeta_p, \eta_1, \ldots, \eta_r).$$

Ω_x is called the *retrenchment* of Ω to F_x.

Now observe that a p-form, $\oint_F \Omega$, on B is defined by

$$\left(\oint_F \Omega\right)(x) = \int_{F_x} \Omega_x, \qquad x \in B.$$

Indeed, to see that $\oint_F \Omega$ is smooth, define a bundle map $\Phi_\Omega: \wedge^r V_E \to \wedge^p \tau_B^*$ by setting

$$\Phi_\Omega(z; \eta_1 \wedge \cdots \wedge \eta_r) = \Omega_{\pi z}(\eta_1, \ldots, \eta_r).$$

Then it is immediate from the definitions that $\oint_F \Omega = \oint_F \Phi_\Omega$ and so $\oint_F \Omega$ is smooth. (Compare Proposition VII, sec. 7.11.)

Evidently $\oint_F$ is a linear map from $A_F^{r+p}(E)$ to $A^p(B)$.

We extend $\oint_F$ to forms of degree $< r$ by setting $\oint_F \Omega = 0$ ($\Omega \in A_F^q(E)$, $q < r$).

Now consider a second oriented smooth bundle $(\hat{E}, \hat{\pi}, \hat{B}, \hat{F})$. Assume $\varphi: E \to \hat{E}$ is a smooth fibre preserving map inducing $\psi: B \to \hat{B}$. Suppose each φ_x is an orientation preserving diffeomorphism of F_x onto an open subset of $\hat{F}_{\psi(x)}$.

Proposition VIII: With the notation and hypotheses above, assume $\Omega \in A_{\hat{F}}(\hat{E})$ satisfies

$$\operatorname{carr} \Omega \cap \hat{F}_{\psi(x)} \subset \operatorname{Im} \varphi_x .$$

Then $\varphi^*\Omega \in A_F(E)$ and

$$\oint_F \varphi^*\Omega = \psi^* \oint_{\hat{F}} \Omega.$$

Proof: According to Proposition VI, sec. 7.10, $\varphi^*\Omega \in A_F(E)$. Now fix $x \in B$ and denote the linear map

$$\wedge(d\psi)_x^*: \wedge T_x(B)^* \leftarrow \wedge T_{\psi(x)}(\hat{B})^*$$

by β. β induces (cf. sec. 4.14) a linear map

$$\beta_*: A(F_x ; \wedge T_x(B)^*) \leftarrow A(F_x ; \wedge T_{\psi(x)}(\hat{B})^*)$$

and it is clear from the definitions that

$$(\varphi^*\Omega)_x = (\beta_* \circ \varphi_x^*)(\Omega_{\psi(x)}).$$

It follows that (cf. Equation (4.9), sec. 4.14)

$$\left(\oint_F \varphi^*\Omega\right)(x) = \int_{F_x} (\beta_* \circ \varphi_x^*)(\Omega_{\psi(x)}) = \beta\left(\int_{F_x} \varphi_x^* \Omega_{\psi(x)}\right).$$

Since φ_x is an orientation preserving diffeomorphism onto an open subset of $\hat{F}_{\psi(x)}$ which contains $\operatorname{carr} \Omega_{\psi(x)}$, we obtain

$$\int_{F_x} \varphi_x^*\Omega_{\psi(x)} = \int_{\hat{F}_{\psi(x)}} \Omega_{\psi(x)} .$$

Hence

$$\left(\oint_F \varphi^*\Omega\right)(x) = \beta\left(\int_{\hat{F}_{\psi(x)}} \Omega_{\psi(x)}\right) = \beta\left[\left(\oint_{\hat{F}} \Omega\right)(\psi(x))\right].$$

Finally, note that for any $\Phi \in A(\hat{B})$,

$$\beta[\Phi(\psi(x))] = (\psi^*\Phi)(x).$$

Q.E.D.

Corollary I: If each φ_x is a diffeomorphism, then φ^* restricts to a homomorphism $\varphi_F^*\colon A_F(E) \leftarrow A_{\hat{F}}(\hat{E})$ and

$$\oint_F \circ \varphi_F^* = \psi^* \circ \oint_{\hat{F}}.$$

Corollary II: If F is compact and each φ_x is a diffeomorphism, then $\varphi_F^* = \varphi^*$ and

$$\oint_F \circ \varphi^* = \psi^* \circ \oint_{\hat{F}}.$$

Examples: **1.** Suppose $E = F$, $B =$ (point). Then F is an oriented manifold, and $\oint_F$ is the linear map

$$\int_F \colon A_c^r(F) \to \mathbb{R}$$

extended to $A_c(F)$ by the rule $\oint_F \Omega = 0$ if $\deg \Omega < r$.

2. Products: Suppose $E = B \times F$. If $\Omega \in A^{p+r}(B \times F)$ then $\Omega_x \in A^r(F; \wedge^p T_x(B)^*)$ is given by

$$\langle \Omega_x(y; \eta_1, \dots, \eta_r), \xi_1 \wedge \cdots \wedge \xi_p \rangle = \Omega(x, y; \xi_1, \dots, \xi_p, \eta_1, \dots, \eta_r)$$
$$x \in B, \quad \xi_i \in T_x(B), \quad y \in F, \quad \eta_i \in T_y(F).$$

Recall from Example 2, sec. 3.20, the bigradation of $A(B \times F)$. The equation above shows that, if $\Omega \in A^{s,q}(B \times F)$, then $\Omega_x = 0$ unless $q = r$.

Now assume $\Delta_F \in A^r(F)$ orients F. Then $1 \times \Delta_F$ orients the bundle (cf. the example, sec. 7.4). Moreover, if $\Psi \in A_c(F)$, $\Phi \in A(B)$, then $\Phi \times \Psi \in A_F(B \times F)$ and

$$\oint_F (\Phi \times \Psi) = \left(\int_F \Psi\right) \Phi. \tag{7.3}$$

In fact, $(\Phi \times \Psi)_x \in A_c(F; \wedge T_x(B)^*)$ is given by

$$(\Phi \times \Psi)_x = \Psi \otimes \Phi(x), \qquad x \in B,$$

(cf. sec. 4.7). It follows that (cf. sec. 4.14)

$$\left(\oint_F (\Phi \times \Psi)\right)(x) = \left(\int_F \Psi\right) \Phi(x),$$

whence (7.3).

7.13. Properties of $\oint_F$. In this section (E, π, B, F) denotes a fixed smooth oriented bundle.

Proposition IX: $\oint_F : A_F(E) \to A(B)$ is a surjective linear map satisfying

$$\oint_F \pi^*\Phi \wedge \Psi = \Phi \wedge \oint_F \Psi, \qquad \Phi \in A(B), \quad \Psi \in A_F(E).$$

Proof: First we establish the equation. Fix $x \in B$. Let σ be the linear map in $\wedge T_x(B)^*$ given by

$$\sigma(z) = \Phi(x) \wedge z, \qquad z \in \wedge T_x(B)^*.$$

σ induces a linear map σ_* in $A(F_x; \wedge T_x(B)^*)$, and

$$(\pi^*\Phi \wedge \Psi)_x = \sigma_*(\Psi_x)$$

(cf. sec. 4.14). It follows via Equation (4.9), sec. 4.14, that

$$\left(\oint_F \pi^*\Phi \wedge \Psi\right)(x) = \int_{F_x} \sigma_*(\Psi_x) = \sigma\left(\int_{F_x} \Psi_x\right)$$

$$= \Phi(x) \wedge \left(\oint_F \Psi\right)(x) = \left(\Phi \wedge \oint_F \Psi\right)(x).$$

It remains to show that $\oint_F$ is surjective. Fix $\Phi \in A(B)$ and let $\{U_\alpha\}$ be a locally finite open cover of B such that the bundle is trivial over each U_α. Denote the restriction of Φ to U_α by Φ_α. Combining Proposition VIII and Example 2 of sec. 7.12, we find $\Omega_\alpha \in A_F(\pi^{-1}(U_\alpha))$ such that $\oint_F \Omega_\alpha = \Phi_\alpha$. Let $\{p_\alpha\}$ be a partition of unity for B subordinate to the open cover. If

$$\Omega = \sum_\alpha (\pi^* p_\alpha) \cdot \Omega_\alpha \in A_F(E), \qquad \text{then} \qquad \oint_F \Omega = \sum_\alpha p_\alpha \cdot \oint_F \Omega_\alpha = \Phi.$$

Q.E.D.

Corollary: $\oint_F$ restricts to a surjective linear map

$$\oint_F : A_c(E) \to A_c(B).$$

Proof: Evidently, for $\Omega \in A_F(E)$,

$$\operatorname{carr} \oint_F \Omega \subset \pi(\operatorname{carr} \Omega).$$

Thus $\oint_F$ restricts to a linear map $A_c(E) \to A_c(B)$. To show that the restriction is surjective, fix $\Phi \in A_c(B)$. Then (Proposition IX) choose $\Omega \in A_F(E)$ so that

$$\oint_F \Omega = \Phi.$$

Since Φ has compact carrier, for some $f \in \mathscr{S}_c(B)$, $f \cdot \Phi = \Phi$. Since $\Omega \in A_F(E)$, $\pi^*f \cdot \Omega \in A_c(E)$. But

$$\oint_F \pi^*f \cdot \Omega = f \cdot \oint_F \Omega = f \cdot \Phi = \Phi.$$

Q.E.D.

Proposition X: Integration over the fibre satisfies the relations:

(1) $i(X) \circ \oint_F = \oint_F \circ\, i(Z)$
(2) $\theta(X) \circ \oint_F = \oint_F \circ\, \theta(Z)$

and

(3) $\delta \circ \oint_F = \oint_F \circ\, \delta$,

where $Z \in \mathscr{X}(E)$, $X \in \mathscr{X}(B)$ are π-related.

Proof: (1) Fix $x \in B$ and set $\xi = X(x)$. The operator $i(\xi)$ in $\wedge T_x(B)^*$ induces a linear map

$$i(\xi)_*: A(F_x\,; \wedge T_x(B)^*) \to A(F_x\,; \wedge T_x(B)^*).$$

Moreover, since $Z \underset{\pi}{\sim} X$ it follows that (cf. sec. 7.12)

$$(i(Z)\Omega)_x = i(\xi)_*\Omega_x\,, \qquad \Omega \in A_F(E).$$

Hence (using Equation (4.9), sec. 4.14), we find

$$\left(\oint_F i(Z)\Omega\right)(x) = \int_{F_x} i(\xi)_*\Omega_x = i(\xi) \int_{F_x} \Omega_x$$

$$= i(\xi) \left(\oint_F \Omega\right)(x) = \left(i(X) \oint_F \Omega\right)(x).$$

(3) Fix $\Omega \in A_F(E)$. We wish to show that $\delta \oint_F \Omega = \oint_F \delta\Omega$. First, apply Proposition VIII, sec. 7.12, to a suitable coordinate representation for the bundle, to reduce to the following case:

(i) $B = \mathbb{R}^n$
(ii) $E = B \times F$

and

(iii) $\operatorname{carr} \Omega \subset B \times K$ (K compact, $K \subset F$).

Next, choose finitely many charts $(U_i, u_i, \mathbb{R}^r)$ $(i = 1, \dots, m)$ on F so that the U_i cover K. Write $\Omega = \Omega_1 + \cdots + \Omega_m$, where

$$\operatorname{carr} \Omega_i \subset B \times K_i \subset B \times U_i, \qquad K_i \text{ compact.}$$

Because $\oint_F$ is linear it is sufficient to establish (3) for each Ω_i. But $\operatorname{carr} \delta\Omega_i \subset B \times K_i$. Thus we can apply Proposition VIII to the maps

$$\iota \times u_i^{-1}: \quad B \times \mathbb{R}^r \to B \times U_i$$

and reduce further to the case

(iv) $F = \mathbb{R}^r$.

Finally, since δ and $\oint_F$ are linear (over $\mathbb{R}$) and homogeneous we may also reduce to the case

(v) $\Omega \in A_F^{p,q}(B \times F)$.

Now consider $\Omega \in A_F^{p,q}(B \times F)$ and assume conditions (i)–(iv) hold. A simple computation using Example 2, sec. 7.12, and sec. 4.6 shows that

$$(\delta_F \Omega)_x = (-1)^p\, \delta(\Omega_x).$$

It follows from Proposition XIV, sec. 4.13, that

$$\left(\oint_F \delta_F \Omega\right)(x) = (-1)^p \int_F \delta(\Omega_x) = 0$$

(because Ω_x has compact carrier). Since, in $A(B \times F)$, $\delta = \delta_B + \delta_F$ (cf. sec. 4.6), we are reduced to proving

$$\oint_F \delta_B \Omega = \delta \oint_F \Omega. \tag{7.4}$$

But $\delta_B\Omega \in A^{p+1,q}(B \times F)$; hence, if $q \neq r$, both sides of (7.4) are zero (cf. Example 2, sec. 7.12). On the other hand, assume $q = r$. Let $e_1, \ldots, e_n$ be a basis for B and let Δ_F be a positive determinant function in F. Then

$$\Omega = \sum_{i_1 < \cdots < i_p} f_{i_1 \cdots i_p} \cdot (e^{*i_1} \wedge \cdots \wedge e^{*i_p} \times \Delta_F),$$

where each $f_{i_1 \cdots i_p} \in \mathscr{S}(B \times F)$ has carrier in $B \times K$. Thus we may restrict ourselves to the case

$$\Omega = f \cdot (e^{*1} \wedge \cdots \wedge e^{*p} \times \Delta_F), \qquad f \in \mathscr{S}(B \times F)$$

and $\operatorname{carr} f \subset B \times K$.

Then (cf. the example of sec. 4.3)

$$\delta_B\Omega = \sum_{i=1}^{n} \frac{\partial f}{\partial e_i} (e^{*i} \wedge e^{*1} \wedge \cdots \wedge e^{*p} \times \Delta_F).$$

Hence, for suitable coordinate functions y_i in F:

$$\left(\oint_F \delta_B\Omega\right)(x) = \sum_{i=1}^{n} \left(\int_F \frac{\partial f}{\partial e_i}(x, y)\, dy_1 \cdots dy_r\right) e^{*i} \wedge e^{*1} \wedge \cdots \wedge e^{*p}$$

$$= \sum_{i=1}^{n} \frac{\partial}{\partial e_i} \left\{\int_F f(x, y)\, dy_1 \cdots dy_r\right\} e^{*i} \wedge e^{*1} \wedge \cdots \wedge e^{*p}$$

$$= (\delta\Phi)(x).$$

Here $\Phi \in A^p(B)$ is given by

$$\Phi(x) = \left[\int_F f(x, y)\, dy_1 \cdots dy_r\right] e^{*1} \wedge \cdots \wedge e^{*p}$$

$$= \left(\oint_F \Omega\right)(x), \qquad x \in B.$$

Thus $\oint_F \delta_B\Omega = \delta \oint_F \Omega$. This finishes the proof of (7.4), and hence completes the proof of (3).

(2) Immediate from (1), (3), and the formula $\theta = i \circ \delta + \delta \circ i$ of Proposition II, sec. 4.3.

Q.E.D.

Proposition X and the corollary to Proposition IX show that $\oint_F$ induces linear maps

$$\oint_F^{\#} : H_F(E) \to H(B) \qquad \text{and} \qquad \oint_F^{\#} : H_c(E) \to H_c(B).$$

These maps are homogeneous of degree $-r$. If F is compact, the first is a map $H(E) \to H(B)$. In this case

$$\oint_F^{\#} \pi^{\#}\alpha * \beta = \alpha * \oint_F^{\#} \beta, \qquad \alpha \in H(B), \quad \beta \in H_c(E).$$

If B is compact $\oint_F^{\#} \pi^{\#}\alpha * \beta = \alpha \cdot \oint_F^{\#}\beta, \qquad \alpha \in H(B), \quad \beta \in H_c(E)$.

7.14. The Fubini theorem. Theorem I: Let (E, π, B, F) be an oriented fibre bundle over an oriented n-manifold B, with $\dim F = r$. Let E have the corresponding local product orientation (cf. sec. 7.6). Then

$$\int_E = \int_B \circ \oint_F \; : \; A_c^{n+r}(E) \to \mathbb{R}.$$

Proof: In view of Proposition VIII, sec. 7.12, a partition of unity argument reduces the problem to the case $E = B \times F$. In this case the theorem becomes

Lemma VII: Let B and F be oriented manifolds of dimensions n and r. Let $B \times F$ have the product orientation. Then

$$\int_{B\times F} = \int_B \circ \oint_F .$$

Proof: First, let $\Phi \in A_c^n(B)$, and let $\Psi \in A_c^r(F)$. Then according to Proposition XIII, sec. 4.13

$$\int_{B\times F} \Phi \times \Psi = \int_B \Phi \cdot \int_F \Psi. \tag{7.5}$$

In view of Example 2, sec. 7.12, we obtain from (7.5) that

$$\int_B \oint_F \Phi \times \Psi = \int_B \left(\int_F \Psi\right) \Phi = \int_{B\times F} \Phi \times \Psi.$$

Now let $\Omega \in A_c^{n+r}(B \times F)$. The Künneth theorem for forms with compact carrier (Theorem V, sec. 5.19) shows that, for some $\Phi \in A_c^n(B)$, $\Psi \in A_c^r(F)$, and $\Omega_1 \in A_c^{n+r-1}(B \times F)$,

$$\Omega = \Phi \times \Psi + \delta\Omega_1 .$$

Since $\Omega_1 \in A_c(B \times F)$, it follows that $\oint_F \Omega_1 \in A_c(B)$; hence (cf. Proposition X, sec. 7.13)

$$\int_B \oint_F \delta\Omega_1 = \int_B \delta \oint_F \Omega_1 = 0 = \int_{B\times F} \delta\Omega_1 .$$

Thus

$$\int_B \oint_F \Omega = \int_B \oint_F \Phi \times \Psi = \int_{B\times F} \Phi \times \Psi = \int_{B\times F} \Omega.$$

Q.E.D.

Corollary I: The maps

$$\oint_F^{\#} : H_c(E) \to H_c(B) \qquad \text{and} \qquad \pi^{\#} : H(B) \to H(E)$$

are dual with respect to the Poincaré scalar products (cf. sec. 5.11); i.e.,

$$\mathscr{P}_B\left(\alpha , \oint_F^{\#} \beta\right) = \mathscr{P}_E(\pi^{\#}\alpha , \beta), \qquad \alpha \in H(B), \quad \beta \in H_c(E).$$

Equivalently,

$$\left\langle D_B\alpha , \oint_F^{\#} \beta\right\rangle = \langle D_E\pi^{\#}\alpha , \beta\rangle.$$

In particular $\oint_F^{\#}$ is surjective (resp. injective, a linear isomorphism) if and only if $\pi^{\#}$ is injective (resp. surjective, a linear isomorphism).

Proof: It is sufficient to consider the case $\alpha \in H^p(B)$, $\beta \in H_c^{n+r-p}(E)$. Let $\Phi \in A^p(B)$ and $\Psi \in A_c^{n+r-p}(E)$ be representing cocycles. Then

$$\mathscr{P}_E(\pi^{\#}\alpha , \beta) = \int_E \pi^*\Phi \wedge \Psi = \int_B \left(\Phi \wedge \oint_F \Psi\right) = \mathscr{P}_B\left(\alpha , \oint_F^{\#} \beta\right).$$

Q.E.D.

Corollary II: Assume E is compact, and let $\omega_E \in H^{n+r}(E)$ and $\omega_B \in H^n(B)$ be the orientation classes (cf. sec. 5.13). Then

$$\oint_F^{\#} \omega_E = \omega_B .$$

If F is compact, then π is proper; hence it induces a homomorphism

$$(\pi_c)^{\#} : H_c(B) \to H_c(E).$$

On the other hand, in this case $\oint_F^{\#}$ is a linear map $H(E) \to H(B)$.

Corollary III: If F is compact, then

$$\oint_F^{\#}: H(E) \to H(B) \quad \text{and} \quad (\pi_c)^{\#}: H_c(B) \to H_c(E)$$

are dual with respect to the Poincaré scalar products. Thus $\oint_F^{\#}$ is injective (resp. surjective, a linear isomorphism) if and only if $(\pi_c)^{\#}$ is surjective (resp. injective, a linear isomorphism).

Problems

$\mathscr{B} = (E, \pi, B, F)$ denotes a smooth fibre bundle.

1. Orientations. (i) Show that $\mathscr{B}$ is orientable if and only if there is an orientation of F and a coordinate representation $\{(U_\alpha, \varphi_\alpha)\}$ for $\mathscr{B}$ such that each $\varphi_{\alpha,x}^{-1} \circ \varphi_{\beta,x}$ is orientation preserving.

(ii) Suppose E is connected. Prove that $\mathscr{B}$ admits no orientations, or precisely two orientations. Prove that B admits no orientations, or precisely two orientations.

(iii) Assume E connected and orientable. Prove that F is orientable. Establish a bijection between orientations in $\mathscr{B}$ and orientations in H_E (H_E is any horizontal bundle). Conclude that $\mathscr{B}$ is orientable if B is.

(iv) Construct a nonorientable bundle with connected, orientable total space.

(v) Suppose E connected and orientable, but that B is nonorientable. Let $\rho_B: \tilde{B} \to B$ be the orientable double cover. Show that $\mathscr{B}$ is orientable if and only if there is a smooth bundle $(E, \tilde{\pi}, \tilde{B}, F_1)$ such that $\pi = \rho_B \circ \tilde{\pi}$. Prove that then $F = F_1 \times \mathbb{Z}_2$.

(vi) Suppose E connected and orientable, and that F has a finite, odd, number of components. Use (v) to establish a bijection between orientations in $\mathscr{B}$ and orientations in B.

2. Fibre integration over products. Let $\hat{\mathscr{B}} = (\hat{E}, \hat{\pi}, \hat{B}, \hat{F})$ be a second smooth bundle and let

$$\mathscr{B} \times \hat{\mathscr{B}} = (E \times \hat{E}, \pi \times \hat{\pi}, B \times \hat{B}, F \times \hat{F}).$$

(i) Show that a homomorphism

$$A_F(E) \otimes A_{\hat{F}}(\hat{E}) \to A_{F \times \hat{F}}(E \times \hat{E})$$

is defined by $\Phi \otimes \Psi \mapsto \Phi \times \Psi$. Thus obtain a homomorphism

$$H_F(E) \otimes H_{\hat{F}}(\hat{E}) \to H_{F \times \hat{F}}(E \times \hat{E}).$$

(ii) Show that orientations in $\mathscr{B}$ and $\hat{\mathscr{B}}$ determine an orientation in $\mathscr{B} \times \hat{\mathscr{B}}$. Prove that

$$\oint_{F\times\hat{F}} (\Omega \times \hat{\Omega}) = \oint_F \Omega \times \oint_{\hat{F}} \hat{\Omega}.$$

(iii) Assume that $\hat{B} = B$. Construct a unique bundle, $\mathscr{B} \oplus \hat{\mathscr{B}}$, with base B and with fibre $F_x \times \hat{F}_x$ at $x \in B$ such that the inclusion map $i\colon \mathscr{B} \oplus \hat{\mathscr{B}} \to \mathscr{B} \times \hat{\mathscr{B}}$ (over the diagonal) is a smooth embedding.

(iv) Suppose $\mathscr{B}$ and $\hat{\mathscr{B}}$ oriented. Obtain an orientation in $\mathscr{B} \oplus \hat{\mathscr{B}}$ and prove that with respect to this orientation

$$\oint_{F\times\hat{F}} i^*(\Phi \times \Psi) = \left(\oint_F \Phi\right) \wedge \left(\oint_{\hat{F}} \Psi\right).$$

3. Composite bundles. Let $\mathscr{B}_1 = (E_1, \pi_1, E, F_1)$ be a smooth bundle over E such that $\hat{\pi} = \pi \circ \pi_1$ is the projection of a third bundle $\hat{\mathscr{B}} = (E_1, \hat{\pi}, B, \hat{F})$ (cf. problem 15, below).

(i) Let V_E, V_{E_1}, $\hat{V}_{E_1}$ be the vertical subbundles for $\mathscr{B}$, $\mathscr{B}_1$, and $\hat{\mathscr{B}}$. Show that $\hat{V}_{E_1} \cong \pi_1^*(V_E) \oplus V_{E_1}$. Conclude that orientations in $\mathscr{B}$ and $\mathscr{B}_1$ determine an orientation in $\hat{\mathscr{B}}$.

(ii) Prove that $\oint_{\hat{F}} = \oint_F \circ \oint_{F_1}$.

4. Stokes' theorem. Let M be a manifold and let $(N, \partial N)$ be a manifold-with-boundary.

(i) Define (the notion of) a smooth bundle $\mathscr{P} = (P, \rho, M, N)$ such that $(P, \partial P)$ is a manifold-with-boundary and $\partial\mathscr{P} = (\partial P, \rho|_{\partial P}, M, \partial N)$ is an ordinary smooth bundle.

(ii) Define the notion of an orientation in $\mathscr{P}$ and show that an orientation in $\mathscr{P}$ induces an orientation in $\partial\mathscr{P}$.

(iii) Show that if $\mathscr{P}$ is oriented and $i\colon \partial P \to P$ is the inclusion, then

$$\oint_N \delta\Phi - \delta \oint_N \Phi = (-1)^{k+r+1} \oint_{\partial N} i^*\Phi, \qquad \Phi \in A_N^k(P), \quad r = \dim N.$$

5. Define a subalgebra, $A(E)_{i=0,\theta=0}$, of $A(E)$ by setting

$$A(E)_{i=0,\theta=0} = \{\Phi \in A(E) \mid i(X)\Phi = 0,\ \theta(X)\Phi = 0,\ X \in \mathscr{X}_V(E)\}.$$

(i) Show that $A(E)_{i=0,\theta=0}$ is stable under δ.

(ii) Assume E is connected. Show that F is of the form $F_0 \times A$ where F_0 is connected and A is discrete. Construct smooth bundles $(E, \tilde{\pi}, \tilde{B}, F_0)$ and $(\tilde{B}, \rho, B, A)$ such that $\rho \circ \tilde{\pi} = \pi$.

(iii) Show that $\tilde{\pi}$ induces an isomorphism

$$A(\tilde{B}) \xrightarrow{\cong} A(E)_{i=0,\theta=0}.$$

6. Partial exterior derivatives. Let M and N be manifolds with $\dim M = m$, $\dim N = n$. Consider the trivial bundle $(M \times N, \pi_M, M, N)$.

(i) Show that $A_N(M \times N)$ is stable under δ_N. Set

$$H_{(N)}(M \times N) = H(A_N(M \times N), \delta_N)$$

and show that $H_{(N)}(M \times N)$ is an $A(M)$-module. Show that δ_M induces an operator $\delta_M^\#$ in $H_{(N)}(M \times N)$ and that this pair is a graded differential algebra.

(ii) Construct an isomorphism

$$A(M) \otimes H_c(N) \xrightarrow{\cong} H_{(N)}(M \times N)$$

of graded differential algebras. Show that this is an $A(M)$-isomorphism.

(iii) Assume N oriented. Regard $A_N(M \times N)$ as an $\mathscr{S}(M)$-module and define an $\mathscr{S}(M)$-bilinear map

$$A(M \times N) \times A_N(M \times N) \to A(M)$$

by

$$(\Phi, \Psi) \mapsto \oint_N (\Phi \wedge \Psi).$$

Thus obtain a bilinear map

$$H(A(M \times N), \delta_N) \times H_{(N)}(M \times N) \to A(M).$$

Interpret this via the isomorphism in (ii).

(iv) Obtain isomorphisms

$$H^{p,q}(A(M \times N), \delta_N) \xrightarrow{\cong} \operatorname{Hom}_M(H_{(N)}^{m-p,n-q}(M \times N); A^m(M)).$$

(v) Let $\Phi \in A^{p,q}(M \times N)$. Show that $\delta_N \Phi = 0$ if and only if the retrenchment, Φ_x, of Φ to each $x \times N$ is closed. Show that $\Phi = \delta_N \Psi$ if and only if each Φ_x is exact. Obtain analogous results for $A_N(M \times N)$.

7. Van Est's theorem. Let M and N be manifolds. Assume $\mathscr{N}$ is the nerve of a simple finite covering of M and establish an isomorphism

$$H(A(M \times N), \delta_M) \cong H(\mathscr{N}; A(N)).$$

8. Vertical cohomology. (i) Mimic the constructions in Chap. IV to define operators $i_V(X)$, $\theta_V(X)$ and δ_V in Sec $\wedge V_E^*$ (for $X \in \mathscr{X}_V(E)$). Show that they satisfy the identities of Chap. IV.

The cohomology algebra, $H(\text{Sec} \wedge V_E^*, \delta_V)$ is denoted by $H_V(E)$ and is called the *vertical cohomology of E*.

(ii) Show that the inclusion $j_x: F_x \to E$ induces a homomorphism Sec $\wedge V_E^* \to A(F_x)$ of graded differential algebras.

(iii) Assume that $\dim H(F) < \infty$. Use a coordinate representation for $\mathscr{B}$ to construct a vector bundle ξ over B with fibre $H(F_x)$ at x. Show that Sec $\xi \cong H_V(E)$.

9. Integral cohomology. Suppose F is compact. Let $\alpha \in H(E)$ be an integral class (cf. problem 13, Chap. V). Assume that $\mathscr{B}$ is oriented and show that $\oint_F^\# \alpha$ is again integral.

10. Submersions. Let $\varphi: M \to Q$ be a submersion.

(i) Show that for $x \in Q$, $\varphi^{-1}(x)$ is a closed submanifold of M with trivial normal bundle.

(ii) Extend the notion of vertical subbundle, bundle orientation, and vertical cohomology to submersions.

(iii) Extend the notion of fibre-compact carrier to $A(M)$. Show that the space of forms with fibre-compact carrier, $A_\varphi(M)$, is a graded ideal in $A(M)$, stable under δ.

(iv) Assume φ is surjective. Define a linear map $\oint: A_\varphi(M) \to A(Q)$ with the same properties as the fibre integral for bundles. In particular, prove a Fubini theorem.

11. Let ξ be an oriented involutive distribution over an n-manifold M. Assume that the maximal integral manifolds, F_α, of ξ are closed r-submanifolds of M. Let $j_\alpha: F_\alpha \to M$ denote the inclusion map.

(i) Let $\Phi \in A_c^r(M)$. Show that $j_\alpha^*\Phi \in A_c^r(F_\alpha)$ and define $f_\Phi: M \to \mathbb{R}$ by

$$f_\Phi(x) = \int_{F_\alpha} j_\alpha^*\Phi, \qquad x \in F_\alpha.$$

(ii) Show that if ξ is the vertical subbundle of the total space of a fibre bundle, then f_Φ is smooth.

(iii) Let $\dot{\mathbb{C}} = \mathbb{C} - \{0\}$. Define a vector field, X, on $\dot{\mathbb{C}} \times \mathbb{C}$ by $X(z, \zeta) = (piz, i\zeta)$. Show that the orbits of X are all circles. Prove that $\dot{\mathbb{C}} \times \mathbb{C}$ is not the total space of a smooth circle bundle whose fibres are the orbits of X. (Here p is an integer.)

(iv) Construct a compact manifold N and a vector field X on N such that the orbits of X are all circles, but such that N is not the total space of a smooth circle bundle whose fibres are the orbits of X.

12. Ehresmann connections. Let V_M be the vertical bundle for a submersion $\varphi: M \to Q$, and suppose H_M is a subbundle of τ_M such that $H_M \oplus V_M = \tau_M$. Let $t \mapsto x(t)$ and $t \mapsto z(t)$ be smooth paths respectively on M and Q. Then $z(t)$ is called a *horizontal lift* of $x(t)$ if

$$\varphi z(t) = x(t) \qquad \text{and} \qquad \dot{z}(t) \in H_{z(t)}(M), \qquad t \in \mathbb{R}.$$

If, for each path $x(t)$ $(t_0 \leqslant t \leqslant t_1)$ and each $z_0 \in \varphi^{-1}(x(t_0))$, there is a horizontal lift $z(t)$ $(t_0 \leqslant t \leqslant t_1)$ such that $z(t_0) = z_0$, then H_M is called an *Ehresmann connection.*

(i) Assume that H_M is an Ehresmann connection. Show that φ is the projection of a smooth bundle (M, φ, Q, N), if Q is connected.

(ii) Show that every proper submersion is the projection of a smooth bundle.

(iii) Show that every smooth bundle admits an Ehresmann connection. *Hint*: If U, V are open subsets of Q over which Ehresmann connections are defined, "piece them together" to obtain an Ehresmann connection in $U \cup V$.

(iv) Assume $E = B \times F$. Let H_1, H_2 be horizontal subbundles of τ_E and let $\rho_1, \rho_2 \in \mathrm{Hom}(\tau_E; \tau_E)$ be the corresponding projection operators with kernel V_E and images H_1, H_2. Fix $\lambda \in \mathbb{R}$. Show that $\mathrm{Im}(\lambda\rho_1 + (1 - \lambda)\rho_2)$ is a horizontal bundle H. Show that even if H_1, H_2 are both Ehresmann connections, H need not be.

13. Homotopy lifting theorem. Assume that H_E is an Ehresmann connection in $\mathscr{B}$.

(i) Let $\varphi: \mathbb{R} \times M \to B$, $\psi_0: M \to E$ be smooth maps such that $\varphi(0, x) = \pi\psi_0(x)$, $x \in M$. Show that there is a unique smooth map

$\psi: \mathbb{R} \times M \to E$ extending ψ_0 and such that for fixed $x \in M$, $\psi(x, t)$ is a horizontal lift for $\varphi(x, t)$.

(ii) Extend the notions of pull-back from vector bundles to general smooth bundles. Let $\varphi_0^*\mathscr{B}$, $\varphi_1^*\mathscr{B}$ denote the pull-backs of $\mathscr{B}$ under smooth maps $\varphi_0, \varphi_1: \hat{B} \to B$. If φ_0 and φ_1 are homotopic, define a fibre preserving diffeomorphism between the total spaces of $\varphi_0^*\mathscr{B}$ and $\varphi_1^*\mathscr{B}$ which induces the identity in $\hat{B}$.

14. (Leray–Hirsch). Assume B connected, and that for some $x \in B$

$$j_x^{\#}: H(E) \to H(F_x)$$

is surjective.

(i) Show that $\mathscr{B}$ is orientable, if F is.

(ii) Show that $H(E) \cong H(B) \otimes H(F)$ (as $H(B)$-modules), if dim $H(F) < \infty$.

15. Let $\mathscr{B}_1 = (E_1, \pi_1, E, F_1)$ be a smooth bundle over E. Show that the projection $\hat{\pi} = \pi \circ \pi_1$ is the projection of a third bundle $\hat{\mathscr{B}} = (E_1, \hat{\pi}, B, \hat{F})$. Show that $(\hat{F}, \pi_1, F, F_1)$ is a smooth bundle.

Chapter VIII

Cohomology of Sphere Bundles

§1. Euler class

8.1. Let $\mathscr{B} = (E, \pi, B, S)$ be an oriented r-sphere bundle ($r \geqslant 1$). Since S is compact, integration over the fibre is a linear surjection

$$\oint_S : A(E) \to A(B)$$

homogeneous of degree $-r$ (cf. Proposition IX, sec. 7.13). Since $r \geqslant 1$ this proposition implies that

$$\oint_S \pi^*\Phi = \Phi \wedge \oint_S 1 = 0, \qquad \Phi \in A(B).$$

Thus π^* can be considered as a linear map

$$\beta\colon A(B) \to \ker \oint_S .$$

In view of Proposition X, sec. 7.13, $\ker \oint_S$ is stable under δ and so β induces a map

$$\beta_{\#}\colon H(B) \to H\left(\ker \oint_S\right).$$

Proposition I: Let $\mathscr{B} = (E, \pi, B, S)$ be an oriented r-sphere bundle. Then the induced map

$$\beta_{\#}\colon H(B) \to H\left(\ker \oint_S\right)$$

is an isomorphism.

Lemma I: The proposition is true if the bundle is trivial: $E = B \times S$.

Proof: Recall that $\Phi \otimes \Psi \mapsto \Phi \times \Psi$ defines a homomorphism of graded differential algebras

$$\kappa\colon A(B) \otimes A(S) \to A(B \times S)$$

(cf. sec. 5.17). It follows from Example 2, sec. 7.12, that

$$\left(\iota \otimes \int_S\right) = \fint_S \circ \kappa.$$

Thus we obtain the row-exact commutative diagram

$$\begin{array}{ccccccccc}
0 & \longrightarrow & A(B) \otimes \ker \int_S & \longrightarrow & A(B) \otimes A(S) & \xrightarrow{\iota \otimes \int_S} & A(B) & \longrightarrow & 0 \\
 & & \downarrow{\scriptstyle \kappa_1} & & \downarrow{\scriptstyle \kappa} & & \downarrow{\scriptstyle \iota} & & \\
0 & \longrightarrow & \ker \fint_S & \longrightarrow & A(B \times S) & \xrightarrow{\fint_S} & A(B) & \longrightarrow & 0,
\end{array}$$

where κ_1 denotes the restriction of κ.

This diagram yields a commutative diagram of long exact cohomology sequences. Since $\iota_{H(B)}$ and $\kappa_\#$ are isomorphisms (cf. Theorem VI, sec. 5.20), we can apply the five-lemma to obtain that $(\kappa_1)_\#$ is an isomorphism,

$$(\kappa_1)_\#\colon H(B) \otimes H\left(\ker \int_S\right) \xrightarrow{\cong} H\left(\ker \fint_S\right).$$

On the other hand, let

$$\gamma\colon A(B) \to A(B) \otimes \ker \int_S$$

be the linear map given by $\gamma(\Phi) = \Phi \otimes 1$. Since κ_1 is the restriction of κ, the diagram

$$\begin{array}{ccc}
A(B) \otimes \ker \int_S & \xrightarrow{\kappa_1} & \ker \fint_S \\
{\scriptstyle \gamma} \nwarrow & & \nearrow {\scriptstyle \beta} \\
 & A(B) &
\end{array}$$

commutes. Thus we have only to show that $\gamma_\#$ is an isomorphism.

In view of Theorem II, sec. 5.13, we have

$$\ker \int_S = \sum_{p=0}^{r-1} A^p(S) \oplus \delta(A^{r-1}(S)),$$

whence

$$H\left(\ker \int_S\right) = \sum_{p=0}^{r-1} H^p(S) = H^0(S) = \mathbb{R},$$

and so

$$H(B) \otimes H\left(\ker \int\right) = H(B).$$

It follows that $\gamma_{\#}$ is an isomorphism.

Q.E.D.

Proof of the proposition: According to Proposition XI, sec. 1.13, we can find a finite cover $U_1, \ldots, U_m$ of B such that the sphere bundles $(\pi^{-1}U_i, \pi, U_i, S)$ are trivial. We proceed by induction on m. The case $m = 1$ is settled in the lemma.

Suppose by induction that the proposition has been proved for decompositions with fewer than m elements. Let

$$U = U_1, \qquad V = \bigcup_{\alpha=2}^{m} U_\alpha.$$

Set

$$E_U = \pi^{-1}(U), \qquad E_V = \pi^{-1}(V), \qquad E_{U \cap V} = \pi^{-1}(U \cap V).$$

Then the sphere bundles

$$(E_U, \pi, U, S), \qquad (E_V, \pi, V, S)$$

and

$$(E_{U \cap V}, \pi, U \cap V, S)$$

satisfy the induction hypothesis.

Consider the row-exact commutative diagram

$$\begin{array}{ccccccccc} 0 & \longrightarrow & A(E) & \longrightarrow & A(E_U) \oplus A(E_V) & \longrightarrow & A(E_{U \cap V}) & \longrightarrow & 0 \\ & & \downarrow{\scriptstyle \oint_S} & & \downarrow{\scriptstyle \oint_S \oplus \oint_S} & & \downarrow{\scriptstyle \oint_S} & & \\ 0 & \longrightarrow & A(B) & \longrightarrow & A(U) \oplus A(V) & \longrightarrow & A(U \cap V) & \longrightarrow & 0 \end{array}$$

(cf. sec. 5.4). Since the vertical maps in this diagram are surjective, it follows from the nine-lemma (cf. sec. 0.6) that the induced sequence

$$0 \to K_B \to K_U \oplus K_V \to K_{U \cap V} \to 0$$

is short exact. Here

$$K_B = \ker \left(\oint_S : A(E) \to A(B)\right), \qquad K_V = \ker \left(\oint_S : A(E_V) \to A(V)\right)$$

$$K_U = \ker \left(\oint_S : A(E_U) \to A(U)\right), \quad K_{U \cap V} = \ker \left(\oint_S : A(E_{U \cap V}) \to A(U \cap V)\right).$$

Now consider the maps

$$\beta_B: A(B) \to K_B, \qquad \beta_V: A(V) \to K_V$$
$$\beta_U: A(U) \to K_U, \qquad \beta_{U \cap V}: A(U \cap V) \to K_{U \cap V}.$$

We have the commutative row-exact diagram

$$
\begin{array}{ccccccccc}
0 & \longrightarrow & A(B) & \longrightarrow & A(U) \oplus A(V) & \longrightarrow & A(U \cap V) & \longrightarrow & 0 \\
& & \downarrow{\scriptstyle \beta_B} & & \downarrow{\scriptstyle \beta_U \oplus \beta_V} & & \downarrow{\scriptstyle \beta_{U \cup V}} & & \\
0 & \longrightarrow & K_B & \longrightarrow & K_U \oplus K_V & \longrightarrow & K_{U \cap V} & \longrightarrow & 0.
\end{array}
\qquad (8.1)
$$

Since $(\beta_U)_\#$, $(\beta_V)_\#$, and $(\beta_{U \cap V})_\#$ are isomorphisms (by induction), the five-lemma can be applied to the diagram of long exact cohomology sequences induced by (8.1) to obtain that $(\beta_B)_\#$ is an isomorphism.

Q.E.D.

8.2. The Gysin sequence. Let $\mathscr{B} = (E, \pi, B, S)$ be an oriented r-sphere bundle ($r \geqslant 1$). Consider the exact sequence of differential spaces

$$0 \longrightarrow \ker \int_S \xrightarrow{\ i\ } A(E) \xrightarrow{\ f_S\ } A(B) \longrightarrow 0.$$

It yields an exact triangle

$$
\begin{array}{ccc}
H\left(\ker \int_S\right) & \xrightarrow{\quad i_\#\quad} & H(E) \\
& & \\
\ {}^{\partial_1}\nwarrow & & \swarrow{\scriptstyle f_S^\#}\ \\
& H(B) &
\end{array}
$$

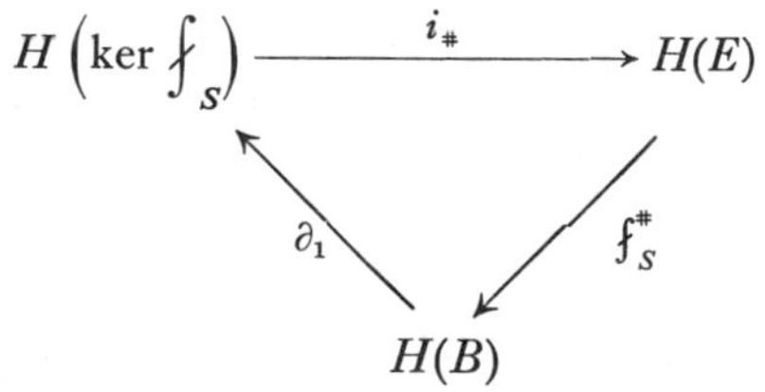

where the connecting homomorphism ∂_1 is homogeneous of degree $r + 1$ (cf. sec. 0.7).

On the other hand, in view of Proposition I, sec. 8.1, we have an isomorphism

$$\beta_\#: H(B) \xrightarrow{\ \cong\ } H\left(\ker \int_S\right)$$

which makes the triangle

$$
\begin{array}{ccc}
& H(B) & \\
{}^{(\beta_\#)^{-1}}\nearrow\ {\scriptstyle \cong} & & \searrow{\scriptstyle \pi^\#} \\
H\left(\ker \int_S\right) & \xrightarrow[\ i_\#\ ]{} & H(E)
\end{array}
$$

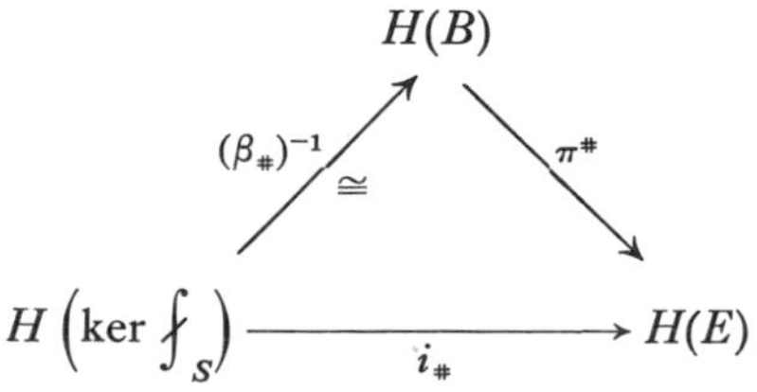

commute. Combining these diagrams and setting $D = (\beta_{\#})^{-1} \circ \partial_1 \circ \omega$, (where $\omega(\alpha) = (-1)^{p+1}\alpha$, $\alpha \in H^p(B)$) we obtain the exact triangle

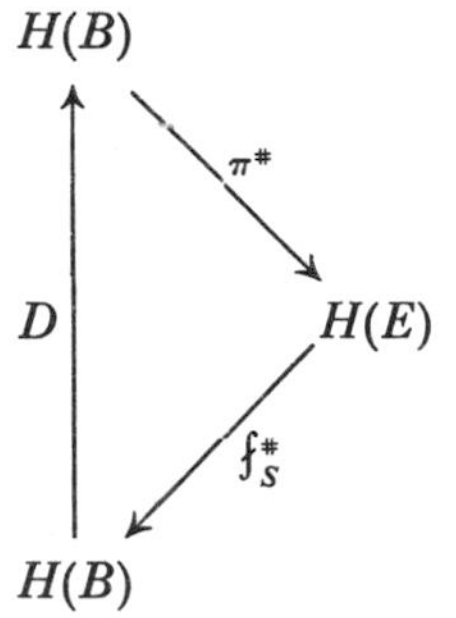

(8.2)

The linear map

$$D\colon H(B) \to H(B),$$

homogeneous of degree $r + 1$, is called the *Gysin map*.

Definition: The long exact sequence

$$\cdots \longrightarrow H^p(B) \xrightarrow{D} H^{p+r+1}(B) \xrightarrow{\pi^*} H^{p+r+1}(E) \xrightarrow{\oint_S^*} H^{p+1}(B) \longrightarrow \cdots$$

corresponding to the triangle (8.2) is called the *Gysin sequence* for the sphere bundle. The element $D(1) \in H^{r+1}(B)$ is called the *Euler class* of the oriented sphere bundle and is denoted by $\chi_{\mathscr{B}}$.

Proposition II: The map $D\colon H(B) \to H(B)$ can be written in the form

$$D\alpha = \alpha \cdot \chi_{\mathscr{B}}, \qquad \alpha \in H^p(B).$$

Proof: This is an immediate consequence of

Lemma II: The map D satisfies the relation

$$D(\alpha \cdot \gamma) = \alpha \cdot D\gamma, \qquad \alpha \in H^p(B), \quad \gamma \in H^q(B).$$

Proof: Let $\Phi \in A^p(B)$ and $\Psi \in A^q(B)$ be cocycles representing α and γ. Choose an r-form Ω on E such that

$$\oint_S \Omega = 1.$$

Then the element $\partial_1(\alpha \cdot \gamma) \in H(\ker \oint_S)$ is represented by the differential form

$$\hat{\Omega} = \delta(\pi^*\Phi \wedge \pi^*\Psi \wedge \Omega) = (-1)^p \pi^*\Phi \wedge \delta(\pi^*\Psi \wedge \Omega).$$

Hence, we obtain

$$\partial_1(\alpha \cdot \gamma) = (-1)^p \beta_\# \alpha \cdot \partial_1 \gamma$$

and so

$$D(\alpha \cdot \gamma) = (-1)^{p+q+1}(\beta_\#)^{-1}\, \partial_1(\alpha \cdot \gamma) = \alpha \cdot D\gamma.$$

Q.E.D.

Example: Assume an oriented r-sphere bundle ($r \geqslant 1$) admits a cross-section σ. Then

$$\sigma^\# \circ \pi^\# = (\pi \circ \sigma)^\# = \iota$$

and so $\pi^\#$ is injective. Thus $\operatorname{Im} D = \ker \pi^\# = 0$; i.e., $D = 0$. In this case the Gysin sequence becomes the short exact sequence

$$0 \longrightarrow H(B) \xrightarrow{\pi^\#} H(E) \xrightarrow{\oint_S^\#} H(B) \longrightarrow 0.$$

It follows that

$$\chi_{\mathscr{B}} = D(1) = 0.$$

In particular the trivial bundle $B \times S$ has zero Euler class.

Proposition III: Let (E, π, B, S) be an oriented r-sphere bundle with r even. Then the Euler class of the bundle is zero.

Proof: Let $\Phi \in A^{r+1}(B)$ represent $\chi_{\mathscr{B}}$. Then there is an r-form Ω on E such that

$$\delta\Omega = \pi^*\Phi \qquad \text{and} \qquad \oint_S \Omega = -1.$$

Since r is even, we have

$$\pi^*\Phi \wedge \Omega = \tfrac{1}{2}\delta(\Omega \wedge \Omega) \tag{8.3}$$

Now Proposition IX, sec. 7.13, and Proposition X, sec. 7.13, yield

$$\Phi = -\oint_S \pi^*\Phi \wedge \Omega = \delta\left(-\tfrac{1}{2}\oint_S \Omega \wedge \Omega\right),$$

whence $\chi_{\mathscr{B}} = 0$.

Q.E.D.

8.3. Naturality of the Gysin sequence. Let $\mathscr{B} = (E, \pi, B, S)$ and $\hat{\mathscr{B}} = (\hat{E}, \hat{\pi}, \hat{B}, S)$ be oriented r-sphere bundles ($r \geqslant 1$) with the same fibre, and let $\varphi: E \to \hat{E}$ be a smooth fibre preserving map which restricts to orientation preserving diffeomorphisms in each fibre and induces $\psi: B \to \hat{B}$. Then Proposition VIII, sec. 7.12 gives the commutative row-exact diagram

$$\begin{array}{ccccccccc} 0 & \longrightarrow & \ker \oint_S & \longrightarrow & A(E) & \xrightarrow{\oint_S} & A(B) & \longrightarrow & 0 \\ & & \uparrow \alpha & & \uparrow \varphi^* & & \uparrow \psi^* & & \\ 0 & \longrightarrow & \ker \oint_S & \longrightarrow & A(\hat{E}) & \xrightarrow{\oint_S} & A(\hat{B}) & \longrightarrow & 0 \end{array}$$

(α is the restriction of φ^*). On the other hand, we have the commutative diagram

$$\begin{array}{ccc} A(B) & \xrightarrow{\beta} & \ker \oint_S \\ \psi^* \uparrow & & \uparrow \alpha \\ A(\hat{B}) & \xrightarrow[\hat{\beta}]{} & \ker \oint_S \end{array}$$

where β, $\hat{\beta}$ are the restrictions of π^* and $\hat{\pi}^*$ (cf. sec. 8.1).

Passing to cohomology we obtain a commutative diagram between the Gysin sequences of the sphere bundles (E, π, B, S) and $(\hat{E}, \hat{\pi}, \hat{B}, S)$,

$$\begin{array}{ccccccccccc} \cdots & \longrightarrow & H^p(B) & \xrightarrow{D} & H^{p+r+1}(B) & \xrightarrow{\pi^\#} & H^{p+r+1}(E) & \xrightarrow{\oint_S^\#} & H^{p+1}(B) & \longrightarrow & \cdots \\ & & \psi^\# \uparrow & & \psi^\# \uparrow & & \varphi^\# \uparrow & & \psi^\# \uparrow & & \\ \cdots & \longrightarrow & H^p(\hat{B}) & \xrightarrow{D} & H^{p+r+1}(\hat{B}) & \xrightarrow{\hat{\pi}^\#} & H^{p+r+1}(\hat{E}) & \xrightarrow{\oint_S^\#} & H^{p+1}(\hat{B}) & \longrightarrow & \cdots \end{array}$$

In particular, it follows that the Euler classes $\chi_{\mathscr{B}}$, $\chi_{\hat{\mathscr{B}}}$ of these bundles are connected by

$$\chi_{\mathscr{B}} = \psi^\# \chi_{\hat{\mathscr{B}}} .$$

8.4. The cohomology of E. Let (E, π, B, S) be an oriented r-sphere bundle with $r \geqslant 1$. Let $\Phi \in A^{r+1}(B)$ be a closed form representing the Euler class. Then we can choose $\Omega \in A^r(E)$ so that

$$\oint_S \Omega = -1 \qquad \text{and} \qquad \delta\Omega = \pi^*\Phi.$$

Let $\wedge\Omega$ denote the exterior algebra over the one-dimensional graded space (homogeneous of degree r) spanned by Ω (Note that, if r is even,

$\wedge \Omega$ need not be a subalgebra of $A(E)$). Form the graded anticommutative algebra $A(B) \otimes \wedge\Omega$. Define

$$d: A(B) \otimes \wedge\Omega \to A(B) \otimes \wedge\Omega$$

by

$$d(\Psi \otimes 1) = \delta\Psi \otimes 1$$

$$d(\Psi \otimes \Omega) = \delta\Psi \otimes \Omega + (-1)^p \Psi \wedge \Phi \otimes 1, \qquad \Psi \in A^p(B).$$

Then d is homogeneous of degree 1, and

$$d^2 = 0.$$

Thus $(A(B) \otimes \wedge\Omega, d)$ is a graded differential space. If r is odd, d is an antiderivation and $(A(B) \otimes \wedge\Omega, d)$ is a graded differential algebra.

Next, define a linear map, homogeneous of degree zero,

$$\mu: A(B) \otimes \wedge\Omega \to A(E)$$

by

$$\mu(\Psi \otimes 1) = \pi^*\Psi, \quad \mu(\Psi \otimes \Omega) = \pi^*\Psi \wedge \Omega, \qquad \Psi \in A(B).$$

Then $\mu \circ d = \delta \circ \mu$ and so μ induces a linear map

$$\mu_\#: H(A(B) \otimes \wedge\Omega) \to H(E)$$

homogeneous of degree zero.

Proposition IV: With the notation and hypotheses above,

$$\mu_\#: H(A(B) \otimes \wedge\Omega) \to H(E)$$

is an isomorphism of graded spaces.

Proof: Let $i: A(B) \to A(B) \otimes \wedge\Omega$ be the inclusion map given by

$$i(\Psi) = \Psi \otimes 1, \qquad \Psi \in A(B).$$

Define $\rho: A(B) \otimes \wedge\Omega \to A(B)$ by

$$\rho(\Psi_1 \otimes \Omega + \Psi_2 \otimes 1) = -\Psi_1 .$$

Then

$$0 \longrightarrow A(B) \xrightarrow{i} A(B) \otimes \wedge\Omega \xrightarrow{\rho} A(B) \longrightarrow 0$$

is a short exact sequence of graded differential spaces. Hence it induces a long exact sequence of cohomology spaces.

On the other hand, the diagram

$$\begin{array}{ccccccccc} 0 & \longrightarrow & A(B) & \xrightarrow{i} & A(B) \otimes \wedge\Omega & \xrightarrow{\rho} & A(B) & \longrightarrow & 0 \\ & & \downarrow{\scriptstyle\beta} & & \downarrow{\scriptstyle\mu} & & \downarrow{\scriptstyle\iota} & & \\ 0 & \longrightarrow & \ker \oint_S & \longrightarrow & A(E) & \xrightarrow{\oint_S} & A(B) & \longrightarrow & 0 \end{array}$$

commutes. Thus it induces a commutative diagram of long exact cohomology sequences. Since (cf. Proposition I, sec. 8.1) $\beta_{\#}$ is an isomorphism, the five-lemma implies that $\mu_{\#}$ is an isomorphism.

Q.E.D.

Corollary I: If r is odd, then $\mu_{\#}$ is an isomorphism of graded algebras.

Proof: Since r is odd, we have $\Omega \wedge \Omega = 0$ (in $A(E)$). It follows that μ is a homomorphism of graded algebras. Hence so is $\mu_{\#}$.

Q.E.D.

Next observe that if the Euler class is zero we may choose Ω so that $\delta\Omega = 0$. With this choice the operator d in $A(B) \otimes \wedge\Omega$ is given by $d = \delta \otimes \iota$ and so $\mu_{\#}$ becomes an isomorphism

$$\mu_{\#}: H(B) \otimes \wedge\Omega \xrightarrow{\cong} H(E).$$

Since (cf. Corollary I to Proposition II, sec. 5.6) $\wedge\Omega \cong H(S)$, we obtain

Corollary II: If the Euler class is zero (in particular if r is even), then there is an isomorphism

$$H(B) \otimes H(S) \xrightarrow{\cong} H(E)$$

of graded spaces. If r is odd, this is an isomorphism of graded algebras.

§2. The difference class

8.5. Pull-back of a cross-section. Suppose that $\mathscr{B} = (E, \pi, B, F)$ and $\hat{\mathscr{B}} = (\hat{E}, \hat{\pi}, \hat{B}, \hat{F})$ are smooth bundles and that $\varphi: E \to \hat{E}$ is a smooth fibre preserving map inducing $\psi: B \to \hat{B}$ and restricting to diffeomorphisms $\varphi_x: F_x \to \hat{F}_{\psi(x)}$ in each fibre.

Given a cross-section $\hat{\sigma}: \hat{B} \to \hat{E}$ define a set map $\sigma: B \to E$ by

$$\sigma(x) = \varphi_x^{-1}\hat{\sigma}\psi(x), \qquad x \in B.$$

Lemma III: σ is a cross-section in $\mathscr{B}$.

Proof: Evidently $\pi \circ \sigma = \iota$. It remains to be shown that σ is smooth. Since smoothness is a local property we can restrict ourselves to the case $E = B \times F$ and $\hat{E} = \hat{B} \times F$. In this case (cf. Lemma VI, sec. 7.10) we can write

$$\varphi = (\psi \times \iota) \circ \chi$$

where $\chi: B \times F \to B \times F$ is a smooth fibre preserving diffeomorphism.

Now define a smooth map $\tau: \hat{B} \to F$ by

$$\hat{\sigma}(y) = (y, \tau(y)), \qquad y \in \hat{B}.$$

Then

$$\sigma(x) = \chi^{-1}(x, \tau\psi(x)), \qquad x \in B$$

and so σ is smooth.

Q.E.D.

The cross-section σ defined above will be denoted by $\varphi^{\#}(\hat{\sigma})$ and is called the *pull-back* of $\hat{\sigma}$ under φ. Observe that in the case of vector bundles this definition agrees with that one given in sec. 2.15. If $\tilde{B} = (\tilde{E}, \tilde{\pi}, \tilde{B}, \tilde{F})$ is a third bundle and $\tilde{\varphi}: \hat{E} \to \tilde{E}$ a smooth fibre preserving map inducing diffeomorphisms in the fibres, then

$$(\tilde{\varphi} \circ \varphi)^{\#} = \varphi^{\#} \circ \tilde{\varphi}^{\#}.$$

8.6. Difference class. Let $\mathscr{B} = (E, \pi, B, S)$ be an oriented r-sphere bundle ($r \geq 1$) which admits a cross-section τ. Then, according to the example of sec. 8.2, the sequence

$$0 \longrightarrow H(B) \xrightarrow{\pi^{\#}} H(E) \xrightarrow{\oint_S^{\#}} H(B) \longrightarrow 0$$

is short exact. Since $\tau^\# \circ \pi^\# = \iota$, it follows that $\oint_S^\#$ maps $\ker \tau^\#$ isomorphically onto $H(B)$. In particular there is a unique class $\omega_\tau \in H^r(E)$ such that

$$\tau^\# \omega_\tau = 0 \qquad \text{and} \qquad \oint_S^\# \omega_\tau = 1.$$

If σ is a second cross-section in $\mathscr{B}$, then

$$\oint_S^\# (\omega_\tau - \omega_\sigma) = 0.$$

Hence, by exactness, there is a unique element $[\tau, \sigma] \in H^r(B)$ such that

$$\pi^\#([\tau, \sigma]) = \omega_\tau - \omega_\sigma .$$

It depends only on $\tau^\#$ and $\sigma^\#$.

Definition: The element $[\tau, \sigma]$ is called the *difference class* for τ and σ.

Lemma IV: Let τ and σ be cross-sections in $\mathscr{B}$. Then

$$\tau^\#\gamma - \sigma^\#\gamma = -\left(\oint_S^\# \gamma\right) \cdot [\tau, \sigma], \qquad \gamma \in H(E).$$

Proof: γ can be written (cf. sec. 8.4)

$$\begin{aligned} \gamma &= \pi^\#\alpha + (\pi^\#\beta) \cdot \omega_\tau \\ &= \pi^\#\alpha + (\pi^\#\beta) \cdot (\omega_\sigma + \pi^\#[\tau, \sigma]), \end{aligned}$$

where $\alpha, \beta \in H(B)$. Then

$$\tau^\#\gamma - \sigma^\#\gamma = \alpha - \alpha - \beta \cdot [\tau, \sigma] = -\beta \cdot [\tau, \sigma].$$

On the other hand,

$$\oint_S^\# \gamma = \beta \cdot \oint_S^\# \omega_\tau = \beta$$

and the lemma follows.

Q.E.D.

Corollary I: $\tau^\# = \sigma^\#$ if and only if $[\tau, \sigma] = 0$, i.e., if and only if $\omega_\tau = \omega_\sigma$.

Corollary II: If $H^r(B) = 0$, then $[\tau, \sigma] = 0$ and $\tau^\# = \sigma^\#$ for every two cross-sections σ and τ.

Now suppose $\hat{\mathscr{B}} = (\hat{E}, \hat{\pi}, \hat{B}, S)$ is a second oriented r-sphere bundle. Assume $\varphi: E \to \hat{E}$ is a smooth fibre preserving map which restricts to diffeomorphisms between the fibres. Let $\psi: B \to \hat{B}$ be the induced map between the base manifolds.

Proposition V: Let $\hat{\tau}$, $\hat{\sigma}$ be cross-sections in $\hat{\mathscr{B}}$ and write $\tau = \varphi^{\#}\hat{\tau}$, $\sigma = \varphi^{\#}\hat{\sigma}$. Then

$$\varphi^{\#}\omega_{\hat{\tau}} = \omega_{\tau}, \qquad \varphi^{\#}\omega_{\hat{\sigma}} = \omega_{\sigma}$$

and

$$\psi^{\#}([\hat{\tau}, \hat{\sigma}]) = [\tau, \sigma].$$

Proof: The equations $\varphi \circ \tau = \hat{\tau} \circ \psi$, $\varphi \circ \sigma = \hat{\sigma} \circ \psi$ yield

$$\tau^{\#} \circ \varphi^{\#} = \psi^{\#} \circ \hat{\tau}^{\#} \qquad \text{and} \qquad \sigma^{\#} \circ \varphi^{\#} = \psi^{\#} \circ \hat{\sigma}^{\#}.$$

Moreover,

$$\oint_S^{\#} \circ \varphi^{\#} = \psi^{\#} \circ \oint_S^{\#}$$

(cf. Corollary II to Proposition VIII, sec. 7.12). The proposition follows.

Q.E.D.

8.7. The main theorem. Consider an oriented sphere bundle $\mathscr{B} = (E, \pi, B, S)$, $r \geqslant 1$. Suppose there are open sets $U \subset B$ and $V \subset B$ such that

$$U \cup V = B$$

and cross-sections $\tau_U: U \to E$, $\sigma_V: V \to E$. Let τ and σ denote the restrictions of τ_U and σ_V to $U \cap V$.

On the other hand, consider the bundles

$$\mathscr{B}_U = (E_U, \pi_U, U, S), \qquad \mathscr{B}_V = (E_V, \pi_V, V, S)$$

and

$$\mathscr{B}_{U \cap V} = (E_{U \cap V}, \pi_{U \cap V}, U \cap V, S).$$

Then τ and σ are cross-sections in $\mathscr{B}_{U \cap V}$ and so their difference class $[\tau, \sigma] \in H^r(U \cap V)$ is defined.

Theorem I: With the notation and hypotheses above, let

$$\partial: H(U \cap V) \to H(B)$$

denote the connecting homomorphism for the Mayer–Vietoris sequence of the triple (B, U, V). Then the Euler class of $\mathscr{B}$ is given by

$$\chi_{\mathscr{B}} = \partial([\tau, \sigma]).$$

Proof: Let $\Phi \in A^{r+1}(B)$ represent the Euler class and choose $\Omega \in A^r(E)$ so that

$$\oint_S \Omega = 1 \qquad \text{and} \qquad \delta\Omega = -\pi^*\Phi.$$

Define Ω_τ and Ω_σ in $A^r(E_{U\cap V})$ by

$$\Omega_\tau = \Omega - \pi^*\tau^*\Omega \qquad \text{and} \qquad \Omega_\sigma = \Omega - \pi^*\sigma^*\Omega.$$

Then Ω_τ and Ω_σ are closed forms and represent respectively ω_τ and ω_σ. Hence $\omega_\tau - \omega_\sigma$ is represented by

$$(\Omega - \pi^*\tau^*\Omega) - (\Omega - \pi^*\sigma^*\Omega) = \pi^*(\sigma^*\Omega - \tau^*\Omega).$$

It follows that $\sigma^*\Omega - \tau^*\Omega$ is closed and represents the difference class $[\tau, \sigma]$.

On the other hand, Φ is closed and represents $\chi_{\mathscr{B}}$. The restrictions of Φ to U and V are given by

$$\Phi_U = \tau_U^*\pi^*\Phi = -\delta\tau_U^*\Omega \qquad \text{and} \qquad \Phi_V = -\delta\sigma_V^*\Omega.$$

Since $\sigma_V^*\Omega$ and $\tau_U^*\Omega$ restrict to $\sigma^*\Omega$ and $\tau^*\Omega$ in $U \cap V$ and since $\sigma^*\Omega - \tau^*\Omega$ represents $[\tau, \sigma]$, we obtain (cf. sec. 5.4),

$$\chi_{\mathscr{B}} = \partial[\tau, \sigma].$$

Q.E.D.

§3. Index of a cross-section at an isolated singularity

8.8. Definition. In this article $\mathscr{B} = (E, \pi, B, S)$ denotes a sphere bundle with $\dim B = n$, $\dim S = n - 1$ and $n \geqslant 2$. A *local cross-section in $\mathscr{B}$ with an isolated singularity at a* is a cross-section $\sigma: \dot{U} \to E$ where U is some neighbourhood of a and $\dot{U} = U - \{a\}$. Now assume that the *manifold E* is oriented. We shall define the index of a local cross-section, σ, at an isolated singularity, a.

Choose a neighbourhood V of a such that

(i) $V \subset U$
(ii) V is diffeomorphic to $\mathbb{R}^n$
(iii) There is a trivializing map $\psi: V \times S \to \pi^{-1}V$ for $\mathscr{B}$.

Let $\mathscr{B}_V = (E_V, \pi_V, V, S)$ denote the restriction of $\mathscr{B}$ to V. Choose orientations in $\mathscr{B}_V$ and V so that the local product orientation in E_V coincides with the orientation induced from that of E. Finally, choose a cross-section τ in $\mathscr{B}_V$ and denote the restriction of τ to $\dot{V} = V - \{a\}$ by $\dot{\tau}$.

Since $\mathscr{B}_V$ is oriented we can form the difference class

$$[\dot{\tau}, \sigma_V] \in H^{n-1}(\dot{V}),$$

where σ_V denotes the restriction of σ to $\dot{V}$. Since V is oriented, we have the canonical isomorphism

$$\alpha_V: H^{n-1}(\dot{V}) \xrightarrow{\cong} \mathbb{R}$$

(cf. sec. 6.8) and hence a real number

$$\alpha_V([\dot{\tau}, \sigma_V])$$

is determined.

Lemma V: With the notations and hypotheses above, $\alpha_V([\dot{\tau}, \sigma_V])$ depends only on σ and the choice of the orientation in E.

Proof: It has to be shown that $\alpha_V([\dot{\tau}, \sigma_V])$ is independent of

(i) the orientations of V and $\mathscr{B}_V$
(ii) the choice of τ
(iii) the choice of V.

(i): Reversing the orientation of V forces us to reverse the orientation of $\mathscr{B}_V$. In this case α_V is replaced by $-\alpha_V$ and $[\dot{\tau}, \sigma_V]$ is replaced by $-[\dot{\tau}, \sigma_V]$. Hence $\alpha_V([\dot{\tau}, \sigma_V])$ remains unchanged.

(ii): Let τ_1 be a second cross-section in $\mathscr{B}_V$. Since $V \cong \mathbb{R}^n$, it follows that $H^{n-1}(V) = 0$. Hence $[\tau, \tau_1] = 0$ and so, by Proposition V, sec. 8.6,

$$[\dot{\tau}, \dot{\tau}_1] = 0.$$

Now Corollary I to Lemma IV, sec. 8.6 shows that $\dot{\tau}^\# = \dot{\tau}_1{}^\#$, whence

$$[\dot{\tau}, \sigma_V] = [\dot{\tau}_1, \sigma_V].$$

(iii): Let W be a second neighbourhood of a satisfying the conditions imposed above on V and let $\tau_1: W \to E_W$ be a cross-section. To show that

$$\alpha_V([\dot{\tau}, \sigma_V]) = \alpha_W([\dot{\tau}_1, \sigma_W])$$

clearly we may assume that $W \subset V$.

In view of (i) we may assume that the inclusion maps

$$i: E_{\dot{W}} \to E_{\dot{V}} \qquad \text{and} \qquad j: \dot{W} \to \dot{V}$$

preserve bundle and base orientations. In view of (ii) we may assume that τ_1 is the restriction of τ to W. Then

$$\sigma_W = i^\# \sigma_V \qquad \text{and} \qquad \dot{\tau}_1 = i^\# \dot{\tau}.$$

Since i preserves the bundle orientations, Proposition V, sec. 8.6, implies that

$$[\dot{\tau}_1, \sigma_W] = j^\#([\dot{\tau}, \sigma_V]).$$

Now, since j preserves orientations, Proposition V, sec. 6.7, gives

$$\alpha_W([\dot{\tau}_1, \sigma_W]) = \alpha_W j^\#([\dot{\tau}, \sigma_V]) = \alpha_V([\dot{\tau}, \sigma_V]).$$

Q.E.D.

Definition: The number $\alpha_V([\dot{\tau}, \sigma_V])$ is called the *index of* σ *at* a and is denoted by $j_a(\sigma)$.

Remark: It follows immediately from the definition that the index changes sign if the orientation of E is reversed.

Next, consider a second $(n-1)$-sphere bundle $\hat{\mathscr{B}} = (\hat{E}, \hat{\pi}, \hat{B}, S)$ over an n-manifold $\hat{B}$ and assume that $\hat{E}$ is oriented. Suppose that $\varphi: \hat{E} \to E$ is a fibre preserving and orientation preserving diffeomorphism of $\hat{E}$ onto an open subset of E. Assume φ induces a diffeomorphism ψ of $\hat{B}$ onto an open subset of B and restricts to diffeomorphisms

$$\varphi_x: S_x \xrightarrow{\cong} S_{\psi(x)}, \qquad x \in \hat{B}.$$

If σ is a local cross-section in $\mathscr{B}$ with an isolated singularity at $a = \psi(b)$, then $\varphi^\# \sigma$ is a local cross-section in $\hat{\mathscr{B}}$ with an isolated singularity at b. Combining Proposition V, sec. 6.7, with Proposition V, sec. 8.6, yields (as in the proof of Lemma V, iii above),

Lemma VI: With the notation and hypotheses above

$$j_b(\varphi^\# \sigma) = j_a(\sigma).$$

8.9. Index and degree. Suppose that E is oriented and that $\sigma: \dot{U} \to E$ is a local cross-section in $\mathscr{B}$ with an isolated singularity at a. Let F be a Euclidean n-space and let

$$\begin{array}{ccc} F \times S & \xrightarrow[\cong]{\varphi} & E_V \\ \downarrow & & \downarrow \pi \\ F & \xrightarrow[\psi]{\cong} & V, \end{array} \qquad E_V = \pi^{-1} V,$$

define a smooth fibre preserving diffeomorphism φ such that

$$\psi(0) = a \qquad \text{and} \qquad V \subset U.$$

Choose orientations in F and S so that φ is orientation preserving. Denote the unit sphere of F by S_F and give it the induced orientation.

The cross-section

$$\varphi^\# \sigma : \dot{F} \to F \times S, \qquad \dot{F} = F - \{0\}$$

determines a smooth map $\hat{\sigma}: \dot{F} \to S$ by the equation

$$\varphi^\# \sigma(x) = (x, \hat{\sigma} x), \qquad x \in \dot{F}.$$

Restricting $\hat{\sigma}$ to S_F we finally obtain a smooth map

$$\hat{\sigma}_S: S_F \to S.$$

Proposition VI: The degree of $\hat{\sigma}_S$ is equal to the index of σ at a,

$$j_a(\sigma) = \deg \hat{\sigma}_S .$$

In particular, the index at an isolated singularity is an integer.

Proof: In view of Lemma VI, sec. 8.8, we lose no generality in assuming that

$$E = F \times S, \qquad B = F, \qquad \varphi = \iota, \qquad \psi = \iota.$$

In this case $\hat{\sigma}: \dot{F} \to S$ is given by

$$\sigma(x) = (x, \hat{\sigma}(x)), \qquad x \in \dot{F}.$$

Now fix a point $e \in S$ and let $\tau: F \to F \times S$ be the cross-section given by

$$\tau(x) = (x, e), \qquad x \in F.$$

If $\Omega \in A^{n-1}(S)$ satisfies $\oint_S \Omega = 1$, then $1 \times \Omega \in A(\dot{F} \times S)$ and

$$\dot{\tau}^*(1 \times \Omega) = 0 \qquad \text{and} \qquad \oint_S (1 \times \Omega) = 1.$$

Hence $1 \times \Omega$ represents $\omega_{\dot{\tau}}$ in $H^{n-1}(\dot{F} \times S)$. It follows that

$$\omega_{\dot{\tau}} = \rho^{\#} \omega_S ,$$

where $\omega_S \in H^{n-1}(S)$ is the orientation class, and $\rho: \dot{F} \times S \to S$ is the projection.

Next, observe that

$$[\dot{\tau}, \sigma] = \sigma^{\#} \omega_{\dot{\tau}} = \sigma^{\#} \rho^{\#} \omega_S ,$$

whence

$$j_0(\sigma) = \alpha_F \sigma^{\#} \rho^{\#} \omega_S = \alpha_F \hat{\sigma}^{\#} \omega_S .$$

Applying Proposition VI, sec. 6.8, yields

$$j_0(\sigma) = \int_{S_F}^{\#} \hat{\sigma}_S^{\#} \omega_S = \deg \hat{\sigma}_S .$$

Q.E.D.

8.10. Cross-sections with index zero. Proposition VII: Assume E is oriented. Let σ be a cross-section with a single singularity at $a \in B$, such that $j_a(\sigma) = 0$. Then there exists a cross-section $\tau: B \to E$ (without singularities) such that $\tau = \sigma$ outside a neighbourhood of a.

Proof: It is clearly sufficient (in view of Lemma VI, sec. 8.8) to consider the case $E = \mathbb{R}^n \times S$ and $a = 0$. Define $\hat{\sigma}: \dot{\mathbb{R}}^n \to S$ by

$$(x, \hat{\sigma}(x)) = \sigma(x), \qquad x \in \mathbb{R}^n.$$

Let $\hat{\sigma}_S$ denote the restriction of $\hat{\sigma}$ to the unit sphere in $\mathbb{R}^n$ (with respect to some Euclidean metric). Then according to Proposition VI, sec. 8.9,

$$\deg \hat{\sigma}_S = j_0(\sigma) = 0.$$

Hence, by Corollary III to Theorem III, sec. 6.15 there is a smooth map (consider S as the unit sphere in $\mathbb{R}^n$)

$$\tilde{\tau}: \mathbb{R}^n \to \dot{\mathbb{R}}^n$$

such that

$$\tilde{\tau}(x) = \hat{\sigma}(x), \qquad |x| \geqslant 1.$$

Set

$$\tau(x) = \left(x, \frac{\tilde{\tau}(x)}{|\tilde{\tau}(x)|}\right), \qquad x \in \mathbb{R}^n.$$

Q.E.D.

§4. Index sum and Euler class

8.11. In this section $\mathscr{B} = (E, \pi, B, S)$ will denote an oriented sphere bundle such that

(1) B is a compact, oriented n-manifold ($n \geqslant 2$)
(2) $\dim S = n - 1$
(3) E is given the local product orientation.

A cross-section, σ, in $\mathscr{B}$ with finitely many singularities $a_1, \ldots, a_k$ is a cross-section

$$\sigma : B - \{a_1, \ldots, a_k\} \to E.$$

The purpose of this section is to prove

Theorem II: Let σ be a cross-section in $\mathscr{B}$ with finitely many singularities $a_1, \ldots, a_k$. Then the Euler class $\chi_{\mathscr{B}}$, of the bundle is given by

$$\chi_{\mathscr{B}} = \left[\sum_{\nu=1}^{k} j_\nu(\sigma)\right] \omega_B$$

where ω_B denotes the orientation class of B (cf. sec. 5.13) and $j_\nu(\sigma)$ is the index of σ at a_ν. Equivalently,

$$\int_B^{\#} \chi_{\mathscr{B}} = \sum_{\nu=1}^{k} j_\nu(\sigma).$$

Proof: Choose, for each ν, a neighborhood U_ν of a_ν diffeomorphic to $\mathbb{R}^n$ such that the U_ν are disjoint and such that the restriction of $\mathscr{B}$ to U_ν is trivial. Set

$$U = \bigcup_{\nu=1}^{k} U_\nu, \qquad V = B - \{a_1, \ldots, a_k\}.$$

Then $U \cup V = B$. Next choose arbitrary cross-sections $\tilde{\tau}_\nu: U_\nu \to E$. These cross-sections determine a cross-section over U. On the other hand, σ is a cross-section over V. Let

$$\hat{\tau}: U \cap V \to E \qquad \text{and} \qquad \hat{\sigma}: U \cap V \to E$$

denote the restrictions of these cross-sections to $U \cap V$.

According to Theorem I, sec. 8.7, we have

$$\chi_{\mathscr{B}} = \partial([\hat{\tau}, \hat{\sigma}]),$$

where ∂ is the connecting homomorphism for the Mayer–Vietoris sequence of the triple (B, U, V).

Next, set $\dot{U}_\nu = U_\nu - \{a_\nu\}$ and let τ_ν and σ_ν denote the restrictions of $\tilde{\tau}_\nu$ and σ to $\dot{U}_\nu$ respectively. Then

$$[\tau_\nu, \sigma_\nu] \in H^{n-1}(\dot{U}_\nu)$$

and, evidently,

$$[\hat{\tau}, \hat{\sigma}] = \bigoplus_{\nu=1}^{k} [\tau_\nu, \sigma_\nu].$$

Finally denote the canonical isomorphism

$$\alpha_{U_\nu}: H^{n-1}(\dot{U}_\nu) \xrightarrow{\cong} \mathbb{R}$$

(cf. the corollary to Proposition VI, sec. 6.8) by α_ν. Applying Proposition VII, sec. 6.9, we obtain

$$\begin{aligned}\int_B^{\#} \chi_{\mathscr{B}} = \int_B^{\#} \partial \bigoplus_{\nu=1}^{k} ([\tau_\nu, \sigma_\nu]) &= \sum_{\nu=1}^{k} \alpha_\nu([\tau_\nu, \sigma_\nu]) \\ &= \sum_{\nu=1}^{k} j_\nu(\sigma).\end{aligned}$$

Q.E.D.

Corollary I: The index sum $\sum_{\nu=1}^{k} j_\nu(\sigma)$ is independent of σ.

Corollary II: If n is odd, then $\sum_{\nu=1}^{k} j_\nu(\sigma) = 0$.

In §5 the following theorem will be established:

Theorem III: Every sphere bundle with fibre dimension $n - 1 \geqslant 1$ and connected base manifold of dimension n admits a cross-section with a single singularity. Moreover, if the base manifold is not compact, then the bundle admits a cross-section without singularities.

With the aid of this and Theorem II we shall now establish:

Theorem IV: An oriented $(n - 1)$-sphere bundle, (E, π, B, S), $(n \geqslant 2)$ over an oriented connected n-manifold B admits a cross-section $\tau: B \to E$ if and only if its Euler class is zero.

Proof of Theorem IV: If B is not compact, Theorem III implies that a cross-section exists. On the other hand, according to Proposition IX sec. 5.15, $H^n(B) = 0$. Hence $\chi_{\mathscr{B}} \in H^n(B)$ is zero.

If B is compact, let $\sigma: B - \{a\} \to E$ be a cross-section with a single singularity at a (cf. Theorem III). Then we have (by Theorem II) that

$$j_a(\sigma) = \int_B^{\#} \chi_{\mathscr{B}} .$$

Thus, if $\chi_{\mathscr{B}} = 0$, then $j_a(\sigma) = 0$. Hence Proposition VII, sec. 8.10, implies that there is a cross-section $\tau: B \to E$.

On the other hand, assume $\tau: B \to E$ is a cross-section. Then the example of sec. 8.2 shows that $\chi_{\mathscr{B}} = 0$.

Q.E.D.

Corollary: If n is odd, the bundle always admits a cross-section.

§5. Existence of cross-sections in a sphere bundle

8.12. It is the purpose of this article to prove the following theorem, which was used in sec. 8.11.

Theorem III: Every sphere bundle with fibre dimension $n - 1$ over a connected base manifold of dimension $n \geqslant 2$ admits a cross-section with a single singularity. If the base is not compact, then the bundle admits a cross-section without singularities.

In sec. 8.13 we construct a cross-section with discrete singularities. This is then modified to give a cross-section without singularities, if the base is not compact. The compact case follows easily.

In this article $\mathscr{B} = (E, \pi, B, S)$ denotes a fixed $(n - 1)$-sphere bundle over a connected n-manifold B, where $n \geqslant 2$.

8.13. Proposition VIII: Let $K \subset A \subset O \subset B$, where K is closed and discrete, A is closed and O is open. Let $\sigma: O - K \to E$ be a cross-section.

Then there is a closed discrete set $L \subset B$ and a cross-section $\tau: B - L \to E$ such that $L \cap A = K$, and τ coincides with σ in a neighbourhood of A.

Lemma VII: The proposition is correct if $E = B \times S$.

Proof: Regard S as the unit sphere of an n-dimensional Euclidean space F. Write

$$\sigma(x) = (x, \sigma_1(x)), \qquad x \in O - K.$$

Consider σ_1 as a smooth map $\sigma_1: O - K \to F$.

Let f be a smooth function on B such that $f = 1$ in some neighbourhood W of A and such that $\operatorname{carr} f \subset O$. Then $\varphi = f \cdot \sigma_1$ is a smooth map of $B - K$ into F. By Sard's theorem (cf. [10, p. 47, Lemma 3.2], and sec. 6.3) there is a regular value $b \in F$ for φ with $|b| < 1$.

Define a closed subset, C, of B by $C = \varphi^{-1}(b)$. Clearly $C \subset B - W$. Moreover, since

$$\dim B = \dim(B - K) = n = \dim F,$$

φ is a local diffeomorphism at each $x \in C$. Hence C is discrete.

Finally, by Theorem III, sec. 1.12, there is a diffeomorphism $\psi: F \to F$ such that

$$\psi(b) = 0 \quad \text{and} \quad \psi(z) = z \quad (|z| \geqslant 1).$$

Set $\tau_1 = \psi \circ \varphi$;

$$\tau_1 : B - (K \cup C) \to F - \{0\}$$

and define τ by

$$\tau(x) = \left(x, \frac{\tau_1(x)}{|\tau_1(x)|}\right), \qquad x \in B - (K \cup C).$$

This proves the lemma, with $L = K \cup C$.

Q.E.D.

Proof of the proposition: Let $\{(U_i, \psi_i)\}$ $(i = 1, 2, \ldots)$ be a coordinate representation for $\mathscr{B}$ such that $\{U_i\}$ is a locally finite open cover for B, and each $\bar{U}_i$ is compact. Choose an open cover V_i of B such that

$$\bar{V}_i \subset U_i .$$

Let V be open in B and assume that $A \subset V \subset \bar{V} \subset O$.
Set

$$A_i = \bigcup_{j \leqslant i} \bar{V}_j \cup \bar{V}$$

We shall construct finite sets $K_i \subset A_i$ and open sets $O_i \supset A_i$ together with cross-sections

$$\sigma_i: O_i - (K \cup K_i) \to E, \qquad i = 0, 1, 2, \ldots$$

satisfying the conditions

(i) $K_i \cap A_{i-1} = K_{i-1}$
(ii) $\sigma_i(x) = \sigma_{i-1}(x), \qquad x \in A_{i-1} - (K \cup K_{i-1}),\ i = 2, 3, \ldots$
(iii) $\sigma_i(x) = \sigma(x), \quad x \in \bar{V} - K.$

In fact, set $K_0 = \varnothing$, $O_0 = O$ and $\sigma_0 = \sigma$. Now assume by induction that $\sigma_0, \ldots, \sigma_i$ have been constructed. Restrict σ_i to a cross-section

$$\sigma_i : (O_i - (K \cup K_i)) \cap U_{i+1} \to E.$$

Note that $A_i \cap U_{i+1}$ is closed in U_{i+1} and that

$$(K \cup K_i) \cap U_{i+1} \subset A_i \cap U_{i+1} \subset O_i \cap U_{i+1} .$$

Now Lemma VII yields a closed discrete set $C \subset U_{i+1} - A_i$ and a cross-section

$$\tau_{i+1} : [U_{i+1} - (C \cup K \cup K_i)] \to E$$

which agrees with σ_i near $A_i \cap U_{i+1}$. Choose a neighbourhood, W, of A_i (in B) so that

$$W \cap C = \varnothing \qquad \text{and} \qquad \tau_{i+1}(x) = \sigma_i(x), \qquad x \in U_{i+1} \cap [W - (K \cup K_i)].$$

Next, observe that $X = C \cap \overline{V}_{i+1}$ is finite, because $\overline{V}_{i+1}$ is a compact subset of U_{i+1}. Choose a neighbourhood, $\tilde{W}$, of $\overline{V}_{i+1}$ in U_{i+1} so that $\tilde{W} \cap C = X$. Then set

$$K_{i+1} = K_i \cup X, \qquad O_{i+1} = W \cup \tilde{W}.$$

Since σ_i and τ_{i+1} agree in $W \cap \tilde{W}$, they define a cross-section

$$\sigma_{i+1} : (O_{i+1} - K \cup K_{i+1}) \to E.$$

This closes the induction.

Finally observe that $\bigcup_i K_i$ has finite intersection with each $\overline{V}_i$. Since the V_i are an open cover of B, $\bigcup_i K_i$ is closed and discrete. Hence $L = K \cup \bigcup_i K_i$ is closed and discrete. Now define $\tau: B - L \to E$ by

$$\tau(x) = \sigma_i(x), \qquad x \in \Big(V \cup \bigcup_{j \leqslant i} V_j\Big) - L.$$

Q.E.D.

Corollary: $\mathscr{B}$ admits a cross-section $\tau: B - L \to E$, where L is closed and discrete. If B is compact, then τ has only finitely many singularities.

8.14. Lemma VIII: Let K be a compact subset of a connected manifold B. Then there exists a compact set $L \subset B$ such that $K \subset L$ and none of the components of the open set $B - L$ has compact closure in B.

Proof: Let $\{O_i\}_{i \in \mathbb{N}}$ denote the components of $B - K$:

$$B - K = \bigcup_{i \in \mathbb{N}} O_i .$$

Choose an open set U such that $U \supset K$ and such that $\overline{U}$ is compact.

Now fix an index i. Then the open sets

$$O_i \qquad \text{and} \qquad U \cup \bigcup_{j \neq i} O_j$$

cover B. Since B is connected, they cannot be disjoint. But by construction we have

$$O_i \cap O_j = \varnothing, \qquad i \neq j,$$

whence

$$O_i \cap U \neq \varnothing, \qquad i \in \mathbb{N}.$$

Next, since

$$\bigcup_{i\in\mathbb{N}} O_i \cup U = B$$

and since $\bar{U}$ is compact, we can choose an integer m such that

$$\bar{U} \subset \bigcup_{i=1}^{m} O_i \cup U.$$

We shall show that $U, O_1, \ldots, O_m$ is an open cover of B.

In fact, since the O_j are disjoint we have, for $j > m$, that

$$O_j = (O_j \cap \bar{U}) \cup (O_j \cap (B - \bar{U})) = (O_j \cap U) \cup (O_j \cap (B - \bar{U})).$$

Since the O_j are connected and $O_j \cap U \neq \varnothing$ (as was proved above), it follows that

$$O_j \subset U, \qquad j > m.$$

Hence

$$B = \bigcup_{j=1}^{m} O_j \cup U.$$

Number the O_j so that $\bar{O}_1, \ldots, \bar{O}_p$ are compact and $\bar{O}_{p+1}, \ldots, \bar{O}_m$ are not compact. Set

$$L = B - \bigcup_{j=p+1}^{m} O_j .$$

Then L is closed. Since

$$L \subset U \cup \bigcup_{j=1}^{p} O_j \subset \bar{U} \cup \bigcup_{j=1}^{p} \bar{O}_j ,$$

it follows that L is compact. Clearly, $L \supset K$. Finally,

$$B - L = \bigcup_{j=p+1}^{m} O_j .$$

Hence the components of $B - L$ are the open sets O_j ($j = p + 1, \ldots, m$) and these components do not have compact closures in B.

Q.E.D.

As an immediate consequence of Lemma VIII, we have

Proposition IX: Let B be a connected manifold. Then there is a sequence of compact subsets $A_i \subset B$, and a sequence of open subsets $O_i \subset B$ such that

(i) $O_i \subset A_i \subset O_{i+1}$
(ii) $B = \bigcup_i O_i$
(iii) none of the components of $B - A_i$ has compact closure.

8.15. Consider the $(n - 1)$-sphere bundle $\mathscr{B} = (E, \pi, B, S)$.

Proposition X: Let O be a connected open subset of B and let $a, b \in O$. Then there is a fibre preserving diffeomorphism $\varphi: E \to E$ inducing $\psi: B \to B$ and such that $\psi(a) = b$ and $\varphi(z) = z$ ($z \notin \pi^{-1}(O)$).

Proof: If the bundle is trivial, $E = B \times S$, use Theorem III, sec. 1.12, to find $\psi: B \to B$ with

$$\psi(a) = b \qquad \text{and} \qquad \psi(x) = x, \qquad x \notin O.$$

Set $\varphi = \psi \times \iota$.

In general, use the local triviality of the bundle, and mimic the proof of Theorem III, sec. 1.12.

Q.E.D.

Corollary: Let $U_1, \ldots, U_m$ be open connected subsets of B such that for each i, j

$$U_i = U_j \qquad \text{or} \qquad U_i \cap U_j = \varnothing.$$

Let $x_i \in U_i$ and $y_i \in U_i$ ($i = 1, \ldots, m$) be two sets of m distinct points. Then there is a fibre preserving diffeomorphism $\varphi: E \to E$ such that

$$\psi(x_i) = y_i \qquad \text{and} \qquad \varphi(z) = z, \qquad z \notin \bigcup_i \pi^{-1}U_i .$$

8.16. Proof of Theorem III: According to the corollary to Proposition VIII, sec. 8.13 there is a closed discrete set K and a cross-section τ: $B - K \to E$.

Assume first that B is not compact. Let A_i ($i = 1, 2, \ldots$) be a sequence of compact subsets of B and let O_i ($i = 1, 2, \ldots$) be a sequence of open subsets of B satisfying conditions (i), (ii), and (iii) of Proposition IX, sec. 8.14. We shall construct a family of closed discrete subsets K_i of B and a sequence of cross-sections $\sigma_i: B - K_i \to E$ such that

(1) $K_i \cap A_i = \varnothing$
(2) $\sigma_i(x) = \sigma_{i-1}(x) \quad (x \in O_{i-1})$.

In fact, set $O_1 = A_1 = \varnothing$ and $\sigma_1 = \tau$. Now suppose that K_p, σ_p have been constructed. Since A_{p+1} is compact, $K_p \cap A_{p+1}$ is a finite set, say

$$K_p \cap A_{p+1} = \{x_1, \ldots, x_m\}.$$

Let U_i ($i = 1, \ldots, m$) denote the component of $B - A_p$ which contains x_i (possibly $U_i = U_j$). Since no U_i has compact closure, each U_i meets $B - A_{p+1}$. Now set

$$C_p = K_p - \{x_1, \ldots, x_m\}.$$

Then C_p is discrete and closed, and

$$C_p \subset B - A_{p+1}.$$

Set

$$V_i = U_i - C_p.$$

Then V_i is connected (because $\dim V_i \geqslant 2$) and meets $B - A_{p+1}$. Thus there are distinct points $y_1, \ldots, y_m$ such that

$$y_i \in V_i, \qquad y_i \notin A_{p+1}.$$

The corollary to Proposition X, sec. 8.15, now yields a fibre preserving diffeomorphism $\varphi: E \xrightarrow{\cong} E$ inducing $\psi: B \xrightarrow{\cong} B$ and such that

$$\psi(x_i) = y_i \qquad \text{and} \qquad \varphi(z) = z \qquad \Big(z \notin \bigcup_i \pi^{-1}V_i\Big).$$

Set $K_{p+1} = \psi(K_p)$. Then

$$K_{p+1} = \{y_1, \ldots, y_m\} \cup C_p,$$

whence

$$K_{p+1} \cap A_{p+1} = \varnothing.$$

Define a cross-section $\sigma_{p+1}: B - K_{p+1} \to E$ by

$$\sigma_{p+1} = \varphi \circ \sigma_p \circ \psi^{-1}.$$

Since $O_p \subset A_p$, and $A_p \cap V_i = \varnothing$, for each i, we have

$$\varphi(z) = z, \qquad z \in \pi^{-1}O_p .$$

It follows that

$$\sigma_{p+1}(x) = \sigma_p(x), \qquad x \in O_p .$$

This closes the induction.

Now define $\sigma: B \to E$ by

$$\sigma(x) = \lim_{p\to\infty} \sigma_p(x).$$

Since B is *not* compact, the sequence $\{O_p\}$ is infinite. Hence for all x, choosing p so that $x \in O_p$, we see that $\sigma_j(x)$ is defined when $j > p$. Thus σ is well defined. Obviously, σ is a cross-section in $\mathscr{B}$. This completes the proof of Theorem III, in the case that B is not compact.

Finally, assume that B is compact. Fix a point $a \in B$. Then $B - \{a\}$ is connected and not compact. Hence there is a cross-section $\sigma: B - \{a\} \to E$. Thus σ is a cross-section in $\mathscr{B}$ with a single singularity at a.

Q.E.D.

Problems

$\mathscr{B} = (E, \pi, B, S)$ is an oriented r-sphere bundle.

1. Cohomology of the base. Let $\mathbb{R}[t]$ denote the graded algebra of polynomials in an indeterminate, t, with t homogeneous of degree $r + 1$.

(i) Define a linear operator d in $\mathbb{R}[t] \otimes A(E)$ by

$$d(z \otimes \Psi) = z \otimes \delta\Psi + (-1)^p z \cdot t \otimes \pi^* \oint_S \Psi, \qquad p = \deg \psi.$$

Show that d is homogeneous of degree $+1$ and that $d^2 = 0$.

(ii) Show that π^* induces a linear map $\lambda: A(B) \to \mathbb{R}[t] \otimes A(E)$. Prove that $d \circ \lambda = \lambda \circ \delta$ and that $\lambda_\#(\chi_{\mathscr{B}})$ is represented by $t \otimes 1$.

(iii) Prove that $\lambda_\#: H(B) \to H(\mathbb{R}[t] \otimes A(E))$ is an isomorphism and interpret the Gysin sequence in terms of $H(\mathbb{R}[t] \otimes A(E))$. *Hint*: Consider $\mathbb{R}[t] \otimes A(B) \otimes \wedge\Omega$.

2. Cohomology of the total manifold. (i) With the notation of sec. 8.4 define ∇ in $H(B) \otimes \wedge\Omega$ by

$$\nabla(\alpha \otimes \Omega + \beta \otimes 1) = (-1)^p \alpha \cdot \chi_{\mathscr{B}} \otimes 1, \qquad \alpha \in H^p(B), \quad \beta \in H(B).$$

Show that $\nabla^2 = 0$ and that $H(E) \cong H(H(B) \otimes \wedge\Omega, \nabla)$.

(ii) Let μ_B denote the multiplication by $\chi_{\mathscr{B}}$ on the right. From the Gysin sequence, obtain a short exact sequence

$$0 \to \operatorname{coker} \mu_B \to H(E) \to \ker \mu_B \to 0$$

($\operatorname{coker} \mu_B = H(B)/\operatorname{Im} \mu_B$).

(iii) If $\dim H(B) < \infty$, show that there are homogeneous classes $\alpha_0, \alpha_1, \ldots, \alpha_p$ in $H(B)$ and nonnegative integers $m_0, \ldots, m_p$ such that

(a) $\alpha_0 = 1, \alpha_1, \ldots, \alpha_p$ represent a basis of $\operatorname{coker} \mu_B$.

(b) $\chi_{\mathscr{B}}^{m_0}, \alpha_1 \cdot \chi_{\mathscr{B}}^{m_1}, \ldots, \alpha_p \cdot \chi_{\mathscr{B}}^{m_p}$ is a basis of $\ker \mu_B$. Conclude that the elements $\alpha_i \cdot \chi_{\mathscr{B}}^j$ ($0 \leqslant j \leqslant m_i$, $i = 0, \ldots, p$) form a basis of $H(B)$.

(iv) With the notation of (iii), show that $H(E)$ has a basis of elements $\beta_0, \dots, \beta_p, \gamma_0, \dots, \gamma_p$, where

$$\beta_i = \pi^\# \alpha_i \qquad \text{and} \qquad \oint_S^\# \gamma_i = \alpha_i \cdot \chi_{\mathscr{B}}^{m_i} \qquad (i = 0, 1, \dots, p).$$

(v) Show that the Poincaré polynomials f_B and f_E are given by

$$f_B(t) = \sum_{i=0}^{p} \frac{t^{q_i}(1 - s^{m_i+1})}{1 - s}, \qquad f_E(t) = \sum_{i=0}^{p} t^{q_i} \left(1 + \frac{s^{m_i+1}}{t}\right),$$

where $q_i = \deg \alpha_i$ and $s = t^{r+1}$.

3. Let ξ and η be oriented Riemannian vector bundles with associated sphere bundles ξ_S and η_S. Let $(\xi \oplus \eta)_S$ denote the sphere bundle associated with $\xi \oplus \eta$. Give $(\xi \oplus \eta)_S$ the induced orientation. Prove that

$$\chi_{(\xi \oplus \eta)_S} = \chi_{\xi_S} \cdot \chi_{\eta_S}.$$

4. Hopf fibering. Let S^{2n+1} denote the unit sphere in $\mathbb{C}^{n+1}$ with respect to a Hermitian inner product. Let $\mathscr{B} = (S^{2n+1}, \rho, \mathbb{C}P^n, S^1)$ be the Hopf fibering (cf. problem 10, Chap I).

(i) Define $\omega \in A^1(\mathbb{C}^{n+1})$ by $\omega(z; \zeta) = (1/2\pi) \operatorname{Im}\langle z, \zeta \rangle$. Let ω also denote the restriction of ω to S^{2n+1} and prove that $\oint_{S^1} \omega = -1$.

(ii) Show $\delta\omega = \rho^*\Psi$ for some $\Psi \in A^2(\mathbb{C}P^n)$.

(iii) Prove that $\oint_{S^{2n+1}} \omega \wedge (\delta\omega)^n = 1$. Conclude that $\chi_{\mathscr{B}}^n$ is an orientation class of $\mathbb{C}P^n$. *Hint*: Integrate $(\delta\omega)^{n+1}$ over the unit ball.

(iv) Repeat (i)–(iii) for the fibering $(S^{4n+3}, p, \mathbb{H}P^n, S^3)$.

5. (i) Assume that the total space of the r-sphere bundle is an $(n + r)$-sphere. Show that r is odd, that $n/(r + 1)$ is an integer, q, and that $H(B) \cong \mathbb{R}[t]/t^{q+1}$, where $\deg t = r + 1$ (cf. problem 2).

(ii) Compute the algebras $H(\mathbb{C}P^n)$ and $H(\mathbb{H}P^n)$ (cf. problem 4).

(iii) Show that the inclusions $\mathbb{C}^{k+1} \to \mathbb{C}^{n+1}$ and $\mathbb{H}^{k+1} \to \mathbb{H}^{n+1}$ induce smooth maps $i: \mathbb{C}P^k \to \mathbb{C}P^n$ and $j: \mathbb{H}P^k \to \mathbb{H}P^n$. Compute $i^\#$ and $j^\#$; in particular show that these homomorphisms are surjective.

(iv) Show that $\mathbb{C}P^{2n}$ and $\mathbb{H}P^{2n}$ are irreversible (cf. sec. 3.22).

(v) Regard S^{2n+1} as the unit sphere in $\mathbb{C}^{n+1}$. Let $\varphi: S^{2n+1} \to S^{2n+1}$ satisfy $\varphi(e^{i\theta}z) = e^{i\theta}\varphi(z)$. Show that $\deg \varphi = +1$.

(vi) Construct a commutative diagram

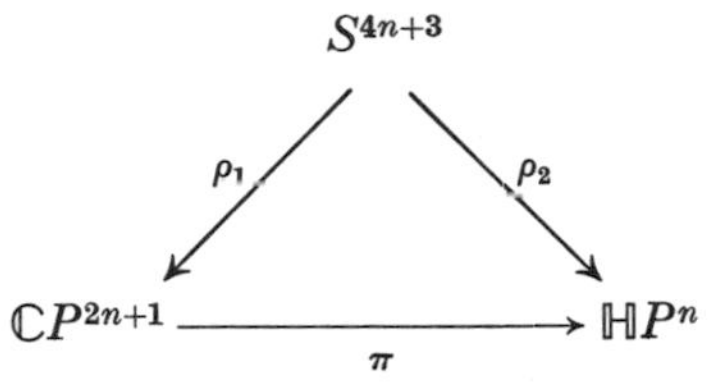

and show that π is the projection of an S^2-sphere bundle. Let χ_1 and χ_2 denote the Euler classes (with respect to ρ_1 and ρ_2). Show that an isomorphism

$$H(\mathbb{H}P^n) \otimes H(S^2) \xrightarrow{\cong} H(\mathbb{C}P^{2n+1})$$

is given by

$$(\alpha \otimes 1 + \beta \otimes \omega_{S^2}) \mapsto \pi^\#\alpha + (\pi^\#\beta)\,\chi_1\,.$$

Prove that $\chi_1^2 = \pi^\#\chi_2$. (*Hint*: Use problem 4.)

6. Assume that r is even.

(i) Show that there are unique classes $\alpha \in H^r(E)$ and $\beta \in H^{2r}(B)$ such that

$$\alpha^2 = \pi^\#\beta \qquad \text{and} \qquad \oint_S^\# \alpha = -1.$$

Show that β does not depend on the choice of the orientation. Conclude that $H(E) \cong H(B) \otimes H(S)$ (as algebras) if and only if $\beta = 0$.

(ii) Suppose $\Omega \in A^r(E)$ satisfies $\oint_S \Omega = -1$ and $\delta\Omega = \pi^*\Psi$. Set $\Omega_1 = \Omega + \frac{1}{2}\pi^*(\oint_S \Omega \wedge \Omega)$. Show that Ω_1 is closed and that

$$(\Omega + \pi^*\Phi)_1 = \Omega_1\,, \qquad \Phi \in A^r(B).$$

Conclude that $\oint_S \Omega_1 \wedge \Omega_1 = 0$.

(iii) Let Ω be as in (ii). Show that $\Omega + \frac{1}{2}\pi^*(\oint_S \Omega \wedge \Omega)$ represents the class α of (i).

7. Orthonormal 2-frames. Let $\mathscr{B} = (V_{n+1,2}\,, \pi, S^n, S^{n-1})$ be the sphere bundle associated with the tangent bundle of S^n.

(i) Identify $V_{n+1,2}$ with the set of ordered pairs of orthonormal vectors in $\mathbb{R}^{n+1}$.

(ii) Fix $e \in S^n$ and define $\sigma\colon S^n - \{e, -e\} \to V_{n+1,2}$ by

$$\sigma(x) = \frac{e - \langle x, e\rangle x}{|e - \langle x, e\rangle x|}.$$

Compute $j_e(\sigma)$ and $j_{-e}(\sigma)$. Conclude that $\chi_{\mathscr{B}} = 2\omega_{S^n}$ (n even) and $\chi_{\mathscr{B}} = 0$ (n odd).

(iii) Compute the algebra $H(V_{n+1,2})$ and show that

$$H(V_{n+1,2}) \cong H(S^{2n-1}) \qquad (n \text{ even})$$

$$H(V_{n+1,2}) \cong H(S^n) \otimes H(S^{n-1}) \qquad (n \text{ odd})$$

(as algebras).

(iv) Let $\hat{\mathscr{B}} = (M, \pi, N, V_{n+1,2})$ be an oriented bundle, with n even, $n = 2m$. Show that π^* induces an isomorphism $H(N) \xrightarrow{\cong} H(\ker \oint_{V_{n+1,2}})$ and obtain a long exact sequence

$$\longrightarrow H^p(N) \longrightarrow H^p(M) \longrightarrow H^{p-2n+1}(N) \xrightarrow{\partial} H^{p+1}(N) \longrightarrow.$$

Interpret ∂ as multiplication by an element $\chi_{\hat{\mathscr{B}}}$ of $H^{2n}(N)$.

8. Let $\mathscr{B}_1 = (E_1, \pi_1, E, S^{r-1})$ be the sphere bundle associated with the vector bundle V_E. Assume r is even, $r = 2m$.

(i) Orient $\mathscr{B}_1$.

(ii) Show that $\hat{\mathscr{B}} = (E_1, \pi \circ \pi_1, B, V_{r+1,2})$ is a smooth oriented bundle.

(iii) Prove that $\oint_{S^r} \chi_{\mathscr{B}^1} = 2$. Conclude that the map

$$H(B) \otimes H(S^r) \to H(E)$$

given by

$$\alpha \otimes 1 + \beta \otimes \omega_{S^r} \mapsto \pi^\# \alpha + \tfrac{1}{2}\pi^\# \beta \cdot \chi_{\mathscr{B}_1}$$

is a linear isomorphism.

9. Let $\mathscr{U} = \{U_i\}$ be an open covering of B, and suppose $\tau_i\colon U_1 \to E$ are local cross-sections.

(i) Construct $\Omega \in A^r(E)$ and $\Phi \in A^{r+1}(B)$ so that Φ represents $\chi_{\mathscr{B}}$, and $\Phi|_{U_i} = \delta\tau_i^*\Omega$.

(ii) If $\mathscr{U}$ contains only k open sets, prove that $(\chi_{\mathscr{B}})^k = 0$.

(iii) Conclude that if $\mathbb{C}P^n$ is covered by n open sets U_i then the restriction of the Hopf fibering to one of the U_i does not admit a cross-section (cf. problem 5).

10. Cochains of differential forms. Let $\{(U_i, \varphi_i)\}$ be a coordinate representation for $\mathscr{B}$. Let $C = \sum_{p,q} C^{p,q}$ denote the algebra of cochains of differential forms defined with respect to the open cover $\{U_i\}$. Fix $e \in S$ and define $\psi_{ij}: U_i \cap U_j \to S$ by $\psi_{ij}(x) = \varphi_{i,x}^{-1}\varphi_{j,x}(e)$. Let $\Omega_S \in A^r(S)$ satisfy $\oint_S \Omega = 1$.

(i) Define $f_1 \in C^{r,1}$ by $f_1(i,j) = \psi_{ij}^*\Omega_S$. Show that $\delta f_1 = 0$. If $\alpha_1 \in H(C, \delta)$ is the class represented by f_1, show that $D_\# \alpha_1 = 0$.

(ii) Show that there are cochains $f_p \in C^{r+1-p,p}$ ($p \geqslant 2$) such that (a) $\nabla(\sum_{i \geqslant 1} f_i) = 0$ and (b) the class represented by $\sum_{i \geqslant 1} f_i$ in $H(C, \nabla)$ corresponds to $\chi_\mathscr{B}$ under the isomorphism $H(B) \cong H(C, \nabla)$.

(iii) Construct a homomorphism $H(\mathscr{N}) \to H(M)$ (cf. problem 25, Chap. V). If $\alpha_1 \notin \operatorname{Im} D_\#$, show that $\chi_\mathscr{B}$ is not in the image of this homomorphism.

11. Stiefel manifolds. (i) By considering the sphere bundle associated with the vertical bundle, obtain from a sphere bundle (E_1, π_1, E_0, S^r) a sequence of sphere bundles $(E_i, \pi_i, E_{i-1}, S^{r-i+1})$ ($i = 1, 2, \ldots$).

(ii) Show that an orientation in the first bundle induces orientations in the following bundles.

(iii) If $E_0 =$ (point), identify E_i with the set of all ordered sequences $(x_1, \ldots, x_i)$ of orthonormal vectors in $\mathbb{R}^{r+1}$. The manifold E_i is called the *ith Stiefel manifold* and is denoted by $V_{r+1,i}$.

(iv) In the sequence of (i) construct bundles

$$(E_i, \pi_{ij}, E_j, V_{r+1-j,i-j}), \qquad \text{where} \quad \pi_{ij} = \pi_{j+1} \circ \cdots \circ \pi_i .$$

(v) Compute the algebra $H(V_{r+1,i})$ and identify the homomorphisms induced by the inclusions $V_{k,i} \to V_{l,i}$ and projections $V_{k,i} \to V_{k,j}$ ($j < i \leqslant k < l$).

(vi) Identify $V_{n,n}$ with the set O^n of isometries of $\mathbb{R}^n$. Thus make O^n into a manifold. Show that the maps $O^n \times O^n \to O^n$ and $O^n \to O^n$ given by $(\varphi, \psi) \mapsto \varphi \circ \psi$ and $\varphi \mapsto \varphi^{-1}$ are smooth.

12. Let $\xi = (M, \rho, N, F)$ be a complex vector bundle with Hermitian metric. Let $\xi_S = (S_M, \rho, N, S)$ denote the associated sphere bundle.

(i) Show that the vertical bundle V_M is a complex bundle with Hermitian metric. Use this to construct a complex bundle η over S_M such that $\eta_{\mathbb{R}} \oplus \epsilon = V_{S_M}$.

(ii) Modify the construction in problem 11 to obtain complex Stiefel manifolds and compute their cohomology. Identify the group of unitary maps of $\mathbb{C}^n$ as a manifold, and compute its cohomology.

(iii) Repeat (i) and (ii) with $\mathbb{C}$ replaced by $\mathbb{H}$.

13. Cohomology with compact carrier. Let $\mathscr{B} = (E, \pi, B, S)$ be an oriented r-sphere bundle.

(i) Establish a long exact sequence for cohomology with compact carrier.

(ii) Show that the map $D_c: H_c(B) \to H_c(B)$ is given by $D_c\alpha = \alpha * \chi_{\mathscr{B}}$, where $\chi_{\mathscr{B}}$ denotes the Euler class.

14. Relative cohomology. Let $\mathscr{B} = (E, \pi, B, S)$ be an oriented r-sphere bundle. Let M be a closed submanifold of B and assume that there is a fixed trivialization $\varphi: U \times S \xrightarrow{\cong} \pi^{-1}U$ for $\mathscr{B}$, where U is a neighbourhood of M.

(i) Let $A(B, M)$ be the ideal in $A(B)$ of forms with carrier in $B - M$. Denote its cohomology by $H(B, M)$. Establish an exact triangle

$$\begin{array}{ccc} H(B) & \xrightarrow{\quad i^* \quad} & H(M) \\ {\scriptstyle \lambda_\#}\nwarrow & & \swarrow{\scriptstyle \partial} \\ & H(B, M) & \end{array}$$

(cf. problem 9, iii, Chap. V).

(ii) Let $\hat{\Omega} \in A^r(S)$ satisfy $\oint_S \hat{\Omega} = -1$. Show that there are elements $\Omega \in A^r(E)$ and $\Phi \in A^{r+1}(B)$ such that

$$\oint_S \Omega = -1, \qquad \delta\Omega = \pi^*\Phi$$

and

$$\mathrm{carr}(\varphi^*\Omega - 1 \times \hat{\Omega}) \subset (U - M) \times S.$$

Conclude that $\Phi \in A(B, M)$. Show that the class in $H(B, M)$ represented by Φ depends only on $\mathscr{B}$ and φ. It is called the *relative Euler class* and is denoted by $\chi_{\mathscr{B},\varphi}$.

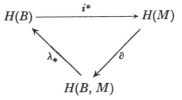

(iii) Show that $\lambda_\# \chi_{\mathscr{B},\varphi} = \chi_{\mathscr{B}}$.

(iv) Construct an example to show that $\chi_{\mathscr{B},\varphi}$ is *not* independent of φ.

(v) Let σ be a cross-section of $\mathscr{B}$ defined in a neighbourhood of M. Construct $\Omega \in A^r(E)$, $\Phi \in A^{r+1}(B)$ so that

$$\oint_S \Omega = -1, \qquad \delta\Omega = \pi^*\Phi$$

and in some neighbourhood V, of M, Ω is closed and represents $-\omega_{\sigma_V}$. (σ_V denotes the restriction of σ to V.)

Conclude that Φ is closed and $\Phi \in A(B, M)$. Show that the class $\chi_{\mathscr{B},\sigma} \in H(B, M)$, represented by Φ, depends only on $\mathscr{B}$ and σ. Show that $\lambda_\# \chi_{\mathscr{B},\sigma} = \chi_{\mathscr{B}}$.

(vi) If τ is another cross-section defined in a neighbourhood of M, let τ_M, σ_M be the restrictions of τ, σ to M. Then $[\tau_M, \sigma_M] \in H^r(M)$. Show that

$$\partial([\tau_M, \sigma_M]) = \chi_{\mathscr{B},\sigma} - \chi_{\mathscr{B},\tau}.$$

15. Manifolds-with-boundary. (i) Define (the notion of) an oriented r-sphere bundle $\mathscr{B} = (E, \pi, B, S)$ over a manifold-with-boundary $(B, \partial B)$. In particular E is a manifold-with-boundary. Show that $\mathscr{B}$ restricts to an oriented smooth bundle $\partial\mathscr{B} = (\partial E, \pi, \partial B, S)$.

(ii) Suppose $(B, \partial B)$ is a compact, oriented n-manifold-with-boundary (and $\partial B \neq \varnothing$). Let $\mathscr{B} = (E, \pi, B, S)$ be an oriented $(n-1)$-sphere bundle over B. Show that $\mathscr{B}$ admits a cross-section $\tau: B \to E$.

(iii) With the notation of (ii) let $\sigma: \partial B \to \partial E$ be a cross-section in $\partial\mathscr{B}$. Show that $\chi_{\mathscr{B},\sigma} \in H_c^n(B)$. The number $\oint_B^\# \chi_{\mathscr{B},\sigma}$ is denoted by $j_{\mathscr{B}}(\sigma)$ and is called the *index* of σ with respect to $\mathscr{B}$. Show that the "index at an isolated singularity" is a special case of this index.

(iv) If the cross-section, σ, of (iii) extends to a cross-section in $\mathscr{B}$, show that $j_{\mathscr{B}}(\sigma) = 0$.

(v) Let τ, σ be the cross-sections of (ii), (iii) and let $\hat{\tau}$ be the restriction of τ to ∂B. With the notation of problem 14, show that

$$\partial[\hat{\tau}, \sigma] = \chi_{\mathscr{B},\sigma}$$

and conclude that

$$\int_{\partial B}^\# \sigma^\#(\omega_{\hat{\tau}}) = \int_{\partial B}^\# [\hat{\tau}, \sigma] = j_{\mathscr{B}}(\sigma).$$

(vi) Let $a_1, ..., a_q \in B - \partial B$. Show that σ extends to a cross-section $\sigma_B: B - \{a_1, ..., a_q\} \to E$. Prove that

$$j_{\mathscr{B}}(\sigma) = \sum_{j=1}^{q} j_{a_i}(\sigma_B).$$

Conclude that σ extends to a global cross-section if and only if $j_{\mathscr{B}}(\sigma) = 0$.

(vii) Let $\Omega \in A^{n-1}(E)$ satisfy $\oint_S \Omega = -1$ and $\delta\Omega = \pi^*\Phi$. Show that

$$j_{\mathscr{B}}(\sigma) = \int_B \Phi - \int_{\partial B} \sigma^*\Omega.$$

16. Two manifolds-with-boundary. Let $\hat{\mathscr{B}} = (\hat{E}, \hat{\pi}, \hat{B}, S)$ be an oriented sphere bundle. Suppose $\hat{B}$ is obtained from two manifolds-with-boundary, $(B, \partial B)$ and $(B_1, \partial B_1)$ by identifying ∂B and ∂B_1 via some diffeomorphism.

Regard ∂B and ∂B_1 as equal to a manifold M which is a submanifold of $\hat{B}$.

(i) Show that $\hat{\mathscr{B}}$ restricts to smooth bundles $\mathscr{B} = (E, \pi, B, S)$, $\mathscr{B}_1 = (E_1, \pi_1, B_1, S)$, and $\mathscr{B}_M = (E_M, \pi_M, M, S)$. Show that $\partial\mathscr{B} = \mathscr{B}_M = \partial\mathscr{B}_1$.

(ii) Suppose σ is a cross-section in $\mathscr{B}_M$. Then $\chi_{\mathscr{B},\sigma} \in H(B, M)$ and $\chi_{\mathscr{B}_1,\sigma} \in H(B_1, M)$. Construct a map

$$+ : H(B, M) \times H(B_1, M) \to H(\hat{B})$$

and show that

$$\chi_{\mathscr{B},\sigma} + \chi_{\mathscr{B}_1,\sigma} = \chi_{\hat{\mathscr{B}}}.$$

(iii) Assume $\dim S = \dim \hat{B} - 1$, and that $\hat{B}$ is compact and oriented. Show that for suitable orientations

$$j_{\mathscr{B}}(\sigma) + j_{\mathscr{B}_1}(\sigma) = \int_{\hat{\mathscr{B}}}^{\#} \chi_{\hat{\mathscr{B}}}.$$

(iv) Apply this to the case that $\hat{B} = S^n$, $M = S^{n-1}$.

Chapter IX

Cohomology of Vector Bundles

§1. The Thom isomorphism

9.1. The main theorem. Let $\xi = (E, \pi, B, F)$ be an oriented fibre bundle. Recall from sec. 7.13 that integration over the fibre induces a linear map

$$\oint_F^{\#} : H_F(E) \to H(B)$$

homogeneous of degree $-r$ where r is the dimension of F. The purpose of this section is to prove:

Theorem I: If the manifold F is contractible, then the map $\oint_F^{\#}$ is a linear isomorphism.

Remark: If B is compact, connected, and oriented, and ξ is a vector bundle, Theorem I follows immediately from Corollary I to Theorem I, sec. 7.14. In fact, in this case, $A_F(E) = A_c(E)$ and so $\oint_F^{\#}$ becomes a map

$$\oint_F^{\#} : H_c(E) \to H(B).$$

In view of the corollary, this map is dual (with respect to the Poincaré scalar products) to the isomorphism

$$\pi^{\#} : H(E) \xleftarrow{\cong} H(B)$$

of Example 3, sec. 5.5. Hence $\oint_F^{\#}$ is an isomorphism.

The proof of Theorem I is preceded by four lemmas.

Lemma I: The theorem holds for the trivial bundle $\xi = (F, \pi, p, F)$ where p is a single point.

Proof: In this case (because F is contractible) $H^0(F) = \mathbb{R}$ and $H^+(F) = 0$. Hence it follows from Theorem I, sec. 5.12, that

$$H_c(F) = H_c^r(F).$$

Thus Theorem II, sec. 5.13, implies that $\oint_F^\#$ is an isomorphism,

$$\oint_F^\# = \int_F^\# : H_c(F) \xrightarrow{\cong} \mathbb{R}.$$

Q.E.D.

Lemma II: The theorem holds for the trivial bundle

$$\xi = (\mathbb{R}^n \times F, \pi, \mathbb{R}^n, F).$$

Proof: Let i: $\{0\} \to \mathbb{R}^n$ and ρ: $\mathbb{R}^n \to \{0\}$ be the inclusion and projection maps. Clearly

$$i \times \iota : \{0\} \times F \to \mathbb{R}^n \times F$$

and

$$\rho \times \iota : \mathbb{R}^n \times F \to \{0\} \times F$$

are bundle maps restricting to diffeomorphisms in the fibres. Hence (cf. the corollary to Proposition VI, sec. 7.10) they induce linear maps

$$(i \times \iota)_F^\# : H_F(\mathbb{R}^n \times F) \to H_F(\{0\} \times F)$$

and

$$(\rho \times \iota)_F^\# : H_F(\{0\} \times F) \to H_F(\mathbb{R}^n \times F).$$

We show first that these maps are inverse isomorphisms. Clearly,

$$(i \times \iota)_F^\# \circ (\rho \times \iota)_F^\# = \iota.$$

On the other hand, let H: $\mathbb{R} \times \mathbb{R}^n \to \mathbb{R}^n$ be a homotopy connecting $i \circ \rho$ and $\iota_{\mathbb{R}^n}$. Then the map

$$(H \times \iota)^* : A(\mathbb{R} \times \mathbb{R}^n \times F) \leftarrow A(\mathbb{R}^n \times F)$$

restricts to a map

$$(H \times \iota)_F^* : A_F(\mathbb{R} \times \mathbb{R}^n \times F) \leftarrow A_F(\mathbb{R}^n \times F).$$

It follows from the remark of sec. 5.2 that the homotopy operator associated with $H \times \iota$ restricts to a linear map

$$h_F : A_F(\mathbb{R}^n \times F) \to A_F(\mathbb{R}^n \times F).$$

Moreover,

$$\iota - (\rho \times \iota)_F^* \circ (i \times \iota)_F^* = \delta \circ h_F + h_F \circ \delta$$

and so

$$(\rho \times \iota)_F^\# \circ (i \times \iota)_F^\# = \iota.$$

Now apply Corollary I of Proposition VIII, sec. 7.12, to obtain the commutative diagram (note that $H_c(F) = H_F(\{0\}) \times F)$)

$$\begin{array}{ccc} H_c(F) & \xleftarrow[\cong]{(i\times\iota)_F^\#} & H_F(\mathbb{R}^n \times F) \\ {\scriptstyle \int_F^\#}\big\downarrow & & \big\downarrow{\scriptstyle \oint_F^\#} \\ \mathbb{R} = H(\{0\}) & \xleftarrow[i^\#]{} & H(\mathbb{R}^n). \end{array}$$

Here $i^\#$ is an isomorphism (cf. Example 1, sec. 5.5). $\int_F^\#$ is an isomorphism by Lemma I. Hence $\oint_F^\#$ is an isomorphism.

Q.E.D.

Lemma III: Let U, V be an open cover of B. Then the short exact sequence (cf. sec. 5.4)

$$0 \to A(E) \to A(\pi^{-1}(U)) \oplus A(\pi^{-1}(V)) \to A(\pi^{-1}(U \cap V)) \to 0$$

restricts to a short exact sequence

$$0 \to A_F(E) \to A_F(\pi^{-1}(U)) \oplus A_F(\pi^{-1}(V)) \to A_F(\pi^{-1}(U \cap V)) \to 0.$$

Moreover, the diagram

$$\begin{array}{ccccccccc} 0 & \longrightarrow & A_F(E) & \longrightarrow & A_F(\pi^{-1}(U)) \oplus A_F(\pi^{-1}(V)) & \longrightarrow & A_F(\pi^{-1}(U \cap V)) & \longrightarrow & 0 \\ & & \big\downarrow{\scriptstyle \oint_F} & & \big\downarrow{\scriptstyle \oint_F \oplus \oint_F} & & \big\downarrow{\scriptstyle \oint_F} & & \\ 0 & \longrightarrow & A(B) & \longrightarrow & A(U) \oplus A(V) & \longrightarrow & A(U \cap V) & \longrightarrow & 0 \end{array}$$

is commutative.

Proof: The same argument is used as in the proof of Lemma I, sec. 5.4. (The commutativity follows from Corollary I to Proposition VIII, sec. 7.12).

Q.E.D.

The following lemma is obvious:

Lemma IV: Suppose B is a disjoint union of open sets,

$$B = \bigcup_\alpha B_\alpha .$$

Then the diagram

$$\begin{array}{ccc} A_F(E) & \xrightarrow[\cong]{\varphi} & \prod_\alpha A_F(\pi^{-1}(B_\alpha)) \\ \oint_F \downarrow & & \downarrow \Pi_\alpha \oint_F \\ A(B) & \xrightarrow[\psi]{\cong} & \prod_\alpha A(B_\alpha) \end{array}$$

commutes, where φ and ψ are the isomorphisms defined by

$$(\varphi\Phi)_\alpha = \Phi|_{\pi^{-1}(B_\alpha)}, \qquad \Phi \in A_F(E)$$

and

$$(\psi\Psi)_\alpha = \Psi|_{B_\alpha}, \qquad \Psi \in A(B).$$

Proof of Theorem I: Consider first the case that ξ is the product bundle, $E = B \times F$. Then the theorem follows from Lemmas II, III, and IV in precisely the same way as Theorem I in sec. 5.12 and Theorem VI in sec. 5.20 are proved.

In the general case cover B by finitely many trivializing neighbourhoods $U_1, \dots, U_m$ (cf. sec. 1.13). Then induction on m (via a Mayer–Vietoris argument using Lemma III) completes the proof.

Q.E.D.

Definition: The isomorphism

$$Th: H(B) \xrightarrow{\cong} H_F(E)$$

inverse to $\oint_F^\#$ is called the *Thom isomorphism* of ξ. It is homogeneous of degree r. The cohomology class $\theta_\xi \in H_F^r(E)$ given by

$$\theta_\xi = Th(1)$$

is called the *Thom class of* ξ.

Next recall that $A_F(E)$ is an ideal in $A(E)$. Thus multiplication in $A(E)$ restricts to bilinear maps

$$A(E) \times A_F(E) \to A_F(E), \qquad A_F(E) \times A(E) \to A_F(E).$$

They induce bilinear maps

$$H(E) \times H_F(E) \to H_F(E), \qquad H_F(E) \times H(E) \to H_F(E)$$

which make $H_F(E)$ into a graded left and right $H(E)$ module. These maps will be denoted by

$$(\alpha, \beta) \mapsto \alpha * \beta, \qquad (\beta, \alpha) \mapsto \beta * \alpha, \qquad \alpha \in H(E), \quad \beta \in H_F(E)$$

(cf. sec. 5.9).

Proposition I: The Thom isomorphism satisfies

$$Th(\alpha \cdot \beta) = \pi^{\#}\alpha * Th(\beta), \qquad \alpha, \beta \in H(B).$$

Proof: Use Proposition IX, sec. 7.13, to obtain

$$\oint_F^{\#} (\pi^{\#}\alpha * Th\,\beta) = \alpha \cdot \oint_F^{\#} Th\,\beta = \alpha \cdot \beta.$$

Now apply Th to both sides of this relation.

Q.E.D.

Corollary: The Thom isomorphism is given by

$$Th(\alpha) = \pi^{\#}\alpha * \theta_{\xi}, \qquad \alpha \in H(B).$$

Next, let $\hat{\xi} = (\hat{E}, \hat{\pi}, \hat{B}, \hat{F})$ be a second oriented bundle with contractible fibre. Suppose $\varphi: E \to \hat{E}$ is a smooth fibre preserving map, inducing $\psi: B \to \hat{B}$ and restricting to orientation preserving diffeomorphisms

$$\varphi_x: F_x \xrightarrow{\cong} \hat{F}_{\psi(x)}, \qquad x \in B.$$

Then (cf. sec. 7.12)

$$\psi^* \circ \oint_{\hat{F}} = \oint_F \circ \varphi_F^* .$$

This yields

Proposition II: With the hypotheses above, the diagram

$$\begin{array}{ccc} H_F(E) & \xleftarrow{\varphi_F^{\#}} & H_{\hat{F}}(\hat{E}) \\ {\scriptstyle Th_{\xi}}\uparrow\cong & & \cong\uparrow{\scriptstyle Th_{\hat{\xi}}} \\ H(B) & \xleftarrow[\psi^{\#}]{} & H(\hat{B}) \end{array}$$

commutes. In particular

$$\varphi_F^{\#}(\theta_\xi) = \theta_\xi .$$

9.2. Corollaries of Theorem I. Corollary I: Let $\Omega \in A_F(E)$ be a closed form such that

$$\oint_F \Omega = \delta\Phi$$

for some $\Phi \in A(B)$. Then there exists a differential form $\Omega_1 \in A_F(E)$ such that $\Omega = \delta\Omega_1$.

Corollary II: Let $\Phi \in A_F^r(E)$ be closed. Then the function $\oint_F \Phi \in \mathscr{S}(B)$ is constant on each component of B. Moreover,

$$\oint_F \Phi = 1$$

if and only if Φ represents the Thom class.

Proof: Observe that

$$\delta \oint_F \Phi = \oint_F \delta\Phi = 0.$$

Q.E.D.

Corollary III: Let $\Phi \in A_F^r(E)$ be a closed r-form such that for every $x \in B$

$$j_x^* \Phi \in \delta(A_c^{r-1}(F_x))$$

($j_x : F_x \to E$ denotes the inclusion map). Then there exists an $(r-1)$-form $\Psi \in A_F^{r-1}(E)$ such that

$$\Phi = \delta\Psi.$$

Proof: It follows from the hypothesis that

$$\left(\oint_F \Phi\right)(x) = \int_{F_x} j_x^* \Phi = 0, \qquad x \in B.$$

Thus Corollary I implies that

$$\Phi = \delta\Psi$$

for some $\Psi \in A_F^{r-1}(E)$.

Q.E.D.

Example: Consider the trivial bundle $\xi = (B \times F, \pi, B, F)$. Define an orientation in the bundle by orienting F. Choose an r-form, Ω, on F with compact carrier and satisfying

$$\int_F \Omega = 1.$$

Then $1 \times \Omega$ represents the Thom class of ξ.

In fact, $1 \times \Omega$ is closed, and satisfies

$$\left(\oint_F (1 \times \Omega)\right)(x) = \left(\int_F \Omega\right)(x) = 1, \qquad x \in B.$$

Hence, by Corollary II, it represents the Thom class.

§2. The Thom class of a vector bundle

The notation established in sec. 9.3 is fixed throughout this article.

9.3. Notation. Let $\xi = (E, \pi, B, F)$ be an oriented vector bundle of rank $r (r \geqslant 2)$. Since F is contractible, Theorem I implies that the map

$$\oint_F^{\#} : H_F(E) \to H(B)$$

is an isomorphism. Hence the Thom isomorphism, $Th = (\oint_F^{\#})^{-1}$ is defined (cf. sec. 9.1). In particular we have the Thom class

$$\theta_\xi = Th(1).$$

Now introduce a Riemannian metric g in ξ. Recall from sec. 3.10 (Examples 5 and 6) the definition of the deleted bundle and associated sphere bundle

$$\dot{\xi} = (\dot{E}, \dot{\pi}, B, \dot{F}) \qquad \text{and} \qquad \xi_S = (E_S, \pi_S, B, S).$$

We have the inclusion and projection:

$$i: E_S \to E \qquad \text{and} \qquad \rho: \dot{E} \to E_S .$$

Finally, every $\epsilon > 0$ determines the open subset $E_\epsilon \subset E$ given by

$$E_\epsilon = \{z \in E \mid |z| < \epsilon\}.$$

Let F_ϵ denote the open ϵ-ball in F with respect to an inner product, and let π_ϵ denote the restriction of π to E_ϵ. Then

$$\xi_\epsilon = (E_\epsilon, \pi_\epsilon, B, F_\epsilon)$$

is a smooth oriented bundle with contractible fibre (prove this via a Riemannian coordinate representation for ξ).

The inclusion $k: E_\epsilon \to E$ is a smooth fibre preserving map; hence (cf. the end of sec. 7.10)

$$(k_{F_\epsilon})_*: A_{F_\epsilon}(E_\epsilon) \to A_F(E)$$

is defined.

Lemma V: For each $\epsilon > 0$ there exists a representative Φ of the Thom class θ_ξ such that $\operatorname{carr} \Phi \subset E_\epsilon$. Two such representatives Φ, Ψ satisfy

$$\Phi - \Psi = \delta\Omega,$$

where $\Omega \in A_F^{r-1}(E)$ and $\operatorname{carr} \Omega \subset E_\epsilon$.

Proof: Since k is orientation preserving, it follows from the definitions that the diagram

$$\begin{array}{ccc} A_{F_\epsilon}(E_\epsilon) & \xrightarrow{(k_{F_\epsilon})_*} & A_F(E) \\ & {\scriptstyle \oint_{F_\epsilon}} \searrow \quad \swarrow {\scriptstyle \oint_F} & \\ & A(B) & \end{array}$$

commutes. Pass to cohomology, and apply Theorem I, sec. 9.1, to both ξ and ξ_ϵ to conclude that $(k_{F_\epsilon})_\#$ is an isomorphism. The lemma follows.

Q.E.D.

9.4. The canonical map α_ξ. In this section we define a canonical linear map

$$\alpha_\xi \colon H(E_S) \to H_F(E)$$

homogeneous of degree 1. Choose a smooth function f on E such that

$$\operatorname{carr} f \subset \dot{E} \qquad \text{and} \qquad f - 1 \in \mathscr{S}_F(E).$$

Then $\delta f \in A_{\dot{F}}(\dot{E})$. Hence a linear map, homogeneous of degree 1,

$$A(E_S) \to A_F(E)$$

is given by

$$\Phi \mapsto (-1)^{p+r-1}\, \delta f \wedge \rho^*\Phi, \qquad \Phi \in A^p(E_S).$$

This map commutes with the exterior derivative and so it induces a linear map

$$\alpha_\xi \colon H(E_S) \to H_F(E).$$

The map α_ξ is independent of the choice of f. In fact, assume that $g \in \mathscr{S}(E)$ is another function satisfying the conditions above. Then $f - g \in \mathscr{S}_F(\dot{E})$. Thus for every closed form Φ on E_S

$$\delta f \wedge \rho^*\Phi - \delta g \wedge \rho^*\Phi = \delta\big((f-g) \wedge \rho^*\Phi\big) \in \delta(A_F(E)).$$

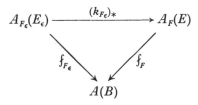

Example: Consider the bundle $\xi = (F, \pi, p, F)$ where p is a single point. Then $E_S = S$, $E = F$ and $H_F(E) = H_c(F)$. Moreover, the diagram

$$\begin{array}{ccc} H^{r-1}(S) & \xrightarrow{\alpha_\xi} & H_c^r(F) \\ {\scriptstyle \rho^\#}\big\downarrow{\scriptstyle\cong} & & {\scriptstyle\cong}\big\downarrow{\scriptstyle \oint_F^\#} \\ H^{r-1}(\dot{F}) & \xrightarrow[\alpha_F]{\cong} & \mathbb{R} \end{array}$$

commutes (cf. secs. 6.7 and 6.8). In particular, α_ξ is an isomorphism in this case.

Proposition III: The diagram

$$\begin{array}{ccc} & & H_F(E) \\ & {\scriptstyle \alpha_\xi}\nearrow & \\ H(E_S) & & {\scriptstyle\cong}\big\uparrow{\scriptstyle Th} \\ & {\scriptstyle \oint_S^\#}\searrow & \\ & & H(B) \end{array}$$

commutes.

Remark: If B is a single point, the proposition reduces to Proposition VI, sec. 6.8. (Use the example above noting that $i^\# = (\rho^\#)^{-1}$.)

Lemma VI: Let $h \in \mathscr{S}(\mathbb{R})$ be a nondecreasing function such that

$$h(t) = \begin{cases} 0, & t < \frac{1}{4} \\ 1, & t > \frac{1}{2} \end{cases}$$

and define $f \in \mathscr{S}(E)$ by

$$f(z) = h(|\, z\,|), \qquad z \in E.$$

Then, for $\Phi \in A^m(E_S)$,

$$(-1)^{m+r-1} \oint_F \delta f \wedge \rho^* \Phi = \oint_S \Phi.$$

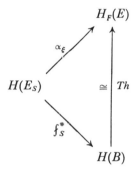

Proof: Use a Riemannian coordinate representation for the bundle to reduce to the case $E = B \times F$ (as Riemannian bundles). Then

$$f(x, y) = h(|y|), \qquad x \in B, \quad y \in F.$$

In particular, for every vector field X on B,

$$i(X)\,\delta f = 0. \tag{9.1}$$

Now let $\Phi \in A^{p,q}(B \times S)$ $(p + q = m)$. It is sufficient to show that

$$i(X_p) \cdots i(X_1)\left[(-1)^{m+r-1} \oint_F \delta f \wedge \rho^*\Phi\right] = i(X_p) \cdots i(X_1) \oint_S \Phi,$$

$$X_1, \ldots, X_p \in \mathscr{X}(B).$$

Consider the X_j as vector fields on $B \times F$ and $B \times S$. In view of (9.1) and Proposition X, sec. 7.13, the relation above is equivalent to

$$(-1)^{q+r-1} \oint_F \delta f \wedge \rho^*(i(X_p) \cdots i(X_1)\Phi) = \oint_S i(X_p) \cdots i(X_1)\Phi.$$

Thus we may assume that $p = 0$. Moreover, we may assume that $q = r - 1$ because otherwise both sides are zero. Then, for $x \in B$, Φ and $\rho^*\Phi$ restrict to $(r-1)$-forms $\Phi_x \in A^{r-1}(S_x)$ and $(\rho^*\Phi)_x \in A^{r-1}(\dot{F}_x)$. Clearly,

$$\rho_x^*\Phi_x = (\rho^*\Phi)_x,$$

where $\rho_x \colon \dot{F}_x \to S_x$ denotes the projection. Finally, observe that f restricts to a function $f_x \in \mathscr{S}(F_x)$ and that

$$(\delta f \wedge \rho^*\Phi)_x = \delta f_x \wedge \rho_x^*\Phi_x.$$

Now, as in the proof of Proposition VI, sec. 6.8, obtain

$$\left(\oint_F \delta f \wedge \rho^*\Phi\right)(x) = \int_{F_x} \delta f_x \wedge \rho_x^*\Phi_x = \int_{S_x} \Phi_x = \left(\oint_S \Phi\right)(x).$$

Q.E.D.

Proof of the proposition: Apply Lemma VI, recalling that

$$Th = \left(\oint_F^{\#}\right)^{-1}.$$

Q.E.D.

9.5. The naturality of α_ξ. Let $\hat{\xi} = (\hat{E}, \hat{\pi}, \hat{B}, F)$ be a second Riemannian bundle with the same typical fibre and let $\varphi: E \to \hat{E}$ be a smooth bundle map which restricts to orientation preserving isomorphisms on the fibres. Then φ determines a fibre preserving map

$$\varphi_S: E_S \to \hat{E}_S$$

given by

$$\varphi_S(z) = \frac{\varphi(z)}{|\varphi(z)|}, \qquad z \in E_S.$$

Proposition IV: With the notation and hypotheses above, the diagram

$$\begin{array}{ccc} H(E_S) & \xleftarrow{\varphi_S^\#} & H(\hat{E}_S) \\ {\scriptstyle \alpha_\xi}\downarrow & & \downarrow{\scriptstyle \alpha_{\hat{\xi}}} \\ H_F(E) & \xleftarrow[\varphi_F^\#]{} & H_F(\hat{E}) \end{array}$$

commutes.

Proof: Let $\hat{f} \in \mathscr{S}(\hat{E})$ satisfy

$$\operatorname{carr} \hat{f} \subset \dot{\hat{E}} \qquad \text{and} \qquad \hat{f} - 1 \in \mathscr{S}_F(\hat{E})$$

and set $f = \varphi^* \hat{f}$. Then f satisfies

$$\operatorname{carr} f \subset \dot{E} \qquad \text{and} \qquad f - 1 \in \mathscr{S}_F(E).$$

Now let $\Phi \in A(\hat{E}_S)$ be closed. Let $\hat{\rho}: \dot{\hat{E}} \to \hat{E}_S$ be the projection. Then

$$\varphi_F^*(\delta \hat{f} \wedge \hat{\rho}^* \Phi) = \delta f \wedge \varphi^* \hat{\rho}^* \Phi.$$

Since

$$\varphi_S \circ \rho = \hat{\rho} \circ \varphi,$$

it follows that

$$\rho^* \circ \varphi_S^* = \varphi^* \circ \hat{\rho}^*$$

and so

$$\varphi_F^*(\delta \hat{f} \wedge \hat{\rho}^* \Phi) = \delta f \wedge \rho^* \varphi_S^* \Phi.$$

Passing to cohomology we obtain the proposition.

Q.E.D.

Corollary: The diagram

$$\begin{array}{ccc} H(E_S) & \xleftarrow{\varphi_S^\#} & H(\hat{E}_S) \\ {\scriptstyle f_S^\#}\downarrow & & \downarrow{\scriptstyle \hat{f}_S^\#} \\ H(B) & \xleftarrow[\psi^\#]{} & H(\hat{B}) \end{array}$$

commutes. (ψ: $B \to \hat{B}$ denotes the map induced by φ.)

Proof: Combine the commutative diagrams of Propositions III and IV, and Corollary I to Proposition VIII, sec. 7.12.

Q.E.D.

9.6. Euler class and Thom class. **Proposition V:** The Thom class θ_ξ, of ξ and the Euler class, χ_S, of ξ_S are connected by the relation

$$\lambda_\# \theta_\xi = \pi^\# \chi_S ,$$

where λ: $A_F(E) \to A(E)$ denotes the inclusion map.

Proof: Let $\Phi \in A^r(B)$ be a closed form representing χ_S. Then, for some $\Omega \in A^{r-1}(E_S)$,

$$\pi_S^* \Phi = \delta \Omega \qquad \text{and} \qquad \oint_S \Omega = -1.$$

Next, choose $f \in \mathscr{S}(E)$ as in Lemma VI, sec. 9.4. Consider the closed form $\Psi \in A^r(E)$ given by

$$\Psi = \pi^* \Phi - \delta(f \cdot \rho^* \Omega).$$

We shall show that

(i) $$\Psi \in A_F(E)$$

and

(ii) $$\oint_F \Psi = 1.$$

In fact, since, in $\dot{E}$,

$$\rho^* \, \delta\Omega = \rho^* \pi_S^* \Phi = \pi^* \Phi,$$

we have

$$\Psi = \pi^* \Phi - \delta f \wedge \rho^* \Omega - f \cdot \rho^* \, \delta\Omega = (1 - f) \cdot \pi^* \Phi - \delta f \wedge \rho^* \Omega.$$

But $f(z) = 1$ for $|z| > \frac{1}{2}$ and so

$$\Psi(z) = 0, \qquad |z| > \tfrac{1}{2};$$

i.e., $\Psi \in A_F(E)$.

Next, using Lemma VI we obtain

$$\oint_F \Psi = \Phi \wedge \oint_F (1-f) - \oint_F \delta f \wedge \rho^*\Omega$$
$$= -\oint_S \Omega = 1.$$

Now Corollary II of Theorem I, sec. 9.2, implies that Ψ represents the Thom class θ_ξ. Thus it follows at once from the definition of Ψ that $\lambda_\#\theta_\xi$ is represented by $\pi^*\Phi$; i.e.

$$\lambda_\#\theta_\xi = \pi^\#\chi_S .$$

Q.E.D.

Corollary I: If σ is a cross-section in ξ, then

$$\chi_S = \sigma^\#\lambda_\#\theta_\xi .$$

In particular, the Euler class of the associated sphere bundle is independent of the Riemannian metric.

Corollary II: Assume that B is a compact and oriented r-manifold. Let σ be a cross-section in ξ. Then

$$\int_B^\# \sigma^\#\lambda_\#\theta_\xi = \int_B^\# \chi_S .$$

Denote by D the connecting homomorphism of the Gysin sequence of the sphere bundle ξ_S.

Theorem II: The diagram

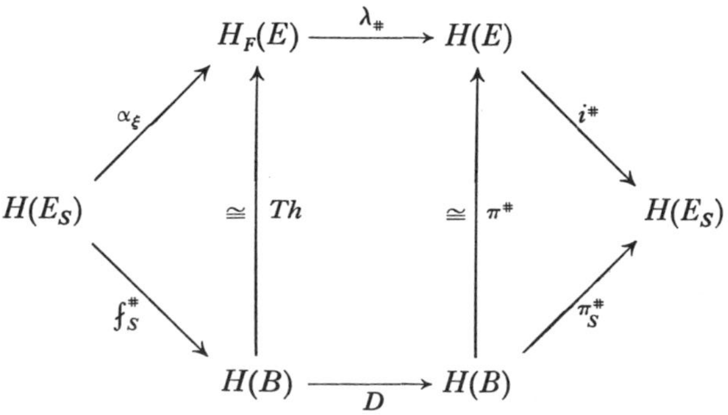

commutes.

Proof: Proposition III, sec. 9.4, states that the left-hand triangle commutes. The commutativity of the right-hand triangle follows from the relation $\pi \circ i = \pi_S$. To show that the square commutes recall from Proposition II, sec. 8.2, and the corollary to Proposition I, sec. 9.1, that

$$D\alpha = \alpha \cdot \chi_S,$$

and

$$Th(\alpha) = \pi^{\#}\alpha * \theta_\xi, \qquad \alpha \in H(B).$$

Now Proposition V yields

$$\lambda_{\#} Th\, \alpha = \pi^{\#}\alpha \cdot \pi^{\#}\chi_S = \pi^{\#} D\alpha.$$

Q.E.D.

Corollary: The triangle

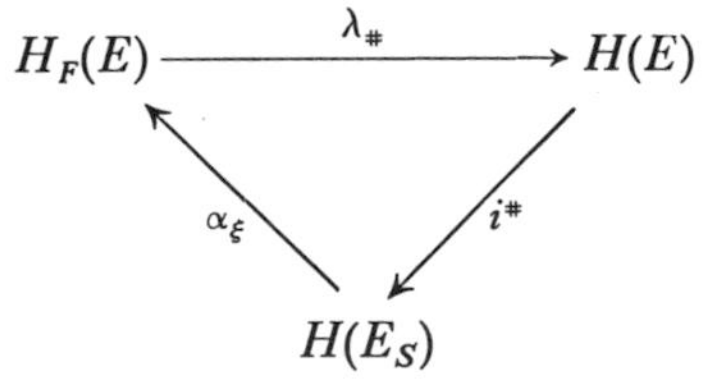

is exact.

Proof: In fact, this triangle corresponds to triangle (8.2) of sec. 8.2 under the isomorphisms of the theorem.

Q.E.D.

§3. Index of a cross-section at an isolated zero

9.7. Index of a cross-section at an isolated zero. Let $\sigma: B \to E$ be a cross-section in a vector bundle $\xi = (E, \pi, B, F)$. A point $a \in B$ is called an *isolated zero* of σ, if $\sigma(a) = 0$ and, for some neighbourhood U of a,

$$\sigma(x) \neq 0, \qquad x \in \dot{U}, \quad \dot{U} = U - \{a\}.$$

Now assume that

(1) The total manifold E is oriented and
(2) $\dim F = \dim B = n \geqslant 2$.

We shall define the index of a cross-section, σ, at an isolated zero a.

Let $\psi: V \times F \xrightarrow{\cong} \pi^{-1}(V)$ be a trivializing map for ξ such that

$$a \in V \subset U \qquad \text{and} \qquad V \cong \mathbb{R}^n.$$

Let $\tilde{\sigma}: V \to F$ be the smooth map satisfying

$$\psi(x, \tilde{\sigma}(x)) = \sigma(x), \qquad x \in V.$$

Orient V and F so that ψ preserves orientations when $V \times F$ is given the product orientation.

Now note that $\tilde{\sigma}^{-1}(0) = a$ and so the local degree

$$\deg_a \tilde{\sigma}$$

of $\tilde{\sigma}$ at a is defined.

On the other hand, introduce a Riemannian metric, g, in ξ, and let $\xi_S = (E_S, \pi_S, B, S)$ be the associated sphere bundle. Give E_S the orientation induced from that of E (cf. Lemma IV, sec. 7.9). Define $\sigma_S: \dot{U} \to E_S$ by

$$\sigma_S(x) = \frac{\sigma(x)}{|\sigma(x)|_x}, \qquad x \in \dot{U}.$$

Then σ_S is a local cross-section in ξ_S with an isolated singularity at a. Thus (cf. sec. 8.8) the index $j_a(\sigma_S)$ of σ_S at a is defined.

Lemma VII: With the notation above

$$\deg_a \tilde{\sigma} = j_a(\sigma_S).$$

In particular,

(1) $\deg_a \tilde{\sigma}$ is independent of the trivializing map
(2) $j_a(\sigma_S)$ is independent of the Riemannian metric.

Proof: Give the trivial bundle, $U \times F$, the Riemannian metric which makes ψ an isometry. Then $(\psi^{\#}\sigma)_S = (\psi_S)^{\#} \sigma_S$, where ψ_S denotes the restriction of ψ to the unit sphere bundle of $U \times F$. Hence by Lemma VI, sec. 8.8, $j_a((\psi^{\#}\sigma)_S) = j_a(\sigma_S)$. Thus we may replace σ by $\psi^{\#}\sigma$; i.e., we may assume that

(i) $V = U = B$ is an oriented Euclidean space with origin a.
(ii) $E = B \times F, \quad \psi = \iota$.

Then $\tilde{\sigma}$ restricts to a smooth map $\dot{\tilde{\sigma}}: \dot{B} \to \dot{F}$ and

$$\deg_a \tilde{\sigma} = \alpha_B \circ \dot{\tilde{\sigma}}^{\#} \circ \alpha_F^{-1}(1).$$

On the other hand, let $\Omega \in A^{n-1}(\dot{F})$ represent $\alpha_F^{-1}(1)$. Let $S_x \subset \dot{F}$ be the unit sphere with respect to the inner product $g(x)$ $(x \in B)$ and let $i : E_S \to \dot{E}$ be the inclusion. Then according to Proposition VI, sec. 6.8,

$$\int_{S_x} i_x^*\Omega = 1, \qquad x \in B.$$

Hence

$$\oint_S i^*(1 \times \Omega) = 1.$$

Moreover, by Proposition VI, sec. 2.17, E_S is diffeomorphic to $B \times S$ (S the unit sphere of F with respect to a fixed inner product). Hence

$$H^{n-1}(E_S) \cong H^{n-1}(S) \cong \mathbb{R}.$$

It follows that $i^*(1 \times \Omega)$ represents the *unique* cohomology class in $H^{n-1}(E_S)$ whose fibre integral is 1. In particular, if $\tau: B \to E_S$ is any cross-section, then $i^*(1 \times \Omega)$ represents ω_τ (cf. sec. 8.6).

Finally, it follows from the definition that

$$j_a(\sigma_S) = \alpha_B \circ \sigma_S^{\#}(\omega_\tau).$$

Let $\tilde{\sigma}_S: \dot{B} \to \dot{F}$ be defined by

$$(x, \tilde{\sigma}_S(x)) = \sigma_S(x).$$

Then $\sigma_S^{\#}(\omega_\tau)$ is represented by $\tilde{\sigma}_S^{*}(\Omega)$; i.e.,

$$\sigma_S^{\#}(\omega_\tau) = \tilde{\sigma}_S^{\#}(\alpha_F^{-1}(1)).$$

Since $\tilde{\sigma}_S$ is homotopic to $\dot{\tilde{\sigma}}$, we obtain

$$\deg_a \tilde{\sigma} = \alpha_B \circ \tilde{\sigma}_S^{\#} \circ \alpha_F^{-1}(1) = j_a(\sigma_S).$$

Q.E.D.

Definition: The integer

$$\deg_a \tilde{\sigma} = j_a(\sigma_S)$$

is called the *index* of σ at a and is denoted by $j_a(\sigma)$.

If a cross-section σ has only finitely many zeros $a_1, ..., a_m$ and if $j_\nu(\sigma)$ denotes the index of σ at a_ν ($\nu = 1,..., m$) then the integer

$$j(\sigma) = \sum_{\nu=1}^{m} j_\nu(\sigma)$$

is called the *index sum of* σ.

Remark: According to Example 4, sec. 3.22, the total manifold, T_M, of the tangent bundle of a manifold M is always orientable. Thus the index of an isolated zero of a vector field on a manifold M is always defined, even if M is nonorientable.

Finally let $\hat{\xi} = (\hat{E}, \hat{\pi}, \hat{B}, \hat{F})$ be a second vector bundle of rank r and assume that $\hat{E}$ is oriented. Let $\varphi: \hat{\xi} \to \xi$ be a bundle map such that

(1) φ restricts to a diffeomorphism on each fibre
(2) the induced map $\psi: \hat{B} \to B$ is a local diffeomorphism.

Then φ is a local diffeomorphism. From Lemma VII we obtain:

Proposition VI: Let σ be a cross-section in ξ with an isolated zero at b and let $\hat{b} \in \hat{B}$ be a point such that $\psi(\hat{b}) = b$. Then $\varphi^{\#}\sigma$ has an isolated zero at $\hat{b}$ and

$$j_{\hat{b}}(\varphi^{\#}\sigma) = \epsilon j_b(\sigma),$$

where $\epsilon = \pm 1$, depending on whether φ preserves or reverses the orientations.

9.8. Index sum and Thom class. In this section $\xi = (E, \pi, B, F)$ denotes an oriented Riemannian vector bundle over an oriented base B, and E is given the local product orientation. It will be assumed that

$$\dim B = \dim F = n \geqslant 2.$$

Now suppose $\sigma \in \operatorname{Sec} \xi$ has only finitely many zeros $a_1, \dots, a_m$. Assume further that for some compact set $K \subset B$ and some $\epsilon_0 > 0$

$$|\sigma(x)| \geqslant \epsilon_0, \qquad x \in B - K. \tag{9.2}$$

Proposition VII: With the notation and hypotheses above let Ω be a representative of the Thom class such that $\operatorname{carr} \Omega \subset E_{\epsilon_0}$. Then

$$\int_B \sigma^*\Omega = j(\sigma).$$

Proof: We show first that the integral is independent of the choice of Ω. In fact, if Ω_1 is another representative of θ_ξ satisfying $\operatorname{carr} \Omega_1 \subset E_{\epsilon_0}$, we can write

$$\Omega_1 - \Omega = \delta\Phi, \qquad \Phi \in A^{n-1}(E), \quad \operatorname{carr} \Phi \subset E_{\epsilon_0}$$

(cf. Lemma V, sec. 9.3). It follows that $\operatorname{carr} \sigma^*\Phi \subset K$. Thus

$$\sigma^*\Omega_1 - \sigma^*\Omega = \delta\sigma^*\Phi \qquad \text{and} \qquad \sigma^*\Phi \subset A_c^{n-1}(B).$$

Hence

$$\int_B \sigma^*\Omega_1 = \int_B \sigma^*\Omega.$$

Next choose trivializing maps (U_ν, ψ_ν) $(\nu = 1, \dots, m)$ so that

(i) $a_\nu \in U_\nu$
(ii) $U_\nu \cong \mathbb{R}^n$
(iii) $U_\nu \cap U_\mu = \varnothing$ $(\nu \neq \mu)$.

In view of equation (9.2) there exists an ϵ $(0 < \epsilon < \epsilon_0)$ such that $|\sigma(x)| \geqslant \epsilon$ $(x \in B - \bigcup_{\nu=1}^m U_\nu)$. Now choose a representative of the Thom class satisfying $\operatorname{carr} \Omega \subset E_\epsilon$. Then $\operatorname{carr} \sigma^*\Omega \subset \bigcup_{\nu=1}^m U_\nu$ and so

$$\int_B \sigma^*\Omega = \sum_{\nu=1}^m \int_{U_\nu} \sigma^*\Omega.$$

Hence it is sufficient to show that

$$\int_{U_\nu} \sigma^*\Omega = j_\nu(\sigma), \qquad \nu = 1, \ldots, m.$$

In other words in addition to the hypotheses above, we may assume that

(iv) ξ is a trivial bundle, $E = B \times F$
(v) the Riemannian metric is given by a fixed inner product in F
(vi) B is an oriented n-dimensional Euclidean space
(vii) σ has a single zero at the origin, a, of B.

Let $\tau: B \to F$ be the map given by

$$\sigma(x) = (x, \tau(x)), \qquad x \in B$$

and let $\dot{\tau}: \dot{B} \to \dot{F}$ denote the restriction of τ to $\dot{B}$. Then

$$j(\sigma) = \alpha_B \circ \dot{\tau}^\# \circ \alpha_F^{-1}(1).$$

Now let $\Phi \in A^{n-1}(\dot{F})$ represent the element $\alpha_F^{-1}(1)$ and choose $f \in \mathscr{S}(F)$ to be zero near the origin and to satisfy $\operatorname{carr}(1 - f) \subset F_\epsilon$. Then τ^*f is zero near a and $1 - \tau^*f$ has compact carrier. Hence,

$$j(\sigma) = \int_B \delta(\tau^*f) \wedge \dot{\tau}^*\Phi = \int_B \tau^*(\delta f \wedge \Phi) = \int_B \sigma^*\big(1 \times (\delta f \wedge \Phi)\big).$$

On the other hand, since Φ represents $\alpha_F^{-1}(1)$, we have

$$⨍_F \big(1 \times (\delta f \wedge \Phi)\big) = \int_F \delta f \wedge \Phi = 1.$$

Thus, according to Corollary II, to Theorem I, sec. 9.2, $1 \times (\delta f \wedge \Phi)$ represents the Thom class. Since it has carrier in $B \times F_\epsilon = E_\epsilon$, the proposition follows.

Q.E.D.

Corollary: Assume in addition that B is compact and let σ be an arbitrary cross-section with finitely many zeros. Then (cf. sec. 9.6)

$$\int_B^\# \sigma^\# \lambda_\# \theta_\xi = j(\sigma).$$

Example: Let M, B be oriented n-manifolds and let F be an oriented Euclidean n-space ($n \geqslant 2$). Assume

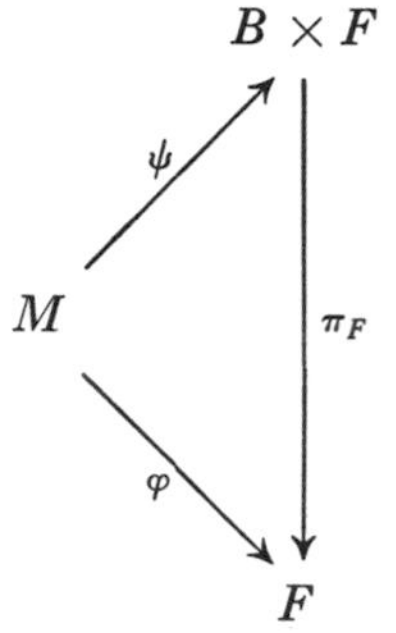

is a commutative diagram of smooth maps such that

$$\varphi^{-1}(0) = a \in M \qquad \text{and} \qquad |\varphi(x)| > \epsilon, \qquad x \in M - K$$

(K a compact subset of M). Choose any representative, Ω, of the Thom class of $B \times F$, subject to the condition: carr $\Omega \subset B \times F_\epsilon$. Then $\psi^*\Omega \in A_c^n(M)$ and

$$\int_M \psi^*\Omega = \deg_a \varphi.$$

In fact, use Lemma V, sec. 9.3, to show that the integral is independent of the choice of Ω (as in the first step of the proof of Proposition VII). Thus, if $\Phi \in A_c^n(F_\epsilon)$ and satisfies $\int_F \Phi = 1$, we can put $\Omega = \pi_F^*\Phi$ and then

$$\int_M \psi^*\Omega = \int_M \psi^*\pi_F^*\Phi = \int_M \varphi^*\Phi.$$

Let $\sigma\colon M \to M \times F$ be the cross-section given by $\sigma(x) = (x, \varphi(x))$. Then, using Proposition VII, we see that

$$\int_M \varphi^*\Phi = \int_M \sigma^*(1 \times \Phi) = j_a(\sigma) = \deg_a \varphi.$$

9.9. Index sum and Euler class. In this section $\xi = (E, \pi, B, F)$ denotes a Riemannian vector bundle such that

(1) B is a connected *compact* n-manifold, $n \geqslant 2$, but not necessarily orientable

(2) $\dim F = \dim B$

(3) E is oriented.

Let σ be a cross-section with finitely many zeros $a_1, \dots, a_m$. We wish to show that the index sum of σ, $j(\sigma)$, is independent of σ.

Consider first the case that B is orientable. Choose an orientation in B. Then it is easy to see that there is a unique orientation in ξ such that the given orientation in E coincides with the local product orientation. This orientation in ξ will be called the *induced bundle orientation.*

Theorem III: Assume that B is oriented and give ξ the induced bundle orientation. Then

$$\int_B^{\#} \chi_S = j(\sigma)$$

where χ_S denotes the Euler class of the associated sphere bundle.

We give two proofs.

Proof I: Consider the associated sphere bundle ξ_S and set

$$\sigma_S(x) = \frac{\sigma(x)}{|\sigma(x)|}, \qquad x \in B - \{a_1, \dots, a_m\}.$$

Then σ_S is a cross-section in ξ_S with finitely many singularities and so Theorem II, sec. 8.11, yields

$$\int_B^{\#} \chi_S = \sum_\nu j_\nu(\sigma_S).$$

But it follows from the definition of $j(\sigma)$ that

$$\sum_\nu j_\nu(\sigma_S) = \sum_\nu j_\nu(\sigma) = j(\sigma).$$

Q.E.D.

Proof II: Combine Corollary II of Proposition V, sec. 9.6, and the corollary to Proposition VII, sec. 9.8.

Q.E.D.

Next assume that B (and hence ξ) is nonorientable. Let $p: \tilde{B} \to B$ be the double covering of B (cf. sec 2.20) and let $\tilde{\xi} = (\tilde{E}, \tilde{\pi}, \tilde{B}, F)$ denote the pull-back of ξ to $\tilde{B}$ via p. Then we have a bundle map

$$\begin{array}{ccc} \tilde{E} & \xrightarrow{p_E} & E \\ {\scriptstyle\tilde{\pi}}\downarrow & & \downarrow{\scriptstyle\pi} \\ \tilde{B} & \xrightarrow[p]{} & B \end{array}$$

restricting to isomorphisms in the fibres. In particular, p_E is a local diffeomorphism. Hence there is a unique orientation in $\tilde{E}$ such that p_E is orientation preserving.

Now give $\tilde{\xi}$ the bundle orientation induced by those of $\tilde{B}$ and $\tilde{E}$. Denote by $\tilde{\chi}_S \in H^n(\tilde{B})$ the Euler class of the sphere bundle associated with $\tilde{\xi}$.

Theorem IV: With the notation and hypotheses above let $\sigma \in \operatorname{Sec} \xi$ have finitely many zeros. Then

$$\int_{\tilde{B}}^{\#} \tilde{\chi}_S = 2j(\sigma).$$

In particular, $j(\sigma)$ is independent of σ.

Proof: Set $p^{-1}(a_\nu) = \{b_\nu, c_\nu\}$. Then $b_1, \ldots, b_m, c_1, \ldots, c_m$ are the zeros of $p_E^{\#}\sigma$. Since p_E is orientation preserving, Proposition VI, sec. 9.7, shows that

$$j_{b_\nu}(p_E^{\#}\sigma) = j_{a_\nu}(\sigma) \qquad \text{and} \qquad j_{c_\nu}(p_E^{\#}\sigma) = j_{a_\nu}(\sigma),$$

whence

$$j(p_E^{\#}\sigma) = 2j(\sigma).$$

Now Theorem III yields

$$\int_{\tilde{B}}^{\#} \tilde{\chi}_S = j(p_E^{\#}\sigma) = 2j(\sigma).$$

Q.E.D.

Theorem V: A vector bundle of rank n over a connected n-manifold M, with oriented total space, admits a cross-section σ with finitely many zeros. It admits a cross-section with no zeros if and only if

(i) M is not compact,

or

(ii) M is compact and $j(\sigma) = 0$.

Proof: Apply Theorem III, sec. 8.11, to obtain σ. If M is not compact, Theorem III, sec. 8.11 shows that there is a cross-section with no zeros; if M is compact the same theorem gives a cross-section, τ, with a single zero, a. Then Theorems III and IV, above show that

$$j(\sigma) = j(\tau) = j_a(\tau).$$

Hence if $j(\sigma) = 0$ Proposition VII, sec. 8.10, applied to τ, gives a cross-section without zeros.

The "only if" part of the theorem is obvious.

Q.E.D.

9.10. Examples: 1. *Vector fields on* S^n: Let E be a Euclidean space of dimension $n+1$ ($n \geqslant 1$). Fix a unit vector $e \in E$ and set $F = e^\perp$. Let S^n be the unit sphere of E. Define maps $\varphi: F \to S^n$, $\psi: F \to S^n$ by

$$\varphi(x) = \frac{x - (1 - |x|^2)e}{|x - (1 - |x|^2)e|}, \qquad x \in F$$

and

$$\psi(x) = \frac{x + (1 - |x|^2)e}{|x + (1 - |x|^2)e|}, \qquad x \in F.$$

Then φ and ψ are diffeomorphisms,

$$\varphi: F \xrightarrow{\cong} S^n - \{e\}, \qquad \psi: F \xrightarrow{\cong} S^n - \{-e\}.$$

A straightforward computation shows that the corresponding diffeomorphism $\chi = \psi^{-1} \circ \varphi$ of $\dot{F} = F - \{0\}$ onto itself is given by

$$\chi(x) = x/|x|^2, \qquad x \in \dot{F}.$$

Now define vector fields X and Y on F by

$$X(x) = a, \qquad x \in F$$

and

$$Y(x) = |x|^2 a - 2\langle a, x\rangle x, \qquad x \in F,$$

where a is a fixed unit vector in F. Evidently,

$$Y(\chi(x)) = \chi'(x; X(x)), \qquad x \in \dot{F};$$

i.e.,

$$\chi_*(X) = Y.$$

It follows that the vector fields $\varphi_* X$ on $S^n - \{e\}$ and $\psi_* Y$ on $S^n - \{-e\}$ agree in $S^n - \{e, -e\}$. Thus they determine a vector field Z on S^n.

Clearly, this vector field has a single zero at $\psi(0) = e$. Moreover, the index of Z at e is given by

$$j_e(Z) = \deg_0 Y,$$

where Y is considered as a smooth map $F \to F$. But according to sec. 6.12, Example 5 (with a replaced by $-a$), $\deg_0 Y = 1 + (-1)^n$. It follows that

$$j(Z) = j_e(Z) = 1 + (-1)^n.$$

Now let τ denote the tangent bundle of S^n and let τ_S be the associated sphere bundle. Then Theorem III implies that

$$\int_{S^n}^{\#} \chi(\tau_S) = 1 + (-1)^n.$$

2. *Vector fields on* $\mathbb{R}P^n$: Consider S^n as the unit sphere in an $(n + 1)$-dimensional Euclidean space E $(n \geqslant 1)$. Let $\varphi\colon E \to E$ be a skew linear map; i.e.,

$$\langle x, \varphi(x)\rangle = 0, \qquad x \in E.$$

Then we have

$$\varphi(x) \in x^\perp = T_x(S^n), \qquad x \in S^n,$$

and so a vector field X on S^n is defined by

$$X(x) = \varphi(x), \qquad x \in S^n.$$

Since $\varphi(-x) = -\varphi(x)$, it follows that

$$\sigma_* X = X$$

where $\sigma\colon S^n \to S^n$ denotes the map given by $\sigma(x) = -x$. Thus X determines a vector field Y on $\mathbb{R}P^n$. Clearly,

$$X \underset{\pi}{\sim} Y,$$

where $\pi\colon S^n \to \mathbb{R}P^n$ denotes the canonical projection.

Now we distinguish two cases:

I. *n odd.* Then $n + 1$ is even and φ can be chosen to be a linear isomorphism. The corresponding vector fields X and Y have then no zeros.

If τ denotes the tangent bundle of $\mathbb{R}P^n$ and if τ_S denote the associated sphere bundle, then the above result together with Theorem III implies that

$$\chi(\tau_S) = 0 \qquad (n \text{ odd}).$$

II. *n even.* Then $n + 1$ is odd and so every skew map has a nonzero kernel. Now fix a unit vector $e \in E$ and choose the skew map φ so that its kernel is the one-dimensional subspace generated by e. Then φ restricts to a linear automorphism of the orthogonal complement F of e.

The corresponding vector field X on S^n has two zeros, at e and $-e$. Since the restriction φ_F of φ to F is a skew linear automorphism, we have

$$\det \varphi_F > 0.$$

Using this it is easy to show that

$$j_e(X) = 1 \qquad \text{and} \qquad j_{-e}(X) = 1.$$

It follows that the corresponding vector field Y on $\mathbb{R}P^n$ has a single zero at the point $\bar{e} = \pi(e)$ and that

$$j(Y) = j_{\bar{e}}(Y) = 1.$$

3. *The Hopf index formula*: Let B be a compact connected oriented n-manifold (n even) and let $\psi: B \to S^n$ be a smooth map.

Consider the pull-back, $\xi = (E, \pi, B, \mathbb{R}^n)$, of the tangent bundle τ_{S^n} of S^n via ψ. Let $\varphi: E \to T_{S^n}$ be the corresponding bundle map.

Since φ restricts to linear isomorphisms on the fibres, the standard metric on S^n induces a Riemannian metric in ξ. If χ_ξ and χ_{S^n} denote the Euler class of the associated sphere bundles of ξ and τ_{S^n}, we have, in view of Example I, that

$$\int_B^\# \chi_\xi = \int_B^\# \psi^\# \chi_{S^n} = \deg \psi \cdot \int_{S^n}^\# \chi_{S^n} = 2 \deg \psi;$$

i.e.,

$$\int_B^\# \chi_\xi = 2 \deg \psi.$$

Since, for each integer p, there is a smooth map $B \to S^n$ of degree p (cf. Example 4, sec. 6.12), we can obtain in this way an infinite number of nonisomorphic vector bundles of rank n over the n-manifold B.

Finally, if σ is a cross-section in ξ with finitely many zeros, Theorem III yields

$$j(\sigma) = 2 \deg \psi.$$

This relation is called the *Hopf index formula*.

Problems

1. Vector fields in $\mathbb{C}$. Find the index of the following vector fields at zero:

(i) $Z(z) = z^n \qquad (n \in \mathbb{Z})$
(ii) $Z(z) = \bar{z}^n \qquad (n \in \mathbb{Z})$
(iii) $Z(z) = f(z)\, e^{1/z}$, where $f(z) = \exp(-|z|^{-2})/|\exp z^{-1}|$,
(iv) $Z(z) = \sin z$.

2. Cartesian products. Let

$$\xi = (E_\xi, \pi_\xi, B_\xi, F_\xi) \qquad \text{and} \qquad \eta = (E_\eta, \pi_\eta, B_\eta, F_\eta)$$

be oriented vector bundles.

(i) Show that

$$(\kappa_E)_\# \circ (Th_\xi \otimes Th_\eta) = Th_{\xi \times \eta} \circ (\kappa_B)_\#,$$

where

$$(\kappa_E)_\# : H_F(E_\xi) \otimes H_F(E_\eta) \to H_F(E_\xi \times E_\eta)$$

and

$$(\kappa_B)_\# : H(B_\xi) \otimes H(B_\eta) \to H(B_\xi \times B_\eta)$$

denote the Künneth homomorphisms. Conclude that

$$\theta_{\xi \times \eta} = (\kappa_E)_\#(\theta_\xi \otimes \theta_\eta).$$

(ii) Assume that $B_\xi = B_\eta$. Show that the Thom class of $\xi \oplus \eta$ is given by

$$\theta_{\xi \oplus \eta} = j^\#(\kappa_E)_\#(\theta_\xi \otimes \theta_\eta),$$

where $j\colon E_{\xi \oplus \eta} \to E_\xi \times E_\eta$ is the inclusion map.

(iii) Conclude that the Euler classes of the sphere bundles associated with $\xi \times \eta$ and $\xi \oplus \eta$ are, respectively, given by

$$\chi_{\xi \times \eta} = (\kappa_B)_\#(\chi_\xi \otimes \chi_\eta)$$

and

$$\chi_{\xi \oplus \eta} = \chi_\xi \cdot \chi_\eta .$$

3. Let ξ, η be vector bundles with oriented total spaces over manifolds B_ξ, B_η. Assume that $\dim B_\xi = \operatorname{rank} \xi$ and $\dim B_\eta = \operatorname{rank} \eta$.

Suppose that $\sigma \in \operatorname{Sec} \xi$ and $\tau \in \operatorname{Sec} \eta$ have isolated zeros at $a \in B_\xi$, $b \in B_\eta$. Show that $\sigma \times \tau$ has an isolated zero at (a, b), and that

$$j_{(a,b)}(\sigma \times \tau) = j_a(\sigma) \cdot j_b(\tau).$$

4. Let X be a vector field on M. Recall from problem 9, Chap. III, that X determines a vector field $Y = \omega_M \circ dX$ on T_M.

(i) Find necessary and sufficient conditions on X for Y to have an isolated zero.

(ii) If Y has an isolated zero at $h \in T_M$, find all possible values for the index of Y at h.

5. Let $f \in \mathscr{S}(M)$. Suppose that $a \in M$ is a point such that $(\delta f)(a) = 0$ and the Hessian of f at a is nondegenerate. Use a Riemannian metric to convert δf into a vector field, X, with an isolated zero at a. Regard the Hessian of f as an indefinite metric on $T_a(M)$. Show that $j_a(X) = (-1)^q$, where q is the dimension of a maximal subspace of $T_a(M)$ on which the Hessian is negative definite.

6. Let $\xi = (E, \pi, B, F)$ be a vector bundle with $\operatorname{rank} \xi = \dim B = n$ and let o denote the zero cross-section. Assume that $\sigma \in \operatorname{Sec} \xi$ has an isolated zero at a.

(i) Show that there is a unique linear map $\alpha\colon T_a(B) \to V_{o(a)}(E)$ such that

$$(d\sigma)_a h = (do)_a h + \alpha(h), \qquad h \in T_a(B).$$

(ii) Show that the following conditions are equivalent:

(a) α is a linear isomorphism

(b) $\operatorname{Im}(do)_a \oplus \operatorname{Im} \alpha = T_{o(a)}(E)$

(c) with respect to an appropriate trivializing map, $\sigma(x) = (x, \sigma_1(x))$ and $\sigma_1'(a)$ is an isomorphism.

(iii) Assume that the conditions of (ii) hold. Assume further that ξ is Riemannian and oriented. Identify a neighbourhood U of a with an oriented Euclidean space; let $S(r)$ denote the sphere of radius r about a. Finally, consider the associated sphere bundle (E_S, π_S, B, S) and set $S_a = \pi_S^{-1}(a)$.

Show that if $\Omega \in A^{n-1}(E_S)$ satisfies $(\oint_S \Omega)(u) = 1$, then

$$j_a(\sigma) = \frac{\det \alpha}{|\det \alpha|} = \lim_{r\to 0} \int_{S(r)} \tau^*\Omega,$$

where $\tau(x) = \sigma(x)/|\sigma(x)|$, $x \in U - \{a\}$. *Hint*: Use problem 7, below.

7. Let F be an oriented Euclidean n-space. Denote its unit sphere (resp. sphere of radius r) by S (resp. $S(r)$).

(i) Show that

$$\lim_{r\to 0} \int_{S(r)} f \cdot \Psi = 0,$$

where $\Psi \in A^{n-1}(F)$ and $f \in \mathscr{S}(\dot{F})$ satisfies $|f(x)| \leqslant K |x|^{-p}$ (K a constant, $p < n - 1$).

(ii) Suppose $\varphi_1, \varphi_2 : F \to F$ are smooth maps. Assume that $\varphi_2^{-1}(0) = 0$ and that $\varphi_2'(0)$ is a linear isomorphism. Define $\psi: \dot{F} \to F \times S$ by

$$\psi(x) = \left(\varphi_1(x), \frac{\varphi_2(x)}{|\varphi_2(x)|}\right), \qquad x \in \dot{F}.$$

Show that

$$\lim_{r\to 0} \int_{S(r)} \psi^*\Phi = 0,$$

if $\Phi \in A^{n-1}(F \times S)$ satisfies $\oint_S \Phi = 0$. *Hint*: Compare problem 6, v, Chap. VII.

8. Thom isomorphism with compact supports. Let $\xi = (E, \pi, B, F)$ be an oriented vector bundle of rank r. Establish a Thom isomorphism Th_c: $H_c(B) \xrightarrow{\cong} H_c(E)$ homogeneous of degree r.

If $\theta_\xi = Th(1)$ is represented by Φ and $\Psi \in A_c(B)$ represents a class $\alpha \in H_c(B)$, show that $(\pi^*\Psi) \wedge \Phi$ represents $Th_c(\alpha)$.

9. Normal bundle. Let M be a closed connected oriented submanifold of a connected, oriented n-manifold N and let $\dim M = m$. Let $\xi = (E, \pi, M, F)$ be the normal bundle of M and regard E as an open subset of N (cf. problem 20, Chap. III).

(i) Orient ξ so that the local product orientation in E is induced from the orientation in N.

(ii) Show that $A_F(E) \subset A(N)$ and $A_c(E) \subset A_c(N)$ are ideals and let γ, γ^c denote the corresponding inclusions.

(iii) Show that the inclusion map $i: M \to N$ is proper. Show that $i^{\#}$ and $\gamma_{\#}^{c} \circ Th_{c}$ (resp. $\gamma_{\#} \circ Th$ and $i_{c}^{\#}$) are dual with respect to the Poincaré scalar products. If M is compact, show that $(i^{\#})^{*}(1) = \gamma_{\#}(\theta_{\xi})$. Conclude that the Euler class of ξ_S (cf. sec. 9.3) is in $\operatorname{Im} i^{\#}$.

(iv) Assume M compact. Let $j: E_S \to N - M$ be the inclusion. Show that, up to sign, $\int_S^{\#} \circ j^{\#}$ is dual to the connecting homomorphism of problem 14, Chap. VIII (with $N = B$). Conclude that the diagram

$$\begin{array}{ccc} H(N) & \xleftarrow{\gamma_{\#} \circ Th} & H(M) \\ & {\scriptstyle k^{\#}} \searrow \quad \nearrow {\scriptstyle \int_S^{\#} \circ j^{\#}} & \\ & H(N - M) & \end{array}$$

is exact, where $k: N - M \to N$ denotes the inclusion map.

(v) Establish an exact triangle

$$\begin{array}{ccc} H(N) & \xrightarrow{k^{\#} \oplus i^{\#}} & H(N - M) \oplus H(M) \\ & {\scriptstyle \partial_1} \nwarrow \quad \swarrow {\scriptstyle (j^{\#},\, -\pi_S^{\#})} & \\ & H(E_S) & \end{array}$$

Find its Poincaré dual.

(vi) Regard $\mathbb{C}P^k$ and $\mathbb{H}P^k$ as submanifolds of $\mathbb{C}P^n$ and $\mathbb{H}P^n$, respectively, with inclusions i, j. Compute the classes $(i^{\#})^{*}(1) \in H^{2(n-k)}(\mathbb{C}P^n)$ and $(j^{\#})^{*}(1) \in H^{4(n-k)}(\mathbb{H}P^n)$ (cf. problems 4 and 5, Chap. VIII). Compute $H(\mathbb{C}P^n - \mathbb{C}P^k)$ and $H(\mathbb{H}P^n - \mathbb{H}P^k)$.

10. Disc bundles. Let $\xi = (E, \pi, B, F)$ be an oriented Riemannian bundle and let $\bar{\xi} = (\bar{E}, \bar{\pi}, B, \bar{F})$ be the bundle whose fibre at x is the subset of F_x whose vectors have length $\leqslant 1$.

(i) Show that $\bar{\xi}$ is a bundle with boundary ξ_S (notation as in sec. 9.3, cf. problem 4, Chap. VII).

(ii) Let $i_S: E_S \to \bar{E}$ be the inclusion. Establish an exact triangle

$$\begin{array}{ccc} H(\ker i_S^{*}) & \longrightarrow & H(\bar{E}) \\ & {\scriptstyle \partial} \nwarrow \quad \swarrow {\scriptstyle (i_S)^{*}} & \\ & H(E_S) & \end{array}$$

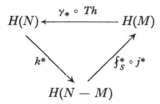

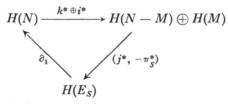

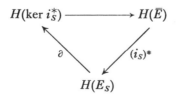

(iii) Construct isomorphisms

$$H(\bar{E}) \cong H(B), \qquad H(\ker i_S^*) \cong H_F(E) \cong H(B)$$

and identify the triangle of (ii) with the Gysin triangle.

(iv) Let (N, ρ, B, S) be a sphere bundle associated with $\xi \oplus \epsilon$ ($\epsilon = B \times \mathbb{R}$). Show that B is a submanifold of N with normal bundle ξ. Identify N with the double of $\bar{E}$. Prove that $H(N) \cong H(B) \otimes H(S)$ and that the sequence

$$0 \longleftarrow H(B) \xleftarrow{i^*} H(N) \xleftarrow{\gamma^* \circ Th} H(B) \longleftarrow 0$$

is exact, where $i : B \to N$, $\gamma : A_F(E) \to A(N)$ are the inclusion maps.

11. Local degree. Let $P_i \subset M_i$ be compact connected oriented p_i-submanifolds of oriented connected n-manifolds M_i ($i = 1, 2$). Let $\xi_i = (E_i, \pi_i, P_i, F_i)$ be the oriented normal bundles with E_i considered as a neighbourhood of P_i in M_i. Write $\dot{M}_i = M_i - P_i$, $\dot{E}_i = E_i - P_i$. Assign ξ_i a Riemannian metric with sphere bundle

$$(\xi_S)_i = ((E_S)_i, \pi_i, P_i, S_i).$$

Finally, suppose $\varphi : M_1 \to M_2$ is a smooth map which restricts to smooth maps $\varphi_P : P_1 \to P_2$ and $\hat{\varphi} : U_1 - P_1 \to U_2 - P_2$ (U_i some neighbourhood of P_i). Then P_1 is called an *isolated manifold* for φ.

(i) Show that the identification of E_i as an open subset of M_i can be chosen so that φ restricts to smooth maps

$$\varphi_E : E_1 \to E_2, \qquad \dot{\varphi}_E : \dot{E}_1 \to \dot{E}_2.$$

(ii) Define $\psi : (E_S)_1 \to (E_S)_2$ by $\psi(z) = \dot{\varphi}_E(z)/|\dot{\varphi}_E(z)|$. Find a neighbourhood, V, of P_2 with the following property: If $\Phi \in A_c(V)$, then $\varphi_E^*\Phi \in A_c(E_1)$. If (for this V) $\Phi \in A_c^n(V)$, show that

$$\int_{E_1} \varphi_E^*\Phi = \deg \psi \cdot \int_{E_2} \Phi.$$

Conclude that the integer $\deg \psi$ is independent of the various choices. It is called the *local degree of* φ *at* P_1 and is written $\deg_{P_1} \varphi$.

(iii) Show that this definition coincides with the definition in the text when P_1 and P_2 are points.

(iv) Suppose M_1, M_2 are compact, and that for an oriented connected closed submanifold $Q \subset M_2$, $\varphi^{-1}(Q)$ consists of finitely many oriented, connected submanifolds. Show that

$$\deg \varphi = \sum_i \deg_{Q_i} \varphi.$$

(v) For some $\epsilon > 0$, show that $(E_2)_\epsilon \subset V$ (cf. sec. 9.3). Hence, from φ_E, obtain a homomorphism $H_{F_1}(E_1) \xleftarrow{\tilde{\varphi}^\#} H_{F_2}(E_2)$. Show that the diagram

$$\begin{array}{ccc} H_{F_1}(E_1) & \xleftarrow{\tilde{\varphi}^\#} & H_{F_2}(E_2) \\ \alpha_{\xi_1}\uparrow & & \uparrow\alpha_{\xi_2} \\ H((E_S)_2) & \xleftarrow[\psi^\#]{} & H((E_S)_1) \end{array}$$

commutes.

12. We retain the notation and hypotheses of problem 11.

(i) Let $\Phi \in A_c^{n-p_2}(V)$ represent θ_{ξ_2}. Show that the class in $H^{p_1-p_2}(P_1)$ represented by $\oint_{F_1} \varphi_E^* \Phi$ is independent of the choice of Φ. It is called the *local fibre degree of* φ *at* P_1 and written $\deg_{P_1}^F \varphi$.

(ii) Let ω be the orientation class of P_2. Then the class $\varphi_P^\# \omega \in H^{p_2}(P_1)$ is called the *local base degree* of φ at P_1 and is written $\deg_{P_1}^B \varphi$. Prove that

$$\deg_{P_1} \varphi = \int_{P_1}^{\#} \deg_{P_1}^B \varphi \cdot \deg_{P_1}^F \varphi.$$

(iii) If $p_1 < p_2$, conclude that $\deg_{P_1} \varphi = 0$.

(iv) If $p_1 = p_2$, note that $\deg_{P_1}^F \varphi \in \mathbb{R}$. In this case obtain $\deg_{P_1} \varphi = \deg \varphi_P \cdot \deg_{P_1}^F \varphi$.

(v) Suppose $p_1 = p_2$. Fix $a \in P_1$ and set $b = \varphi_P a$. Choose a trivializing map χ: $U_b \times F_2 \to \pi^{-1} U_b$ for ξ_2. (U_b, a neighbourhood of b in P_2.) Find a neighbourhood O_a of 0 in $(F_1)_a$ such that $\varphi_E(O_a) \subset \pi^{-1}(U_b)$. Use φ_E to construct a smooth map $\gamma : O_a \to F_2$ such that $\gamma^{-1}(0) = 0$. Show that $\deg_0 \gamma = \deg_{P_1}^F \varphi$ and conclude that the local fibre degree is an integer. Hence conclude that the local degree is an integer.

13. Let $\mathscr{B}_i = (M_i, \rho_i, B_i, N_i)$ be smooth bundles with orientable, compact, connected r-dimensional fibres and n-dimensional bases. Suppose $\varphi : M_1 \to M_2$ is a smooth fibre preserving map inducing $\psi : B_1 \to B_2$. Let a be an isolated point for ψ.

(i) Show that $(N_1)_a$ is an isolated manifold for φ.

(ii) Let $\varphi_a : (N_1)_a \to (N_2)_{\psi(a)}$ be the restriction of φ. Show that, for appropriate orientations,

$$\deg^F_{(N_1)_a} \varphi = \deg_a \psi \qquad \text{and} \qquad \deg^B_{(N_1)_a} \varphi = \deg \varphi_a .$$

Conclude that

$$\deg_{(N_1)_a} \varphi = \deg_a \psi \cdot \deg \varphi_a .$$

(iii) Assume that $\mathscr{B}_i$ and B_i are oriented and that the B_i are connected and compact. Prove that $\deg \varphi_x$ $(x \in B_1)$ is independent of x and that

$$\deg \varphi = \deg \psi \cdot \deg \varphi_x .$$

Conclude that $\varphi^\#$ is an injective if and only if both $\psi^\#$ and $\varphi_x^\#$ are injective.

14. Let $\xi = (E, \pi, B, F)$ be an oriented vector bundle over an oriented base, with $\dim B = \operatorname{rank} \xi = n$. Adopt the notation of sec. 9.3. A submanifold $M \subset B$ is called an *isolated zero manifold for* $\sigma \in \operatorname{Sec} \xi$ if, for some neighbourhood, U, of M, $M = \{x \in U \mid \sigma(x) = 0_x\}$.

Let M be a compact isolated zero manifold for some $\sigma \in \operatorname{Sec} \xi$.

(i) Show that, if Φ is a suitable representative of θ_ξ, then, for sufficiently "small" U, the restriction of $\sigma^*\Phi$ to U has compact support. Prove that $\int_U \sigma^*\Phi$ is independent of the choice of Φ. It is called the *index of* σ *at* M and is denoted by $j_M(\sigma)$.

(ii) Choose a tubular neighbourhood, V, of M whose boundary is a sphere bundle, V_S, over M. Show that σ determines a cross-section, σ_S, in the restriction of ξ_S to V_S. Let $\hat{\xi}_S$ denote the restriction of ξ_S to V. Prove that

$$j_M(\sigma) = \int_V^\# \chi_{\hat{\xi}_S, \sigma_S}$$

(cf. problems 14 and 15, Chap. VIII). Conclude that $j_M(\sigma)$ is an integer.

(iii) Suppose that the set of zeros of σ consists of finitely many submanifolds $M_1, \ldots, M_q$. Assume B compact and show that

$$\int_B^\# \chi_{\xi_S} = \sum_{i=1}^{q} j_{M_i}(\sigma).$$

(Give two different proofs.)

(iv) Suppose the restriction of ξ to V is trivial, $\pi^{-1}(V) = V \times F$. Write $\sigma(x) = (x, \tau(x))$, $x \in V$. Show that M is an isolated submanifold for τ, with image manifold $\{0\}$. Show that $j_M(\sigma) = \deg_M(\tau)$.

15. Tangent bundle. Let B be an oriented manifold containing a compact connected oriented submanifold M with oriented normal bundle $\eta = (V, \rho, M, F)$ and associated sphere bundle $\eta_S = (V_S, \rho, M, S)$. Regard V as an open subset of B. Let $X \in \mathscr{X}(B)$ and suppose that $M = \{x \in V \mid X(x) = 0\}$.

(i) Assume that $X(x)$ is tangent to the submanifold $F_{\rho(x)}$ for $x \in V$. By restricting X obtain on each F_y $(y \in M)$ a vector field X_y with an isolated zero at 0. Prove that $j_0(X_y)$ is constant as y varies through M. Denote this integer by $j_M^F(X)$.

(ii) With X as in (i), let χ_M be the Euler class of the tangent bundle of M. Show that

$$j_M(X) = j_M^F(X) \cdot \int_M^{\#} \chi_M .$$

(iii) Evaluate $j_M^F(X)$ in the following three cases:

(a) X restricts to a vector field tangent to V_S.
(b) X is the radial vertical vector field for η.
(c) X is the negative of the radial vector field for η.

16. Local dashed degree. Let $\xi_i = (E_i, \pi_i, B_i, F)$ $(i = 1, 2)$ be oriented vector bundles of rank n over n-manifolds B_i $(n \geqslant 2)$. Let $\varphi : \xi_1 \to \xi_2$ be a bundle map inducing $\varphi_B : B_1 \to B_2$. A point $a \in B_1$ is called an *isolated singularity of* φ, if for some neighbourhood, U, of a

$$\varphi_B(x) \neq \varphi_B(a), \qquad x \in \dot{U} \quad (\dot{U} = U - \{a\})$$

and φ_x is a linear isomorphism for $x \in \dot{U}$. Let a be a fixed isolated singularity of φ.

(i) Show that in $\dot{U}$ the maps φ_x are all orientation preserving or orientation reversing. Let $\epsilon_a(\varphi) = 1$ in the first case, and let $\epsilon_a(\varphi) = -1$ in the second. (Assume U connected.)

(ii) Let S be a small sphere in B_1 centred at a (use an atlas for B_1). Let $\hat{\varphi} : V_1 \times F \to V_2 \times F$ be a bundle map obtained from φ by trivializations in ξ_1 and ξ_2 $(a \in V_1, \varphi_B(a) \in V_2)$. Fix $h \in S_F$ (S_F the unit sphere of F) and define $\varphi_S : S \to S_F$ by $\varphi_S(x) = \hat{\varphi}_x(h)/|\hat{\varphi}_x(h)|$. Show that $\deg \varphi_S$ depends only on ξ_1, ξ_2, and φ (and not on the other choices). It will be called the *local dashed degree of* φ *at* a and is denoted by $\deg'_a \varphi$.

(iii) If $\varphi_a \neq 0$ show that $\deg'_a \varphi = 0$.

(iv) Let ξ_3 be a third oriented vector bundle of rank n over a third oriented n-manifold. Let $\psi: \xi_2 \to \xi_3$ have an isolated singularity at $b = \varphi_B(a)$. Show that a is an isolated singularity for $\psi \circ \varphi$ and prove the formulae

$$\epsilon_a(\psi \circ \varphi) = \epsilon_a(\varphi) \cdot \epsilon_b(\psi)$$

and

$$\deg'_a(\psi \circ \varphi) = \deg'_b \psi \cdot \deg_a \varphi_B + \epsilon_b(\psi) \cdot \deg'_a \varphi.$$

Simplify this if ψ_b or φ_a is nonzero.

(v) Let $\tau \in \operatorname{Sec} \xi_2$ have an isolated zero at b. Show that $\varphi^{\#}\tau$ has an isolated zero at a and that

$$j_a(\varphi^{\#}\tau) = \deg_a \varphi_B \cdot j_b(\tau) - \deg'_a \varphi.$$

17. Global dashed degree. Let ξ_1 and ξ_2 be as in problem 16 and assume B_1, B_2 compact. Suppose $\varphi : \xi_1 \to \xi_2$ induces $\varphi_B : B_1 \to B_2$. Suppose that $\{a_1, a_2, \dots, a_m\}$ are isolated singularities for φ. Assume that φ_x is an isomorphism for $x \neq a_1, \dots, a_m$. Set $\deg' \varphi = \sum_i \deg'_{a_i} \varphi$.

(i) Let $\chi_i \in H^n(B_i)$ $(i = 1, 2)$ be the Euler classes of the associated sphere bundles. Prove the *Riemann–Hurwitz relation*

$$\int_{B_1}^{\#} \chi_1 = \deg \varphi_B \cdot \int_{B_2}^{\#} \chi_2 - \deg' \varphi.$$

(ii) Consider the case where $\xi_i = \tau_{B_i}$ and $\varphi = d\varphi_B$. In particular suppose f is a polynomial of degree k with complex coefficients. Regard f as a smooth map $S^2 \to S^2$ and use the Riemann–Hurwitz relation to show that the sum of multiplicities of the roots of f is k.

18. Let $\xi = (E, \pi, B, F)$ be a vector bundle over a compact connected base and let $\varphi : E \to E$ be a proper smooth map. Define a map $\varphi_B : B \to B$ by $\varphi_B = \pi \circ \varphi \circ o$.

(i) If ξ is oriented show that $\varphi_C^{\#}\theta_\xi = m(\varphi)\, \theta_\xi$, where $m(\varphi) \in \mathbb{Z}$. If φ is a diffeomorphism show that $m(\varphi) = \pm 1$.

(ii) If B and ξ are oriented and φ is a diffeomorphism show that $\deg \varphi_B = \pm 1$. Prove that φ preserves the orientation of E if and only if

$$\deg \varphi_B \cdot m(\varphi) = 1.$$

(iii) If B and ξ are oriented, if rank $\xi = \dim B$, and if $\chi_B \neq 0$, show that E is irreversible.

19. Let $\xi = (E, \pi, B, F)$ be a bundle of rank $2m$ over a compact n-manifold.

(i) Suppose ξ is orientable and contains a subbundle of odd rank. Prove that the Euler class (of ξ_S) is zero.

(ii) Suppose E is orientable and $n = 2m$. If ξ contains a vector subbundle of odd rank, show that it admits a cross-section without zeros.

(iii) Let η be any vector bundle of rank 1 over B. Show that $\wedge^{2m}(\xi \otimes \eta) \cong \wedge^{2m} \xi$.

(iv) Assume E orientable and $n = 2m$. Let η be a rank 1 vector bundle over B, and let $\hat{E}$ be the total space of $\xi \otimes \eta$. Show that $\hat{E}$ is orientable. If $\sigma \in \operatorname{Sec} \xi$ and $\tau \in \operatorname{Sec}(\xi \otimes \eta)$ have only finitely many zeros, prove that $j(\sigma) = j(\tau)$. In particular, if ξ admits a cross-section without zeros, conclude that every rank 1 vector bundle, η, is a subbundle of ξ.

20. Mod 2 index sum. Let $\xi = (E, \pi, B, F)$ be a vector bundle with $\dim B = \dim F = n \geqslant 2$. Assume E is connected, but not necessarily orientable.

(i) Let $\sigma \in \operatorname{Sec} \xi$ have an isolated zero at a. Choose an orientation of $\pi^{-1}(V)$ (V some neighbourhood of a) and define $j_a(\sigma)$ with respect to it. Let $[j_a(\sigma)]$ denote the (mod 2)-reduction of $j_a(\sigma)$ and show that it does not depend on the choice of orientation.

(ii) Let $\sigma \in \operatorname{Sec} \xi$ and suppose σ has only finitely many zeros, $a_1, \ldots, a_m$. Define $[j(\sigma)] \in \mathbb{Z}_2$ by

$$[j(\sigma)] = \sum_i [j_{a_i}(\sigma)]$$

and call it the (mod 2)-index sum. Show that it is independent of σ.

(iii) Assume E is *not* orientable. Let $\sigma \in \operatorname{Sec} \xi$ have a single zero at a, and define $j_a(\sigma)$ as in (i). Show that for each integer, m, there is a cross-section τ in ξ such that a is the only zero of τ, and $j_a(\tau) = j_a(\sigma) + 2m$. In particular, conclude that ξ admits a cross-section without zeros if and only if the (mod 2)-index sum is zero.

21. Continuous local degree. (i) Extend the notion of local degree to continuous maps $\varphi : \mathbb{R}^n \to \mathbb{R}^n$ with an isolated zero at zero. If

$\psi_1, \psi_2 : \mathbb{R}^n \to \mathbb{R}^n$ are origin preserving homeomorphisms, compare $\deg_0(\psi_1 \circ \varphi \circ \psi_2)$ with $\deg_0 \varphi$.

(ii) Extend the notion of index at an isolated zero to continuous cross-sections.

(iii) Define a continuous vector field, X, on $\mathbb{R}^2$ by $X(x, y) = (\xi, \eta)$, where

$$\xi(x, y) = \begin{cases} x \log |x|, & x \neq 0 \\ 0, & x = 0 \end{cases} \qquad \eta(x, y) = \begin{cases} y \log |y|, & y \neq 0 \\ 0, & y = 0. \end{cases}$$

Find the orbits and zeros of X, and compute the indices at isolated zeros.

22. Let $\sigma : \mathbb{R}^n \to \mathbb{R}^n$ be a smooth map such that $\sigma(0) = 0$ and $\det \sigma'(0) \neq 0$.

(i) Show that 0 is an isolated zero for σ, and that

$$\deg_0 \sigma = \det(\sigma'(0))/|\det \sigma'(0)|$$

(ii) Define $\varphi : \mathbb{R}^n \to \mathbb{R}^n$ by $\varphi(x) = \sigma'(x; x)$. Show that 0 is an isolated zero for φ, and that

$$\deg_0 \varphi = \deg_0 \sigma.$$

(iii) Define $\psi : \mathbb{R}^n \to \mathbb{R}^n$ by $\psi(x) = \sigma'(0; x)$. Show that $\deg_0 \psi = \deg_0 \sigma$.

23. Affine simplices. Let $\sigma = (a_0, ..., a_n)$ be an ordered affine n-simplex in $\mathbb{R}^m$ (cf. problem 20, Chap. V). Let E_σ be the affine n-plane spanned by σ; thus $x \in E_\sigma$ if and only if $x = \sum_0^n \lambda^i a_i$ and $\sum_0^n \lambda^i = 1$. The symbols $+$, $-$ denote addition and subtraction in $\mathbb{R}^m$.

(i) Given $x \in E_\sigma$, find a unique linear structure in E_σ with the same underlying affine structure, and with origin at x. Identify this space with $T_x(E_\sigma)$.

(ii) Show that a vector field, X_σ, on E_σ is defined by

$$X_\sigma(x) = x + \sum_{\nu,\mu=0}^{n} \epsilon_{\nu\mu} \lambda^\nu \lambda^\mu a_\mu,$$

where $x = \sum_\nu \lambda^\nu a_\nu$, $\sum_\nu \lambda^\nu = 1$, and

$$\epsilon_{\nu\mu} = \begin{cases} +1, & \nu < \mu \\ 0, & \nu = \mu \\ -1, & \nu > \mu. \end{cases}$$

With respect to an appropriate trivialization of τ_{E_σ}, show that

$$X_\sigma(x) = \left(x, \sum_{\nu,\mu=0}^{n} \epsilon_{\nu\mu}\lambda^\nu\lambda^\mu a_\mu\right).$$

(iii) Show that $X_\sigma(x) = 0$ if and only if x is of the form

$$x = \sum_{\nu=0}^{2r} (-1)^\nu a_{i_\nu},$$

where $0 \leqslant i_0 < i_1 < \cdots < i_{2r} \leqslant n$. In particular, show that in a neighbourhood, U, of σ the zeros of X_σ are precisely the vertices of σ.

(iv) Show that $j_{a_i}(\sigma) = (-1)^i$. What are the indices of X_σ at the other zeros?

(v) Assume $m = n$ and regard S^n as $\mathbb{R}^n \cup \{x_\infty\}$. Define a vector field, Z, on S^n by

$$Z(x) = \begin{cases} e^{-|x|^2} X_\sigma(x), & x \in \mathbb{R}^n \\ 0, & x = x_\infty . \end{cases}$$

Determine the index of Z at x_∞.

(vi) Find the orbits of X_σ, when $n = m = 2$.

24. Vector fields on an affine complex. Let $|K|$ be a finite affine simplicial complex in $\mathbb{R}^m$ (with corresponding abstract simplicial complex, K) (cf. problem 20, Chap. V). If σ is a simplex of K, and $x \in \sigma$, recall that $T_x(\sigma)$ is the affine space E_σ with x as origin (cf. problem 23). A *continuous vector field* on an open subset $U \subset |K|$ is a continuous map $X : U \to \mathbb{R}^m$ such that $X(x) \in T_x(\sigma)$ if $x \in \sigma$. If $X(x) = x$, then x is a *zero* of X. Let X be a continuous vector field on $|K|$.

(i) Set $\varphi(x) = X(x) - x$. Show that for some strictly positive continuous function f on $|K|$,

$$x + f(x)\,\varphi(x) \in |K|, \qquad x \in |K|.$$

(ii) Assume a is an isolated zero for X, and that $|K|$ is a topological n-manifold. Use a chart $(U, u, \mathbb{R}^n)$ about a, and the linear structure of $\mathbb{R}^n$ to define the index of X at a.

(iii) Show that each vertex of K is a zero of X. Let a be a vertex of K and assume a is an isolated zero for X. Assume further that if $a \in \sigma$, then the restriction of X to σ extends to a smooth map $Y_\sigma : E_\sigma \to E_\sigma$. Thus the derivative of Y_σ at a is a linear map, $Y'_\sigma(a) : T_a(E_\sigma) \to T_a(E_\sigma)$.

Assume $Y'_\sigma(a) - \iota$ is a linear isomorphism for each σ. Show that a continuous vector field $\tilde{X}$ on a neighbourhood of a in $|K|$ is defined by

$$\tilde{X}(x) = Y'_\sigma(a; x), \qquad x \in \sigma.$$

Show that $\tilde{X}$ has an isolated zero at a and that $j_a(\tilde{X}) = j_a(X)$.

(iv) Suppose the vertices of K are given a partial order which converts each simplex, σ, of K into an ordered simplex. Show that a continuous vector field, X, on $|K|$ is defined by $X(x) = X_\sigma(x)$, $x \in \sigma$, where X_σ is the vector field of problem 23. Show that the zeros of X are precisely the vertices of K.

(v) The *barycentre* of a simplex $(a_0, ..., a_p)$ is the point

$$\frac{1}{p+1}(a_0 + \cdots + a_p)$$

Let b_σ be the barycentre of σ ($\sigma \in K$) and write $b_\sigma < b_\tau$ if σ is a face of τ. Show that the ordered simplices of the form $(b_{\sigma_0}, ..., b_{\sigma_p})$ (where $b_{\sigma_0} < \cdots < b_{\sigma_p}$) make up a simplicial complex K'. Identify $|K'|$ with $|K|$. Use the ordering among the vertices of K' to obtain a continuous vector field, X, on $|K'|$ whose zeros are the vertices of K'.

(vi) Assume that $|K|$ (and hence $|K'|$) is a topological manifold and that each simplex of K is a face of a n-simplex. Show that the index of X at b_σ is $(-1)^{\dim \sigma}$.

Chapter X

The Lefschetz Class of a Manifold

§1. The Lefschetz isomorphism

10.1. In this article, M will denote a connected compact oriented n-manifold, $n \geqslant 2$. Recall that the Euler–Poincaré characteristic of M is defined by

$$\chi_M = \sum_{p=0}^{n} (-1)^p b_p = \sum_{p=0}^{n} (-1)^p \dim H^p(M).$$

The purpose of this article is to establish the following theorems.

Theorem I: Let τ_M be the tangent bundle of M. Then the Euler class, χ_S, of the associated sphere bundle and the Euler–Poincaré characteristic of M are related by

$$\chi_M = \int_M^{\#} \chi_S \,.$$

Theorem II: Let N be a compact n-manifold ($n \geqslant 2$). Let X be a vector field on N with finitely many zeros, and index sum $j(X)$. Then

$$j(X) = \chi_N \,.$$

10.2. Notational conventions. $\pi_L: M \times M \to M$ and $\pi_R: M \times M \to M$ will denote the projections given by

$$\pi_L(x, y) = x \qquad \text{and} \qquad \pi_R(x, y) = y, \qquad x, y \in M,$$

while $\Delta: M \to M \times M$ will denote the diagonal map, $\Delta(x) = (x, x)$.

We regard $(M \times M, \pi_L, M, M)$ as an oriented bundle. Consequently, we have the linear maps

$$\oint_M : A(M \times M) \to A(M), \qquad \oint_M^{\#} : H(M \times M) \to H(M).$$

F denotes an oriented Euclidean n-space, and $\tau_M = (T_M, \pi, M, F)$ is the tangent bundle of M. The orientation of M is an orientation of τ_M. The corresponding Thom class is denoted by $\theta_M \in H_F^n(T_M)$.

τ_M is given a fixed Riemannian metric. For each $\epsilon > 0$

$$O_\epsilon = \{\xi \in T_M \mid |\xi| < \epsilon\}$$

and

$$F_\epsilon = \{y \in F \mid |y| < \epsilon\}.$$

π_ϵ is the restriction of π to O_ϵ, and $\{O_\epsilon, \pi_\epsilon, M, F_\epsilon\}$ is an oriented smooth fibre bundle (cf. sec. 9.3). The fibre of this bundle over $x \in M$ will be denoted by

$$T_{\epsilon,x}(M) = O_\epsilon \cap T_x(M).$$

10.3. The Lefschetz isomorphism. Denote by L_M the space of linear transformations $H(M) \to H(M)$ homogeneous of degree zero. Then we can write

$$L_M = \sum_{p=0}^{n} L_{H(M)}^p,$$

where $L_{H(M)}^p$ denotes the space of linear transformations of $H^p(M)$. Since $H^p(M)$ has finite dimension (cf. Theorem III, sec. 5.15), we have, for each p, a canonical isomorphism

$$k_p: H^p(M) \otimes H^p(M)^* \xrightarrow{\cong} L_{H(M)}^p.$$

Define an isomorphism

$$k: \sum_{p=0}^{n} H^p(M) \otimes H^p(M)^* \xrightarrow{\cong} L_M$$

by

$$k = \sum_{p=0}^{n} (-1)^{np} k_p.$$

Next observe that the Poincaré duality isomorphisms (cf. sec. 5.11)

$$D_M^p: H^p(M) \xrightarrow{\cong} H^{n-p}(M)^* \qquad (p = 0, \dots, n)$$

determine an isomorphism

$$\iota \otimes D_M^{-1}: \sum_{p=0}^{n} H^p(M) \otimes H^p(M)^* \xrightarrow{\cong} \sum_{p=0}^{n} H^p(M) \otimes H^{n-p}(M).$$

Finally, we have the Künneth isomorphism

$$\kappa_{\#}: \sum_{p=0}^{n} H^p(M) \otimes H^{n-p}(M) \xrightarrow{\cong} H^n(M \times M)$$

(cf. Theorem VI, sec. 5.20).

Combining these isomorphisms yields the linear isomorphism

$$\lambda_M : L_M \xrightarrow{\cong} H^n(M \times M)$$

given by

$$\lambda_M = \kappa_{\#} \circ (\iota \otimes D_M^{-1}) \circ k^{-1}.$$

Definition: The linear isomorphism λ_M is called the *Lefschetz isomorphism* for M and the class $\Lambda_M \in H^n(M \times M)$ given by

$$\Lambda_M = \lambda_M(\iota)$$

is called the *Lefschetz class* of M.

Proposition I: The Lefschetz isomorphism satisfies the relation

$$\oint_M^{\#} (\pi_R^{\#}\alpha) \cdot (\lambda_M \sigma) = \sigma(\alpha), \qquad \sigma \in L_M, \quad \alpha \in H(M).$$

Proof: Since both sides are linear in σ and in α it is sufficient to consider the case

$$\sigma = k(\beta \otimes D_M \gamma), \qquad \beta \in H^p(M), \quad \gamma \in H^{n-p}(M),$$

and $\alpha \in H^q(M)$. Then (cf. Equation 7.3, sec. 7.12)

$$\sigma(\alpha) = (-1)^{pn} \langle D_M \gamma, \alpha \rangle \beta = (-1)^{pn} \left(\int_M^{\#} \gamma \cdot \alpha \right) \beta$$

$$= (-1)^{pn} \oint_M^{\#} \pi_L^{\#}\beta \cdot \pi_R^{\#}\gamma \cdot \pi_R^{\#}\alpha.$$

Thus both sides are zero unless $p = q$; and in this case

$$\sigma(\alpha) = \oint_M^{\#} (\pi_R^{\#}\alpha) \cdot (\lambda_M \sigma).$$

Q.E.D.

Corollary I: The Lefschetz class is the unique element in $H^n(M \times M)$ satisfying

$$\oint_M^{\#} (\pi_R^{\#}\alpha) \cdot \Lambda_M = \alpha, \qquad \alpha \in H(M).$$

Corollary II:

$$\oint_M^{\#} \Lambda_M = 1.$$

Proposition II: Let $Tr\colon L_M \to \mathbb{R}$ be the linear map given by

$$Tr\,\sigma = \sum_{p=0}^{n} (-1)^p \operatorname{tr} \sigma_p ,$$

where $\sigma = \sum_p \sigma_p$, $\sigma_p \in L^p_{H(M)}$. Then

$$Tr\,\sigma = \int_M^{\#} \Delta^{\#}(\lambda_M \sigma).$$

Proof: Again it is sufficient to consider the case

$$\sigma = k(\beta \otimes D_M\gamma), \qquad \beta \in H^p(M), \quad \gamma \in H^{n-p}(M).$$

Then, by ordinary linear algebra,

$$Tr\,\sigma = (-1)^{np+p}\langle D_M\gamma, \beta\rangle = (-1)^{p(n-p)}\langle D_M\gamma, \beta\rangle.$$

Since

$$\langle D_M\gamma, \beta\rangle = \int_M^{\#} \gamma \cdot \beta = (-1)^{p(n-p)} \int_M^{\#} \beta \cdot \gamma,$$

it follows that

$$Tr\,\sigma = \int_M^{\#} \beta \cdot \gamma.$$

According to Example I, sec. 5.17,

$$\beta \cdot \gamma = \Delta^{\#}\kappa_{\#}(\beta \otimes \gamma) = \Delta^{\#}(\lambda_M \sigma)$$

and so

$$Tr\,\sigma = \int_M^{\#} \Delta^{\#}(\lambda_M \sigma).$$

Q.E.D.

Corollary:

$$\int_M^{\#} \Delta^{\#}(\Lambda_M) = \chi_M \, .$$

Proof: Apply the proposition with $\sigma = \iota$.

Q.E.D.

Example: *The n-sphere S^n:* Since the Künneth isomorphism yields

$$H^n(S^n \times S^n) = (H^n(S^n) \otimes 1) \oplus (1 \otimes H^n(S^n)),$$

it follows that the Lefschetz class for S^n must be of the form

$$\Lambda_{S^n} = a(\omega \otimes 1) + b(1 \otimes \omega), \qquad a, b \in \mathbb{R},$$

where ω denotes the orientation class of S^n. Now a simple calculation, using Corollary I to Proposition I, shows that $a = (-1)^n$, $b = 1$, and so

$$\Lambda_{S^n} = (-1)^n \omega \otimes 1 + 1 \otimes \omega.$$

10.4. Lefschetz class and Thom class. In this section we shall use the Thom class θ_M of τ_M to construct a representative of the Lefschetz class.

In Appendix A we shall construct (for a sufficiently small positive number ϵ) a smooth map

$$\exp\colon O_\epsilon \to M$$

with the following properties (cf. Proposition II, sec. A.3).

(i) $\exp(0_x) = x, \; x \in M$.

(ii) The restriction, $\exp_x\colon T_{\epsilon,x}(M) \to M$, of exp is a diffeomorphism onto an open subset of M. It satisfies

$$(d \exp_x)_{0_x} = \iota : T_x(M) \to T_x(M).$$

(iii) The map $\varphi\colon O_\epsilon \to M \times M$ given by

$$\varphi(\xi) = (\pi\xi, \exp \xi),$$

is a diffeomorphism onto an open subset of $M \times M$.

Observe that φ is a fibre preserving map between the bundles

$\{O_\epsilon, \pi_\epsilon, M, F_\epsilon\}$ and $\{M \times M, \pi_L, M, M\}$. The restriction of φ to a fibre is the smooth map

$$\varphi_x = \exp_x : T_{\epsilon,x}(M) \to M.$$

It follows from condition (ii) above and the connectivity of $T_{\epsilon,x}(M)$ that φ_x is orientation preserving. Hence φ preserves the bundle orientations.

Finally, recall from sec. 5.9 that φ induces a homomorphism

$$(\varphi_c)_*: A_c(O_\epsilon) \to A(M \times M).$$

Proposition III: Let Φ be a representative of the Thom class θ_M of τ_M such that

$$\operatorname{carr} \Phi \subset O_\epsilon$$

(cf. Lemma V, sec. 9.3). Then Φ has compact support and $(\varphi_c)_*\Phi$ represents the Lefschetz class of M.

Proof: Since M is compact, Φ has compact support. Moreover,

$$\operatorname{carr}(\varphi_c)_*\Phi \subset \operatorname{Im} \varphi.$$

Thus, for all $\Psi \in A(M \times M)$,

$$\operatorname{carr}(\Psi \wedge (\varphi_c)_*\Phi) \subset \operatorname{Im} \varphi.$$

Since φ preserves the bundle orientations we can apply Proposition VIII, sec. 7.12, to obtain

$$\oint_{F_\epsilon} \varphi^*(\Psi \wedge (\varphi_c)_*\Phi) = \oint_M \Psi \wedge (\varphi_c)_*\Phi.$$

On the other hand, it follows from the definition of $(\varphi_c)_*$ that $\varphi^* \circ (\varphi_c)_*: A_c(O_\epsilon) \to A(O_\epsilon)$ is the inclusion map. Hence

$$\oint_{F_\epsilon} \varphi^*\Psi \wedge \Phi = \oint_M \Psi \wedge (\varphi_c)_*\Phi.$$

Thus if $(\varphi_c)_*\Phi$ represents $\alpha \in H^n(M \times M)$ and Φ represents $\gamma \in H_c^n(O_\epsilon)$, we have

$$\oint_{F_\epsilon}^{\#} (\varphi^\#\omega) * \gamma = \oint_M^{\#} \omega \cdot \alpha, \qquad \omega \in H(M \times M).$$

Finally, observe that

$$\varphi \circ o = \Delta,$$

where o denotes the zero cross-section in τ_M (cf. Property (i) above). In Example 3, sec. 5.5, it was shown that $o \circ \pi$ is homotopic to the identity in T_M. The same argument shows that $o \circ \pi_\epsilon$ is homotopic to the identity map of O_ϵ. Hence,

$$\varphi \sim \varphi \circ o \circ \pi_\epsilon = \Delta \circ \pi_\epsilon .$$

It follows that for $\beta \in H(M)$

$$\oint_M^{\#} (\pi_R^{\#}\beta) \cdot \alpha = \oint_{F_\epsilon}^{\#} (\varphi^{\#}\pi_R^{\#}\beta) * \gamma = \oint_{F_\epsilon}^{\#} \pi_\epsilon^{\#}(\Delta^{\#}\pi_R^{\#}\beta) * \gamma$$

$$= \beta \cdot \oint_{F_\epsilon}^{\#} \gamma = \beta \cdot \oint_{F_\epsilon} \Phi = \beta$$

(since $\pi_R \circ \Delta = \iota$). Now Corollary I to Proposition I, sec. 10.3, implies that

$$\alpha = \Lambda_M .$$

Q.E.D.

Corollary: For every neighbourhood V of $\Delta(M)$ in $M \times M$, there exists a representative Ψ of the Lefschetz class such that carr $\Psi \subset V$.

Proof: Combine the proposition with Lemma V, sec. 9.3.

Q.E.D.

Now let χ_S denote the Euler class of the sphere bundle associated with τ_M.

Proposition IV: The Lefschetz class, Λ_M, and χ_S are related by the equation

$$\chi_S = \Delta^{\#}(\Lambda_M).$$

Proof: Choose a representative Φ of θ_M such that carr $\Phi \subset O_\epsilon$. Then it follows from Proposition III that $\Delta^{\#}(\Lambda_M)$ is represented by $\Delta^*(\varphi_c)_*\Phi$. Since $\Delta = \varphi \circ o$,

$$\Delta^*(\varphi_c)_*\Phi = o^*\varphi^*(\varphi_c)_*\Phi = o^*\Phi.$$

But Corollary I to Proposition V, sec. 9.6, implies that $o^*\Phi$ represents χ_S.

Q.E.D.

Proof of Theorem I (sec. 10.1): Apply Proposition IV, and the corollary to Proposition II, sec. 10.3, to obtain

$$\chi_M = \int_M^{\#} \Delta^{\#}(\Lambda_M) = \int_M^{\#} \chi_S.$$

Q.E.D.

10.5. Index sum. Proof of Theorem II (sec. 10.1): We may assume that N is connected. If N is orientable, Theorem II is an immediate consequence of Theorem I and Theorem III, sec. 9.9. If N is non-orientable, let $\tilde{N}$ be the oriented double cover (cf. Example 9, sec. 3.21). Then X pulls back to a vector field $\tilde{X}$ on $\tilde{N}$ with finitely many zeros and, since $\tilde{N}$ is oriented,

$$\chi_{\tilde{N}} = j(\tilde{X}).$$

On the other hand, by Theorem IV, sec. 9.9,

$$j(\tilde{X}) = 2j(X).$$

Thus it remains to show that

$$\chi_{\tilde{N}} = 2\chi_N.$$

Recall from sec. 5.7 that

$$H(\tilde{N}) = H_+(\tilde{N}) \oplus H_-(\tilde{N})$$

and $H_+(\tilde{N}) \cong H(N)$. Moreover, if $\omega: \tilde{N} \to \tilde{N}$ is the covering transformation and $\omega^p: H^p(\tilde{N}) \leftarrow H^p(\tilde{N})$ is the induced linear map, then

$$\operatorname{tr} \omega^p = \dim H_+^p(\tilde{N}) - \dim H_-^p(\tilde{N}).$$

Since $\omega(z) \neq z$ ($z \in \tilde{N}$), it will follow from the corollary to Theorem III, sec. 10.8, that

$$\sum_{p=0}^{n} (-1)^p \operatorname{tr} \omega^p = 0.$$

Thus

$$\sum_{p=0}^{n} (-1)^p \dim H_+^p(\tilde{N}) = \sum_{p=0}^{n} (-1)^p \dim H_-^p(\tilde{N}).$$

Recalling that $\dim H_+^p(\tilde{N}) = \dim H^p(N)$ (cf. sec. 5.7), we obtain

$$\chi_{\tilde{N}} = \sum_{p=0}^{n} (-1)^p(\dim H_+^p(\tilde{N}) + \dim H_-^p(\tilde{N})) = 2\chi_N \,.$$

Q.E.D.

Corollary I: If $\dim N$ is odd, then $\chi_N = 0$.

Proof: Apply Theorem IV(1), sec. 5.16.

Q.E.D.

Corollary II: If $\tilde{N}$ is the double cover of N, then

$$\chi_{\tilde{N}} = 2\chi_N \,.$$

Corollary III: Let P be any connected n-manifold ($n \geqslant 2$). Then P admits a vector field without zeros if and only if P is not compact, or P is compact and $\chi_P = 0$.

Proof: Apply Theorem V, sec. 9.9, and Theorem II.

Q.E.D.

Corollary IV: Every odd dimensional manifold admits a vector field without zeros.

§2. Coincidence number

10.6. The Poincaré adjoint. Let M and N be compact oriented connected manifolds of dimensions m and n, respectively, and let $\varphi: M \to N$ be a smooth map. Denote the restriction of $\varphi^\#$ to $H^p(N)$ by φ^p. Let

$$\tilde{\varphi}^{m-p}: H^{m-p}(M) \to H^{n-p}(N)$$

be the unique linear map such that

$$\mathscr{P}_N(\tilde{\varphi}^{m-p}\alpha, \beta) = \mathscr{P}_M(\alpha, \varphi^p\beta), \qquad \alpha \in H^{m-p}(M), \quad \beta \in H^p(N)$$

($\mathscr{P}_N$, $\mathscr{P}_M$ are the Poincaré scalar products—cf. sec. 5.11).

The linear maps $\tilde{\varphi}^q$ define a linear map

$$\tilde{\varphi}: H(M) \to H(N),$$

homogeneous of degree $n - m$; it is called the *Poincaré adjoint of* $\varphi^\#$. If $m = n$, $\tilde{\varphi}$ is homogeneous of degree zero.

If $\psi: N \to Q$ is a map of N into a third compact connected oriented manifold Q, then

$$\widetilde{\psi \circ \varphi} = \tilde{\psi} \circ \tilde{\varphi}.$$

Lemma I: If $m = n$, then

$$\tilde{\varphi} \circ \varphi^\# = \deg \varphi \cdot \iota.$$

In particular, if $M = N$ and $\deg \varphi \neq 0$, then $\varphi^\#$ is a linear isomorphism and

$$\tilde{\varphi} = \deg \varphi \cdot (\varphi^\#)^{-1}.$$

Proof: Observe that, for $\alpha \in H^p(N)$ and $\beta \in H^{m-p}(N)$,

$$\begin{aligned}\mathscr{P}_N(\tilde{\varphi}\varphi^\#\alpha, \beta) &= \mathscr{P}_M(\varphi^\#\alpha, \varphi^\#\beta)\\ &= \int_M^\# \varphi^\#(\alpha \cdot \beta) = \deg \varphi \cdot \mathscr{P}_N(\alpha, \beta).\end{aligned}$$

The lemma follows from the nondegeneracy of $\mathscr{P}_N$.

Q.E.D.

Proposition V: Let M be a compact oriented n-manifold. Then the Lefschetz class of M is given by

$$\Lambda_M = (-1)^n \tilde{\Delta}(1).$$

Proof: Let $\alpha \in H^p(M)$, $\beta \in H^{n-p}(M)$. Then by Corollary I to Theorem I, sec. 7.14,

$$\begin{aligned}\mathscr{P}_M\Big(\beta, \oint_M^{\#} \pi_R^{\#}\alpha \cdot \tilde{\Delta}(1)\Big) &= \mathscr{P}_{M\times M}(\pi_L^{\#}\beta, \pi_R^{\#}\alpha \cdot \tilde{\Delta}(1)) \\ &= (-1)^n \mathscr{P}_{M\times M}(\tilde{\Delta}(1), \pi_L^{\#}\beta \cdot \pi_R^{\#}\alpha) \\ &= (-1)^n \mathscr{P}_M(\beta, \alpha).\end{aligned}$$

It follows that

$$\oint_M^{\#} \pi_R^{\#}\alpha \cdot (-1)^n \tilde{\Delta}(1) = \alpha, \qquad \alpha \in H(M).$$

Now Corollary I to Proposition I, sec. 10.3, implies that

$$\Lambda_M = (-1)^n \tilde{\Delta}(1).$$

Q.E.D.

10.7. Coincidence number and Lefschetz number. The *coincidence number* of two smooth maps $\varphi: M \to N$ and $\psi: M \to N$ between compact, connected, oriented n-manifolds is defined by

$$L(\varphi, \psi) = \sum_{p=0}^{n} (-1)^p \operatorname{tr}(\varphi^p \circ \tilde{\psi}^p).$$

If $N = M$ and ψ is the identity map, this number is denoted by $L(\varphi)$ and called the *Lefschetz number* of φ,

$$L(\varphi) = \sum_{p=0}^{n} (-1)^p \operatorname{tr} \varphi^p.$$

In particular, $L(\iota) = \chi_M$.

Proposition VI: Let $\varphi, \psi: M \to N$ be as above. Then

(1) $L(\varphi, \psi) = (-1)^n L(\psi, \varphi)$.

(2) If $\chi: Q \to M$ is a map of a third connected oriented compact n-manifold into M, then

$$L(\varphi \circ \chi, \psi \circ \chi) = \deg \chi \cdot L(\varphi, \psi).$$

(3) If $\varphi \sim \varphi_1$ and $\psi \sim \psi_1$ then $L(\varphi, \psi) = L(\varphi_1, \psi_1)$.

Proof: (1) Since

$$\mathscr{P}_M(\alpha, \psi^\# \tilde{\varphi} \beta) = \mathscr{P}_M(\varphi^\# \tilde{\psi} \alpha, \beta), \qquad \alpha, \beta \in H(M),$$

it follows that

$$\operatorname{tr}(\psi^{n-p} \circ \tilde{\varphi}^{n-p}) = \operatorname{tr}(\varphi^p \circ \tilde{\psi}^p), \qquad 0 \leqslant p \leqslant n,$$

whence

$$L(\varphi, \psi) = \sum_{p=0}^{n} (-1)^p \operatorname{tr}(\varphi^p \circ \tilde{\psi}^p) = (-1)^n L(\psi, \varphi).$$

(2) In view of Lemma I, sec. 10.6,

$$\begin{aligned} \operatorname{tr}[(\varphi \circ \chi)^p \circ (\widetilde{\psi \circ \chi})^p] &= \operatorname{tr}(\chi^p \circ \varphi^p \circ \tilde{\psi}^p \circ \tilde{\chi}^p) \\ &= \operatorname{tr}(\varphi^p \circ \tilde{\psi}^p \circ \tilde{\chi}^p \circ \chi^p) \\ &= \deg \chi \cdot \operatorname{tr}(\varphi^p \circ \tilde{\psi}^p) \end{aligned}$$

and so (2) follows.

(3) Obvious.

Q.E.D.

Again, suppose $\varphi: M \to N$ and $\psi: M \to N$ are smooth maps between compact, connected, oriented n-manifolds. Recall the Lefschetz isomorphisms (sec. 10.3)

$$\lambda_M: L_M \xrightarrow{\cong} H^n(M \times M), \qquad \lambda_N: L_N \xrightarrow{\cong} H^n(N \times N)$$

and that

$$\Lambda_M = \lambda_M(\iota), \qquad \Lambda_N = \lambda_N(\iota)$$

denote the Lefschetz classes. On the other hand,

$$\varphi^\# \circ \tilde{\psi} = \sum_p \varphi^p \circ \tilde{\psi}^p \in L_M.$$

Proposition VII: With the notations and hypotheses above

$$\lambda_M(\varphi^\# \circ \tilde{\psi}) = (\varphi \times \psi)^\#(\Lambda_N).$$

In particular, if $\Delta_M: M \to M \times M$ denotes the diagonal map, then

$$L(\varphi, \psi) = \int_M^{\#} \Delta_M^{\#} (\varphi \times \psi)^{\#}(\Lambda_N).$$

Lemma II: Let $\sigma \in L_N$. Then

$$(\varphi \times \psi)^{\#}(\lambda_N \sigma) = \lambda_M(\varphi^{\#} \circ \sigma \circ \tilde{\psi}).$$

Proof: As in sec. 10.3 it is sufficient to consider the case

$$\sigma = k(\alpha \otimes D_N \beta), \qquad \alpha \in H^p(N), \quad \beta \in H^{n-p}(N).$$

Then, for $\gamma \in H^p(M)$,

$$\begin{aligned}(\varphi^{\#} \circ \sigma \circ \tilde{\psi})(\gamma) &= (-1)^{np}\langle D_N \beta, \tilde{\psi}\gamma\rangle \cdot \varphi^{\#}\alpha \\ &= (-1)^{np}\mathscr{P}_M(\psi^{\#}\beta, \gamma) \cdot \varphi^{\#}\alpha \\ &= [k(\varphi^{\#}\alpha \otimes D_M \psi^{\#}\beta)](\gamma).\end{aligned}$$

It follows that

$$\varphi^{\#} \circ \sigma \circ \tilde{\psi} = k(\varphi^{\#}\alpha \otimes D_M \psi^{\#}\beta),$$

whence

$$\begin{aligned}\lambda_M(\varphi^{\#} \circ \sigma \circ \tilde{\psi}) &= (\varphi \times \psi)^{\#} \circ \kappa_{\#}(\alpha \otimes \beta) \\ &= (\varphi \times \psi)^{\#}(\lambda_N \sigma).\end{aligned}$$

Q.E.D.

Proof of Proposition VII: Applying Lemma II with σ the identity map of $H(N)$, we obtain

$$\lambda_M(\varphi^{\#} \circ \tilde{\psi}) = (\varphi \times \psi)^{\#}(\lambda_N(\iota)) = (\varphi \times \psi)^{\#}(\Lambda_N).$$

Thus Proposition II, sec. 10.3, yields

$$L(\varphi, \psi) = Tr(\varphi^{\#} \circ \tilde{\psi}) = \int_M^{\#} \Delta_M^{\#}(\lambda_M(\varphi^{\#} \circ \tilde{\psi})) = \int_M^{\#} \Delta_M^{\#}(\varphi \times \psi)^{\#}(\Lambda_N).$$

Q.E.D.

10.8. The weak Lefschetz theorem. Theorem III: Let $\varphi, \psi: M \to N$ be smooth maps between compact, connected, oriented n-manifolds such that $\varphi(x) \neq \psi(x)$ $(x \in M)$. Then $L(\varphi, \psi) = 0$.

Proof: Define $\chi: M \to N \times N$ by

$$\chi(x) = (\varphi(x), \psi(x)), \qquad x \in M.$$

Since M is compact so is $\chi(M)$. By hypothesis,

$$\chi(M) \cap \Delta_N(N) = \varnothing,$$

where $\Delta_N: N \to N \times N$ is the diagonal map. Hence $U = N \times N - \chi(M)$ is a neighbourhood of $\Delta(N)$. According to the corollary of Proposition III, sec. 10.4, there is a representative, Φ, of the Lefschetz class Λ_N such that carr $\Phi \subset U$. Then $\chi^*\Phi = 0$; i.e.,

$$\Delta_M^* \circ (\varphi \times \psi)^*(\Phi) = 0.$$

Passing to cohomology yields

$$\Delta_M^\# \circ (\varphi \times \psi)^\#(\Lambda_N) = 0,$$

whence, by Proposition VII, sec. 10.7,

$$L(\varphi, \psi) = 0.$$

Q.E.D.

Corollary: If the Lefschetz number of a map $\varphi: M \to M$ is different from zero, then φ has at least one fixed point; i.e., for some $a \in M$,

$$\varphi(a) = a.$$

Remark: The rest of this chapter is devoted to proving a strengthened version of Theorem III, which appears in sec. 10.10.

§3. The Lefschetz coincidence theorem

10.9. Local coincidence number. Let $\varphi, \psi: M \to N$ be smooth maps between oriented n-manifolds ($n \geqslant 2$). A point $a \in M$ is called a *coincidence point for* φ *and* ψ if $\varphi(a) = \psi(a)$. A coincidence point, a, is called *isolated*, if there is a neighbourhood O of a such that $\varphi(x) \neq \psi(x)$, $x \in O - \{a\}$. We shall define the *local coincidence number* of φ and ψ at an isolated coincidence point a.

Let $(F, v, \mathbb{R}^n)$ be a chart on N such that $\varphi(a) \in F$ and $v(\varphi(a)) = 0$. Set

$$\dot{F} = F - \{\varphi(a)\} \qquad \text{and} \qquad F \dot{\times} F = F \times F - \Delta(F)$$

(where Δ is the diagonal map). Then we have the inclusion map

$$j : \varphi(a) \times \dot{F} \to F \dot{\times} F.$$

Lemma III: The map j induces an isomorphism of cohomology.

Proof: Use v to give F a linear structure, and define the map $\mu: F \times F \xrightarrow{\cong} F \times F$ by

$$\mu(x, y) = (x, x + y).$$

Since $\varphi(a)$ is the zero of F, the diagram

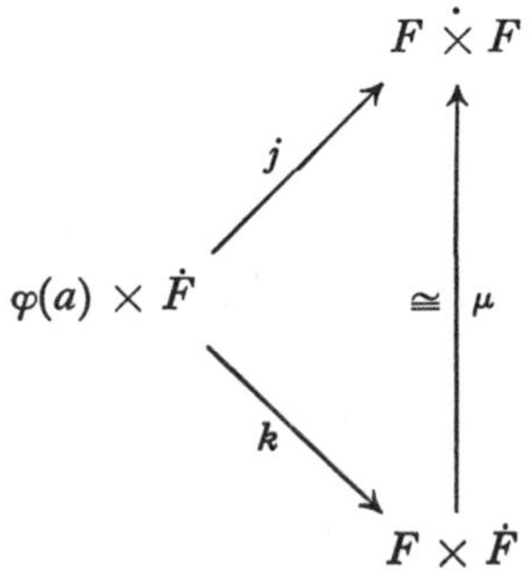

commutes, where k is the obvious inclusion. Since k induces an isomorphism of cohomology, so does j.

Q.E.D.

Next, choose a neighbourhood U of a, diffeomorphic to $\mathbb{R}^n$ and satisfying the conditions (i) $\bar{U}$ is compact, (ii) $\bar{U} \subset \varphi^{-1}(F) \cap \psi^{-1}(F)$, and (iii) $\varphi(x) \neq \psi(x)$, $x \in \bar{U} - \{a\}$. Define $\tau: U \to F \times F$ by

$$\tau(x) = (\varphi(x), \psi(x)).$$

Then τ restricts to a smooth map

$$\dot{\tau}: \dot{U} \to F \dot{\times} F.$$

Thus we can form the real number (cf. sec. 6.7)

$$t = (\alpha_U \circ \dot{\tau}^\# \circ (j^\#)^{-1} \circ \alpha_F^{-1})(1).$$

Lemma IV: The number t is independent of the choice of U and of $(F, v, \mathbb{R}^n)$.

Proof: Suppose $(F_1, v_1, \mathbb{R}^n)$ and U_1 satisfy the conditions above. It is sufficient to consider the case that $F_1 \subset F$ and $U_1 \subset U$. But then the diagram

$$\begin{array}{ccccc} \varphi(a) \times \dot{F} & \xrightarrow{j} & F \dot{\times} F & \xleftarrow{\dot{\tau}} & \dot{U} \\ \uparrow & & \uparrow & & \uparrow \\ \varphi(a) \times \dot{F}_1 & \xrightarrow{j_1} & F_1 \dot{\times} F_1 & \xleftarrow{\dot{\tau}_1} & \dot{U}_1 \end{array}$$

commutes, and the lemma follows.

Q.E.D.

Definition: The number $(\alpha_U \circ \dot{\tau}^\# \circ (j^\#)^{-1} \circ \alpha_F^{-1})(1)$ is called the *local coincidence number* of φ and ψ at a. It is denoted by $L_a(\varphi, \psi)$.

Now (with the notation above), regard φ and ψ as maps of U into the linear space F. Define $\psi - \varphi : U \to F$ by

$$(\psi - \varphi)(x) = \psi(x) - \varphi(x), \qquad x \in U.$$

Then $\psi - \varphi$ has an isolated zero at a. Thus the integer $\deg_a(\psi - \varphi)$ is defined.

Lemma V: $\deg_a(\psi - \varphi) = L_a(\varphi, \psi)$.

Proof: Define $\sigma: U \to F \times F$ by

$$\sigma(x) = (\varphi(x), (\psi - \varphi)(x))$$

and restrict σ to

$$\dot{\sigma} : \dot{U} \to F \times \dot{F}.$$

The diagram

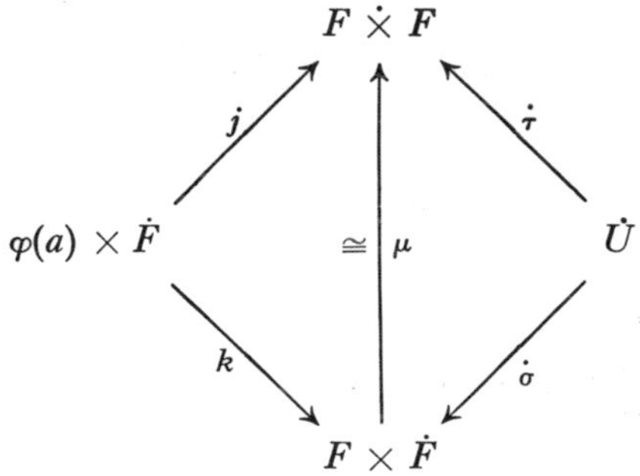

commutes, whence

$$L_a(\varphi, \psi) = \alpha_U \circ \dot{\sigma}^{\#} \circ (k^{\#})^{-1} \circ \alpha_F^{-1}(1).$$

On the other hand, the projection $\rho: F \times \dot{F} \to \dot{F}$ satisfies $\rho \circ k = \iota$, and hence

$$\rho^{\#} = (k^{\#})^{-1}.$$

But

$$\rho \circ \dot{\sigma} = (\psi \mathbin{\dot{-}} \varphi) : \dot{U} \to \dot{F}.$$

It follows that (cf. sec. 6.11)

$$\begin{aligned} L_a(\varphi, \psi) &= \alpha_U \circ (\psi \mathbin{\dot{-}} \varphi)^{\#} \circ \alpha_F^{-1}(1) \\ &= \deg_a(\psi - \varphi). \end{aligned}$$

Q.E.D.

Corollary: $L_a(\varphi, \psi)$ is an integer.

If $N = M$, then the coincidence number $L_a(\varphi, \iota)$ at an isolated fixed point a of φ is called the *index of the fixed point* and is denoted by $L_a(\varphi)$. It is independent of the orientation of M.

10.10. The main theorem. For the rest of this article, M and N denote fixed compact, connected, oriented n-manifolds ($n \geqslant 2$);

$\varphi, \psi: M \to N$ are two smooth maps with only finitely many coincidence points $a_1, \ldots, a_r$. The local coincidence number $L_{a_i}(\varphi, \psi)$ will be denoted by $L_i(\varphi, \psi)$. The rest of this article is devoted to proving the following generalization of Theorem III, sec. 10.8:

Theorem IV: With the notation and hypotheses above, the coincidence number of φ and ψ is the sum of the local coincidence numbers,

$$L(\varphi, \psi) = \sum_{i=1}^{r} L_i(\varphi, \psi).$$

Corollary I: $L(\varphi, \psi)$ is an integer.

Corollary II: Let $\varphi: M \to M$ be a smooth map of a compact, connected, oriented n-manifold, M, into itself with finitely many fixed points $a_1, \ldots, a_r$. Then

$$L(\varphi) = \sum_{p=0}^{n} (-1)^p \operatorname{tr} \varphi^p = \sum_{i=1}^{r} L_i(\varphi).$$

Corollary III: If $\varphi: M \to M$ has finitely many fixed points and $\varphi \sim \iota$, then

$$\chi_M = \sum_{i=1}^{r} L_i(\varphi).$$

Examples: 1. $\varphi: S^n \to S^n$ be a map with finitely many fixed points $a_1, \ldots, a_r$. Since $H^p(S^n) = 0$ $(1 \leqslant p \leqslant n-1)$, we have

$$\operatorname{tr} \varphi^p = 0, \qquad 1 \leqslant p \leqslant n-1.$$

Moreover,

$$\operatorname{tr} \varphi^n = \deg \varphi \qquad \text{and} \qquad \operatorname{tr} \varphi^0 = 1.$$

Thus Corollary II yields

$$1 + (-1)^n \deg \varphi = \sum_{i=1}^{r} L_i(\varphi).$$

2. Let $\varphi, \psi: M \to S^n$ be smooth maps (M a compact, oriented, n-manifold). Then

$$\varphi^p = 0 = \tilde{\psi}^p, \qquad 1 \leqslant p \leqslant n-1.$$

Moreover,

$$\varphi^n \circ \tilde{\psi}^n = \deg \varphi \cdot \iota$$

and

$$\varphi^0 \circ \tilde{\psi}^0 = \deg \psi \cdot \iota.$$

It follows that

$$L(\varphi, \psi) = \deg \psi + (-1)^n \deg \varphi.$$

Thus if φ, ψ have finitely many coincidence points a_i ($i = 1, \ldots, r$), the theorem gives

$$\deg \psi + (-1)^n \deg \varphi = \sum_{i=1}^{r} L_i(\varphi, \psi).$$

3. Again consider $\varphi, \psi: M \to S^n$, but this time assume that $-\varphi$ and ψ have finitely many coincidence points $b_1, \ldots, b_s$, where

$$(-\varphi)(x) = -(\varphi(x)).$$

Since $\deg(-\varphi) = (-1)^{n-1} \deg \varphi$ (cf. Example 1, sec. 6.2), we obtain

$$\deg \psi - \deg \varphi = \sum_{i=1}^{s} L_i(-\varphi, \psi).$$

4. Let $\varphi, \psi: M \to N$ be smooth maps between compact, connected, oriented n-manifolds such that

$$\varphi^{\#} = \psi^{\#}.$$

Then Proposition VI, (2), sec. 10.7, shows that

$$L(\varphi, \psi) = L(\varphi, \varphi) = \deg \varphi \cdot L(\iota, \iota) = \deg \varphi \cdot \chi_M.$$

Thus if φ, ψ have only finitely many coincidence points $a_1, \ldots, a_r$, then

$$\deg \varphi \cdot \chi_M = \sum_{i=1}^{r} L_i(\varphi, \psi).$$

In particular, if $\deg \varphi \neq 0$, $\chi_M \neq 0$, then φ and ψ have at least one coincidence point.

We come now to the proof of Theorem IV. It is broken up into four steps, each occupying a section. (Note that all diagonal maps are denoted by Δ).

10.11. Step I. Preliminaries. Let $\mathbb{R}^n$ have a fixed orientation and Euclidean metric, and choose charts $(F_i, v_i, \mathbb{R}^n)$ for N, subject to the following conditions:

(i) The F_i are disjoint
(ii) v_i is orientation preserving
(iii) $\varphi(a_i) \in F_i$ and $v_i(\varphi(a_i)) = 0$. Assign to each F_i that structure of an oriented Euclidean space for which v_i is an orientation preserving isometry.

Next, choose neighbourhoods V_i of $\varphi(a_i)$ $(i = 1, ..., r)$ so that $\overline{V}_i$ is compact and $\overline{V}_i \subset F_i$. Then there are open sets $W_i, U_i \subset M$ satisfying the conditions

(iv) $a_i \in W_i \subset \overline{W}_i \subset U_i$
(v) U_i is diffeomorphic to $\mathbb{R}^n$
(vi) $\overline{U}_i$ is compact and $\varphi(x) \neq \psi(x)$ $(x \in \overline{U}_i - \{a_i\})$
(vii) $\varphi(\overline{U}_i) \cup \psi(\overline{U}_i) \subset V_i$.

In particular, if $x \in \overline{U}_i$ then $\varphi(x), \psi(x) \in F_i$, and so we can form the difference

$$\psi(x) - \varphi(x) \in F_i .$$

Since $\overline{U}_i - W_i$ is compact, condition (vi) implies that for some $\epsilon > 0$

$$|\psi(x) - \varphi(x)| > \epsilon, \qquad x \in \overline{U}_i - W_i ; \quad i = 1, ..., r. \tag{10.1}$$

Fix such an ϵ and set

$$(F_i)_\epsilon = \{x \in F_i \mid |x| < \epsilon\}.$$

10.12. Step II. Representation of the local coincidence numbers. Let

$$\varphi_i: U_i \to V_i, \qquad \psi_i: U_i \to V_i$$

denote the restrictions of φ and ψ. Then, according to Lemma V, sec. 10.9, the local coincidence number $L_i(\varphi, \psi)$ of φ and ψ at a_i is given by

$$L_i(\varphi, \psi) = \deg_{a_i}(\psi_i - \varphi_i), \tag{10.2}$$

where $\psi_i - \varphi_i: U_i \to F_i$ denotes the difference map.

On the other hand, consider the oriented Riemannian vector bundles $\xi_i = (V_i \times F_i, \pi_i, V_i, F_i)$ and let $\rho_i: V_i \times F_i \to F_i$ denote the projection. Define

$$\sigma_i: U_i \to V_i \times F_i$$

by

$$\sigma_i(x) = (\varphi_i(x), (\psi_i - \varphi_i)(x)), \qquad x \in U_i .$$

Then the diagrams

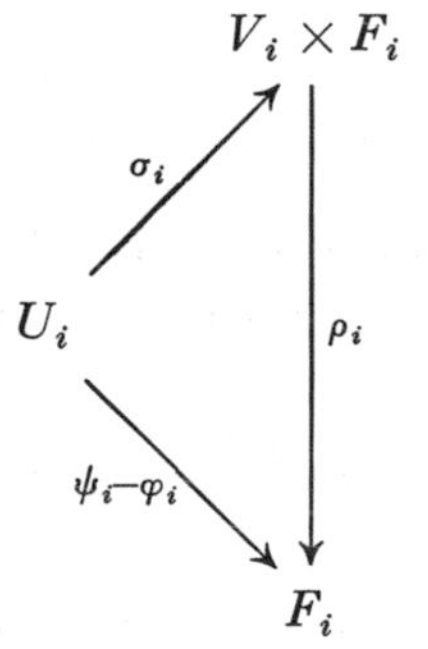

commute.

Now let θ_i be the Thom class for ξ_i and let Ψ_i be a representative of θ_i such that

$$\operatorname{carr} \Psi_i \subset V_i \times (F_i)_\epsilon .$$

In view of formula (10.1) and condition (iv) we can apply the example of sec. 9.8 to obtain

$$\deg_{a_i}(\psi_i - \varphi_i) = \int_{U_i} \sigma_i^* \Psi_i .$$

Hence (cf. formula 10.2)

$$L_i(\varphi, \psi) = \int_{U_i} \sigma_i^* \Psi_i . \tag{10.3}$$

10.13. Step III. Define smooth maps

$$\mu_i: F_i \times F_i \to N \times N$$

by

$$\mu_i(x, y) = (x, x + y), \qquad x, y \in F_i .$$

Each μ_i is a diffeomorphism onto an open subset of $N \times N$. Now consider the open subset $O \subset N \times N$ given by

$$O = \left[\left(N - \bigcup_{i=1}^{r} \bar{V}_i\right) \times N\right] \cup \bigcup_{i=1}^{r} \mu_i(F_i \times (F_i)_\epsilon).$$

O is a neighbourhood of $\Delta(N)$, and satisfies

$$O \cap (\{x\} \times N) = \mu_i(\{x\} \times (F_i)_\epsilon), \qquad x \in V_i . \tag{10.4}$$

Next, observe that the μ_i restrict to smooth maps β_i: $V_i \times F_i \to N \times N$ and that the diagrams

$$\begin{array}{ccc} V_i \times F_i & \xrightarrow{\beta_i} & N \times N \\ \downarrow{\scriptstyle \pi_i} & & \downarrow{\scriptstyle \pi_L} \\ V_i & \xrightarrow{j_i} & N \end{array}$$

commute (j_i denotes the inclusion map). In view of (10.4) we may use Proposition VIII, sec. 7.12, to conclude that if $\Phi \in A(N \times N)$ and carr $\Phi \subset O$, then

$$\operatorname{carr} \beta_i^* \Phi \subset V_i \times (F_i)_\epsilon \tag{10.5}$$

and

$$j_i^* \left(\oint_N \Phi \right) = \oint_{F_i} \beta_i^* \Phi.$$

In particular let Φ be a closed n-form representing the Lefschetz class Λ_N and such that carr $\Phi \subset O$ (cf. the corollary to Proposition III, sec. 10.4). Then (cf. Corollary II to Proposition I, sec. 10.3)

$$\oint_{F_i} \beta_i^* \Phi = \oint_N \Phi = 1.$$

Hence (cf. Corollary II of sec. 9.2) $\beta_i^* \Phi$ represents the Thom class θ_i. In view of (10.5) we can apply formula (10.3) at the end of Step II to obtain

$$L_i(\varphi, \psi) = \int_{U_i} \sigma_i^* \beta_i^* \Phi.$$

10.14. Step IV. Define $\tau : M \to N \times N$ by $\tau = (\varphi \times \psi) \circ \Delta$. Then Proposition VII, sec. 10.7, yields

$$L(\varphi, \psi) = \int_M^{\#} \tau^{\#}(\Lambda_N).$$

But, by hypothesis,

$$\tau(M) \cap \Delta(N) = \{(\varphi(a_i), \psi(a_i))\}_{i=1,\dots,r}.$$

In particular, $\tau(M - \bigcup_i U_i)$ is a compact set disjoint from $\Delta(N)$. Thus, if O is the open set constructed in Step III, then

$$O - \tau\left(M - \bigcup_i U_i\right)$$

is a neighbourhood of $\Delta(N)$.

According to the corollary to Proposition III, sec. 10.4, we may choose a representative, Φ, of the Lefschetz class of N such that

$$\operatorname{carr} \Phi \subset O - \tau\Big(M - \bigcup_i U_i\Big).$$

Then

$$\operatorname{carr} \tau^*\Phi \subset \bigcup_i U_i$$

and so

$$L(\varphi, \psi) = \int_M \tau^*\Phi = \sum_{i=1}^r \int_{U_i} \tau^*\Phi.$$

Finally, observe that the diagrams

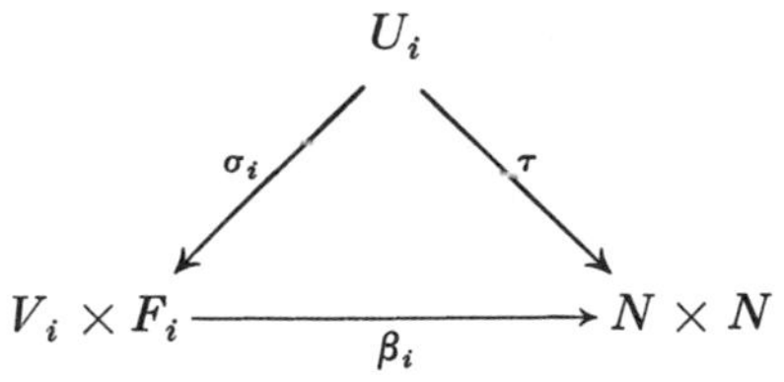

commute. Since $\operatorname{carr} \Phi \subset O$ it follows that (cf. Step III)

$$L(\varphi, \psi) = \sum_{i=1}^r \int_{U_i} \sigma_i^* \beta_i^* \Phi = \sum_{i=1}^r L_i(\varphi, \psi).$$

This completes the proof.

Q.E.D.

Problems

1. Let $\mathscr{B} = (E, \pi, B, F)$ be a smooth fibre bundle with compact base and compact fibre. Show that the Euler–Poincaré characteristics of E, B, F are connected by

$$\chi_E = \chi_B \cdot \chi_F .$$

2. A manifold is said to have a *homogeneous structure*, if it is given an atlas $\{(U_\alpha, u_\alpha)\}$ subject to the following conditions: If $x \in u_\alpha(U_{\alpha\beta})$, then, for some $\epsilon_x > 0$,

$$u_{\beta\alpha}(tx) = tu_{\beta\alpha}(x), \qquad |t - 1| < \epsilon_x .$$

(i) Show that every noncompact manifold admits a homogeneous structure.

(ii) Show that a compact manifold admits a homogeneous structure if and only if its Euler characteristic is nonnegative.

(iii) Let $\{(U_\alpha, u_\alpha)\}$ be a homogeneous structure for M. A point $x \in M$ is called an *origin*, if, for some α, $x \in U_\alpha$ and $u_\alpha(x) = 0$. Show that if x is an origin, then for any U_β containing x, $u_\beta(x) = 0$. Show that the origins form a discrete subset of M. Show that the number of origins of a compact manifold M is χ_M.

3. Let M be a connected compact orientable 4-manifold which admits a vector field, X, without zeros. Show that $H^1(M) \neq 0$ and that dim $H^2(M) = 2(\dim H^1(M) - 1)$.

4. Show that a compact 4-manifold admits a Lorentz metric if and only if its Euler characteristic is zero. (A Lorentz metric is an indefinite metric in τ_M of signature 2.)

5. Let $\varphi : M \to M$ be a smooth map and let $a \in M$ be a fixed point for φ. Assume that the linear transformation $(d\varphi)_a : T_a(M) \to T_a(M)$ satisfies $\det((d\varphi)_a - \iota) \neq 0$. Show that a is an isolated fixed point and that

$$L_a(\varphi) = \begin{cases} -1, & \text{if } \det((d\varphi)_a - \iota) > 0 \\ +1, & \text{if } \det((d\varphi)_a - \iota) < 0. \end{cases}$$

6. Let S^2 be the Riemann sphere. Determine the fixed points and their indices of the map $\varphi : S^2 \to S^2$ given by $\varphi(z) = z^n$ $(n \in \mathbb{Z})$.

7. Projective spaces. (i) Show that every map $\varphi : \mathbb{R}P^n \to \mathbb{R}P^n$ (n even) has a fixed point.

(ii) Construct a map $\mathbb{C}P^2 \to \mathbb{C}P^2$ without fixed points.

8. Let $\varphi : \mathbb{C}P^n \to \mathbb{C}P^n$ be smooth. Show that the restriction of $\varphi^\#$ to $H^2(\mathbb{C}P^n)$ is of the form $\lambda_\varphi \cdot \iota$ $(\lambda_\varphi \in \mathbb{R})$.

(i) Obtain the relations

$$\deg \varphi = \lambda_\varphi^n \qquad \text{and} \qquad L(\varphi) = \sum_{\nu=0}^{n} \lambda_\varphi^\nu$$

(cf. problem 5, Chap. VIII).

(ii) If $\psi : \mathbb{C}P^n \to \mathbb{C}P^n$ is a second smooth map, show that

$$L(\varphi, \psi) = \sum_{p=0}^{n} \lambda_\varphi^p \lambda_\psi^{n-p}.$$

Conclude that $L(\varphi, \psi) \neq 0$ (and so φ and ψ have a coincidence point) unless n is odd and $\deg \psi = -\deg \varphi$.

(iii) Repeat (i) and (ii) for $\mathbb{H}P^n$.

9. Consider two compact oriented n-manifolds-with-boundary $(M_i, \partial M_i)$ $(i = 1, 2)$ and let M be a compact $(n-1)$-manifold. Suppose $\varphi_i : \partial M_i \xrightarrow{\cong} M$ are given diffeomorphisms. Identify ∂M_1 with ∂M_2 via $\varphi_1^{-1} \circ \varphi_2$ to obtain a compact manifold, $M_1 \# M_2$. Let $X_i \in \mathrm{Sec}(\tau_{M_i}|_{\partial M_i})$ $(i = 1, 2)$ denote the outward pointing normal vector fields.

(i) Extend X_i to a vector field $\tilde{X}_i$ on M_i with finitely many zeros. Show that $j(\tilde{X}_i)$ is independent of the extension $\tilde{X}_i$ (cf. problem 15, Chap. VIII).

(ii) Prove that $H(M_i)$ and $H(M_i, \partial M_i)$ have finite dimension (cf. problem 14, Chap. VIII). Let χ_{M_i}, $\chi_{M_i, \partial M_i}$ be the corresponding Euler–Poincaré characteristics.

(iii) Establish the relations

$$j(\tilde{X}_1) + (-1)^n j(\tilde{X}_2) = \chi_{M_1 \# M_2},$$

$$j(\tilde{X}_1) + j(\tilde{X}_2) = \chi_{M_1} + \chi_{M_2},$$

and

$$\chi_{M_1 \# M_2} = \chi_{M_1} + \chi_{M_2} - \chi_M = \chi_{M_1, \partial M_1} + \chi_{M_2, \partial M_2} + \chi_M .$$

(iv) Conclude that

$$j(\tilde{X}_i) = \chi_{M_i} = \chi_{M_i, \partial M_i} + \chi_M .$$

10. Critical points (Morse). Let $f \in \mathscr{S}(M)$ (M a compact n-manifold) have only nondegenerate critical points, $a_1, \ldots, a_k$. Suppose that the Hessian of f at a_i has $n - p_i$ positive directions and p_i negative directions. Show that

$$\chi_M = \sum_{i=1}^{k} (-1)^{p_i}.$$

Generalize this to compact manifolds-with-boundary (cf. problem 9).

11. Let M be a compact manifold. Let $X \in \mathscr{X}(M)$ and suppose the zero-set of X consists of finitely many disjoint connected submanifolds P_i. Assume that X is nowhere normal to the boundaries of tubular neighbourhoods of the P_i. Prove that

$$\chi_M = \sum_i \chi_{P_i} .$$

12. Suppose M is compact and oriented. Assume that two maps $\varphi, \psi : M \to M$ have a single coincidence point, a. Show that ψ is homotopic to a map $\psi_1 : M \to M$ which has no coincidence points with φ, if and only if $L(\varphi, \psi) = 0$.

13. Intersection theory. Let M be a compact submanifold of a manifold N. Suppose $\varphi : P \to N$ is smooth. φ is called *transverse regular to* M, if whenever $\varphi(x) \in M$, then

$$T_{\varphi(x)}(M) + Im(d\varphi)_x = T_{\varphi(x)}(N).$$

(i) If φ is transverse regular to M, prove that $\varphi^{-1}(M)$ is a closed submanifold of P. What is its dimension?

(ii) Use Sard's theorem (cf. problem 13, Chap. III) to prove that any map φ is homotopic to a smooth map φ_1 which is transverse regular to M (Thom's transversality theorem).

(iii) Assume all manifolds oriented. Let $i : M \to N$ be the inclusion and let φ_1 be as in (ii). Denote $\varphi_1^{-1}(M)$ by $M \cap P$ and let $\psi : M \cap P \to N$

be the restriction of φ_1. Orient $M \cap P$. Define cohomology classes in $H(N)$ by

$$[M] = \tilde{\imath}(1), \qquad [P] = \tilde{\varphi}(1), \qquad [M \cap P] = \tilde{\psi}(1).$$

Show that

$$[M \cap P] = [M] \cdot [P]$$

and conclude that $[M \cap P]$ and $\psi^{\#}$ depend only on i and φ.

(iv) Apply (iii) to the case $N = M \times M$, $i = \Delta : M \to M \times M$ and $P = M$. Thus $\varphi(x) = (\sigma(x), \tau(x))$, where $\sigma, \tau : M \to M$. Show that in this case $[M] \cdot [P] = L(\sigma, \tau)\, \omega_{M \times M}$, where $\omega_{M \times M}$ is the orientation class of $M \times M$.

Appendix A

The Exponential Map

A.1. Sprays. Let M be an n-manifold with tangent bundle $\tau_M = (T_M, \pi, M, \mathbb{R}^n)$. The tangent bundle of the manifold T_M will be written $\tau_M^2 = (T_M^2, \pi_2, T_M, \mathbb{R}^{2n})$. Consider the commutative diagram

$$\begin{array}{ccc} T_M^2 & \xrightarrow{d\pi} & T_M \\ {\scriptstyle \pi_2}\downarrow & & \downarrow{\scriptstyle \pi} \\ T_M & \xrightarrow[\pi]{} & M \end{array} .$$

A *spray* for M is a smooth map $Y: T_M \to T_M^2$ which satisfies $d\pi \circ Y = \iota$ and $\pi_2 \circ Y = \iota$. (In particular, a spray is a vector field on T_M.)

Next consider the map $\mu: \mathbb{R} \times T_M \to T_M$ given by

$$\mu(t, \xi) = t\xi, \qquad t \in \mathbb{R}, \quad \xi \in T_M .$$

It determines, for each $t \neq 0$, the diffeomorphism $\mu_t : T_M \to T_M$ given by

$$\mu_t(\xi) = \mu(t, \xi).$$

A spray, Y, for M is called *affine*, if

$$(\mu_t)_* Y = (1/t)Y, \qquad t \neq 0.$$

Example: Assume that the tangent bundle of M is trivial,

$$T_M = M \times \mathbb{R}^n.$$

Then $T_M^2 = (M \times \mathbb{R}^n) \times (\mathbb{R}^n \times \mathbb{R}^n)$ and $d\pi$ and π_2 are given by

$$\begin{aligned} d\pi(x, h; k, l) &= (x, k), \qquad && x \in M \\ \pi_2(x, h; k, l) &= (x, h), \qquad && h, k, l \in \mathbb{R}^n. \end{aligned}$$

In this case, an affine spray is given by

$$Y(x, h) = (x, h; h, 0), \qquad x \in M, \quad h \in \mathbb{R}^n$$

Lemma I: Every manifold M admits an affine spray.

Proof: Cover M by open sets U_α with trivial tangent bundle (e.g., by chart neighborhoods). Let Y_α be an affine spray in U_α (cf. the example above). Let $\{f_\alpha\}$ be a partition of unity in M with $\operatorname{carr} f_\alpha \subset U_\alpha$. Then

$$Y = \sum_\alpha (\pi^* f_\alpha) Y_\alpha$$

is an affine spray in M.

Q.E.D.

A.2. The flow of a spray. Let Y be an affine spray for M. Recall from sec. 3.15 that there is a radial open set $W \subset \mathbb{R} \times T_M$ and a smooth map $\psi: W \to T_M$ such that

$$\dot{\psi}_\xi(t) = Y(\psi_\xi(t)), \qquad (t, \xi) \in W,$$

and

$$\psi_\xi(0) = \xi$$

$(\psi_\xi(t) = \psi(t, \xi))$. In particular,

$$\dot{\psi}_\xi(0) = Y(\xi), \qquad \xi \in T(M).$$

The map ψ is called a *local flow* of the spray Y.

From now on it will be assumed that M is compact. Give τ_M a Riemannian metric and set, for $r > 0$,

$$O_r = \{\xi \in T_M \mid |\xi| < r\}.$$

Since M is compact, we can choose $\delta > 0$ and $\rho > 0$ so that

$$I_\delta \times O_\rho \subset W,$$

where $I_\delta = \{t \in \mathbb{R} \mid |t| < \delta\}$. Then the flow restricts to a map

$$\psi : I_\delta \times O_\rho \to T_M .$$

Lemma II: Let Y be an affine spray on a compact manifold M. Then the corresponding flow satisfies

$$\psi(t, s\xi) = s\psi(st, \xi), \qquad t \in I_\delta, \quad |s| \leqslant 1, \quad \xi \in O_\rho .$$

In particular,

$$\psi(t, 0_x) = 0_x, \qquad x \in M.$$

Proof: Fix $x \in M$, $s \in I_1$ and $\xi \in T_x(M) \cap O_\mu$. Define maps

$$\alpha: I_\delta \to T_M \qquad \text{and} \qquad \beta: I_\delta \to T_M$$

by

$$\alpha(t) = \psi(t, s\xi) \qquad \text{and} \qquad \beta(t) = s\psi(st, \xi).$$

Then

$$\alpha(0) = \beta(0) = s\xi.$$

Moreover, $\dot{\alpha}(t) = Y(\alpha(t))$ while, since Y is affine,

$$\dot{\beta}(t) = d\mu_s(s \cdot \dot{\psi}_\xi(st)) = Y(s \cdot \psi_\xi(st)) = Y(\beta(t)).$$

Thus α and β are orbits for the vector field Y agreeing at $t = 0$. Now Proposition X, sec. 3.15, implies that $\alpha = \beta$.

Q.E.D.

A.3. The exponential map. Again let Y be an affine spray on a compact manifold M and let $\psi: I_\delta \times O_\rho \to T_M$ denote the corresponding flow. Set

$$\chi = \pi \circ \psi : I_\delta \times O_\rho \to M.$$

To each vector $\xi \in O_\rho$ associate the path $\chi_\xi: I_\delta \to M$ given by

$$\chi_\xi(t) = \chi(t, \xi).$$

Then $t \to \dot{\chi}_\xi(t)$ (cf. sec. 3.1) defines a path $\dot{\chi}_\xi$ in T_M and $\ddot{\chi}_\xi = (\dot{\chi}_\xi)^\cdot$ is a path in T_M^2.

Proposition I: With the notation and hypotheses above the map χ satisfies

(1) $\chi(0, \xi) = \pi(\xi)$
(2) $\dot{\chi}_\xi(0) = \xi$
(3) $\ddot{\chi}_\xi(t) = Y(\dot{\chi}_\xi(t)), \quad \xi \in O_\rho,\ t \in I_\delta$

and

(4) $\chi(t, s\xi) = \chi(st, \xi), \quad |s| \leqslant 1.$

Proof: (1) follows immediately from the definition of χ. Since Y is a spray,

$$\dot{\chi}_\xi(t) = (d\pi)\,\dot{\psi}_\xi(t) = (d\pi)\,Y(\psi_\xi(t)) = \psi_\xi(t);$$

i.e.,

$$\dot{\chi}_\xi = \psi_\xi.$$

It follows that

$$\dot{\chi}_\xi(0) = \psi_\xi(0) = \xi$$

and

$$\ddot{\chi}_\xi(t) = \dot{\psi}_\xi(t) = Y(\psi_\xi(t)) = Y(\dot{\chi}_\xi(t)).$$

Thus (2) and (3) hold.

To obtain (4) apply π to both sides of the equation in Lemma II and observe that

$$\pi(s\psi(st, \xi)) = \pi\psi(st, \xi).$$

Q.E.D.

Next set $\epsilon = \frac{1}{2}\rho\delta$ and consider the smooth map

$$\sigma: I_2 \times O_\epsilon \to M$$

given by

$$\sigma(t, \xi) = \chi(\tfrac{1}{2}\delta t, (2/\delta)\xi).$$

Definition: The map exp: $O_\epsilon \to M$ given by

$$\exp(\xi) = \sigma(1, \xi), \qquad \xi \in O_\epsilon$$

is called the *exponential map generated by the affine spray Y*.

The restriction of the exponential map to $T_{\epsilon,x}(M) = O_\epsilon \cap T_x(M)$ will be denoted by $\exp_x$.

Proposition II: The exponential map has the following properties, if ϵ is sufficiently small:

(1) $\exp 0_x = x, \; x \in M$

(2) $\exp_x$ is a diffeomorphism of $T_{\epsilon,x}(M)$ onto an open subset of M and satisfies

$$(d \exp_x)_{0_x} = \iota$$

(3) The map $\varphi: O_\epsilon \to M \times M$ given by

$$\varphi(\xi) = (\pi(\xi), \exp \xi)$$

is a diffeomorphism of O_ϵ onto an open subset of $M \times M$.

Proof: (1) Apply Lemma II, sec. A.2.

(2) Fix $\xi \in T_{\epsilon,x}(M)$ and define $\alpha: (-1, 1) \to M$ by

$$\alpha(t) = \exp(t\xi) = \chi(\tfrac{1}{2}\delta, (2/\delta)t\xi).$$

Then

$$\dot{\alpha}(0) = (d \exp_x)_{0_x}\xi.$$

Now apply Proposition I, (4) and (2), to obtain

$$\dot{\alpha}(0) = \dot{\chi}_\xi(0) = \xi;$$

i.e.,

$$(d \exp_x)_{0_x}\xi = \xi.$$

The rest of (2) is immediate if ϵ is sufficiently small.

(3) In view of (2), φ is injective. It remains to be shown that φ is a local diffeomorphism. In fact, assume that for some $\xi \in T_{\epsilon,x}(M)$ and $\eta \in T_\xi(T_M)$

$$(d\varphi)\eta = 0.$$

Then $(d\pi)\eta = 0$ and so

$$\eta \in V_\xi(T_M) = T_\xi(T_x(M))$$

(cf. sec. 7.1). Thus

$$0 = (d \exp)\eta = (d \exp_x)\eta.$$

Now (2) implies that $\eta = 0$. Hence $d\varphi$ is injective.

Since $\dim T_M = \dim(M \times M)$, each $(d\varphi)_\xi$ must be a linear isomorphism. Thus φ is a local diffeomorphism.

Q.E.D.

References

1. N. Bourbaki, "Éléments de Mathématique, Algèbre I," Hermann, Paris, 1970.
2. C. Chevalley, "Fundamental Concepts of Algebra," Academic Press, New York, 1956.
3. E. Coddington and N. Levinson, "Theory of Ordinary Differential Equations," McGraw-Hill, New York, 1955.
4. J. Dugundji, "Topology," Allyn & Bacon, Rockleigh, New Jersey, 1966.
5. W. H. Greub, "Linear Algebra," Springer-Verlag, Berlin and New York, 1967.
6. W. H. Greub, "Multilinear Algebra," Springer-Verlag, Berlin and New York, 1967.
7. S. Helgason, "Differential Geometry and Symmetric Spaces," Academic Press, New York, 1962.
8. W. Hurewitz and H. Wallman, "Dimension Theory," Princeton Univ. Press, Princeton, New Jersey, 1941.
9. F. Nevanlinna and R. Nevanlinna, "Absolute Analysis," Springer-Verlag, Berlin and New York, 1959.
10. S. Sternberg, "Lectures on Differential Geometry," Prentice-Hall, Englewood Cliffs, New Jersey, 1964.
11. A. H. Wallace, "An Introduction to Algebraic Topology," Pergamon, Oxford, 1963.
12. H. Whitney, "Geometric Integration Theory," Princeton Univ. Press, Princeton, New Jersey, 1957.

Bibliography

Chapter I

Dold, A., Partitions of unity in the theory of fibrations, *Ann. of Math.* (2) **78** (1963), 223–255.

Hopf, H., Introduction à la théorie des espaces fibrés, *Colloq. Topol. (Espaces Fibrés), Bruxelles, 1950*, pp. 107–113. Masson, Paris, 1950.

Chapter II

Greub, W., and Stamm, E., On the multiplication of tensor fields, *Proc. Amer. Math. Soc.* **17** (1966), 1112–1119.

Swan, R. G., Vector bundles and projective modules, *Trans. Amer. Math. Soc.* **105** (1962), 264–277.

Chapter IV

Almgren, Jr., F. J., Mass continuous cochains are differential forms, *Proc. Amer. Math. Soc.* **16** (1965), 1291–1294.

Boothby, W. M., and Wang, H. C., On contact manifolds, *Ann. of Math.* (2) **68** (1958), 721–734.

Bungart, L., Integration on analytic varieties, Stokes' formula, *J. Math. Mech.* **15** (1966), 1047–1054.

Constantinescu, C., Theory of currents on oriented varieties, "Modern Problems Theory of Functions," pp. 151–205. Editura Acad. R.P.R., Bucharest, 1965.

Ehresmann, C., Introduction à la théorie des structures infinitésimales et des pseudogroupes de Lie, *Colloq. Topol. Géom. Différentielle, Strasbourg, 1952*, 1952.

Graeub, W., Liesche Gruppen und affin zusammenhängende Mannigfaltigkeiten, *Acta Math.* **106** (1961), 65–111.

Jonker, L., Natural differential operators, Ph.D. Thesis, Univ. of Toronto, Toronto, Canada, 1967.

Moser, J., On the volume elements on a manifold, *Trans. Amer. Math. Soc.* **120** (1965), 286–294.

Palais, R. S., Natural operations on differential forms, *Trans. Amer. Math. Soc.* **92** (1959), 125–141.

Pohl, W. F., Differential geometry of higher order, *Topology* **1** (1962), 169–211.

Rinehart, G. S., Differential forms on general commutative algebras, *Trans. Amer. Math. Soc.* **108** (1963), 195–222.

Chapter V

Allendoerfer, C. B., Cohomology on real differentiable manifolds, *Proc. Internat. Congr. Math., Cambridge, Massachusetts, 1950*, II, pp. 428–435. American Mathematical Society, 1952.

Allendoerfer, C. B., and Eells, J., Jr., On the cohomology of smooth manifolds, *Comment. Math. Helv.* **32** (1958), 165–179.

Auderset, C., Poincaré duality, Master's Thesis, Univ. of Fribourg, Fribourg, Switzerland, 1968.

de Rham, G., La théorie des formes différentiables extérieurs et l'homologie des variétés différentiables, *Rend. Mat. e Appl.* (5), **20** (1961), 105–146.

de Rham, G., and Bidal, P., Les formes différentielles harmoniques, *Comment. Math. Helv.* **19** (1946), 1–49.

Dolbeaut, P., Formes différentielles et cohomologie à coéfficients entiers (couples d'Allendoerfer-Eells). Séminaire P. Lelong, 1958/59, exp. 1. Fac. Sci., Paris, 1959.

Eilenberg, S., Singular homology in differentiable manifolds, *Ann. of Math.* (2) **48** (1947), 670–681.

Guggenheim, V. K. A. M., and Spencer, D. C., Chain homotopy and the de Rham theory, *Proc. Amer. Math. Soc.* **7** (1956), 144–152.

Hirsch, G., Sur certain opérations dans l'homologie des espaces de Riemann, *Bull. Soc. Math. Belg.* **9** (1957), 115–139.

Hu, S.-T., On singular homology in differentiable spaces, *Ann. of Math.* (2) **50** (1949), 266–269.

Johnson, H. H., An algebraic approach to exterior differential systems, *Pacific J. Math.* 17 (1966), 423–434.

Kervaire, M., Extension d'un théorème de G. de Rham et expression de l'invariant de Hopf par une intégrale, *C. R. Acad. Sci. Paris* **237** (1953), 1486–1448.

Koszul, J. L., "Faisceaux et cohomologie," Inst. de Mat. Pura e Aplicado do C. N. Pq., São Paulo, 1957.

Mayer, W., Duality theorems, *Fund. Math.* **35** (1948), 188–202.

Palais, R. S., Logarithmically exact differential forms, *Proc. Amer. Math. Soc.* **12** (1961), 50–52.

Postnikov, M. M., Homology theory of smooth manifolds and its generalizations, *Uspehi Mat. Nauk* (N. S.), **11**, No. 1 (1956), 115–166.

Samelson, H., On Poincaré duality, *J. Analyse Math.* **14** (1965), 323–336.

Succi, F., Il teorema de de Rham e la dualità per le varietà relative, *Atti Acad. Naz. Lincei Rend. Cl. Sci. Fis. Mat. Natur.* (8), **35** (1963), 496–503.

Uehara, H., and Massey, W. S., The Jacobi identity for Whitehead products, "Algebraic Topology and Geometry," pp. 361–377. Princeton Univ. Press, Princeton, New Jersey, 1957.

Wang, H. C., The homology groups of fibre bundles over a sphere, *Duke Math. J.* **16** (1949), 33–38.

Weier, J., Über gewisse Integralinvarianten bei der Transformation von Differentialformen, *Rend. Circ. Mat. Palermo* (2), **8** (1959), 311–332.

Weil, A., Sur les théorèmes de de Rham, *Comment. Math. Helv.* **26** (1952), 119–145.

Whitney, H., Algebraic topology and integration theory, *Proc. Nat. Acad. Sci. U.S.A.* **33** (1947), 1–6.

Whitney, H., La topologie algébrique et la théorie d'intégration, Topologie algébrique, *Colloq. Internat. Centre Nat. Recherche Sci., Paris*, No. 12, pp. 107–113, CNRS, Paris, 1957.

Chapter VI

Epstein, D. B. A., The degree of a map, *Proc. London Math. Soc.* (3) **16** (1966), 369–383.
Hopf, H., Abbildungsklassen n-dimensionaler Mannigfaltigkeiten, *Math. Ann.* **96** (1927), 209–224.
Hopf, H., Zur Topologie der Abbildungen von Mannigfaltigkeiten I, Neue Darstellung des Abbildungsgrades, *Math. Ann.* **100** (1928), 579–608.
Hopf, H., II Klasseninvarianten, *Math. Ann.* **102** (1930), 562–623.
Hopf, H., Über die Drehung der Tangenten und Sehnen ebener Kurven, *Compositio Math.* **2** (1935), 50–62.
Hopf, H., Über den Defekt stetiger Abbildungen von Mannigfaltigkeiten, *Rend. Mat. e Appl.* **21** (1962), 273–285.
Hopf, H., and Rueff, M., Über faserungstreue Abbildungen der Sphären, *Comment. Math. Helv.* **11** (1938), 49–61.
Olum, P., Mappings of manifolds and the notion of degree, *Ann. of Math.* (2), **58** (1953), 458–480.
Whitney, H., On regular closed curves in the plane, *Compositio Math.* **4** (1937), 276–284.

Chapter VII

Auer, J. W., A spectral sequence for smooth fibre bundles and fibre integration, Ph.D. Thesis, Univ. of Toronto, Toronto, Canada, 1970.
Chern, S.-S., On the characteristic classes of complex sphere bundles and algebraic varieties, *Amer. J. Math.* **75** (1953), 565–597.
Ehresmann, C., Sur les espaces fibrés différentiables, *C. R. Acad. Sci. Paris* **224** (1947), 1611–1612.
Ehresmann, C., Les connexions infinitésimales dans un espaces fibré différentiable, *Colloq. Topol. (Espaces Fibrés), Bruxelles, 1950*, pp. 29–55. Masson, Paris, 1950.
Hattori, A., Spectral sequences in the de Rham cohomology of fibre bundles, *J. Fac. Sci. Univ. Tokyo Sect. I*, **8** (1960), 289–331.
Hermann, R., A sufficient condition that a mapping of Riemannian spaces be a fibre bundle, *Proc. Amer. Math. Soc.* **11** (1960), 236–242.
Hirsch, G., L'anneau de cohomologie d'un espace fibré et les classes caractéristiques, *C. R. Acad. Sci. Paris* **229** (1949), 1297–1299.
Leray, J., L'homologie d'un espace fibré dont la fibré est connexe, *J. Math.* **29** (1950), 169–213.
Reinhart, B. L., Structures transverse to a vector field, *Internat. Symp. Non-linear Differential Equations and Non-linear Mech., Colorado Springs, 1961*, pp. 442–444. Academic Press, New York, 1963.
Spanier, E. H., Homology theory of fibre bundles, *Proc. Internat. Congr. Math., Cambridge, Massachusetts, 1950*, II, pp. 390–396. American Mathematical Society, 1952.
van Est, W. T., A generalization of the Cartan-Leray spectral sequence, *Nederl. Akad. Wetensch. Proc. Ser. A* **61** (1958), 399–413.

Chapter VIII

Boltyanskiĭ, V. G., Second obstructions for cross-sections, *Amer. Math. Soc. Transl.* (2), **21** (1962), 51–86.

Chern, S.-S., Integral formulas for characteristic classes of sphere bundles, *Proc. Nat. Acad. Sci. U.S.A.* **30** (1944), 269–273.

Chern, S.-S., On the multiplication in the characteristic ring of a sphere bundle, *Ann. of Math.* (2), **49** (1948), 362–372.

Chern, S.-S., On the characteristic classes of complex sphere bundles and algebraic varieties, *Amer. J. Math.* **75** (1953), 565–597.

Chern, S.-S., and Spanier, E., The homology structure of sphere bundles, *Proc. Nat. Acad. Sci. U.S.A.* **36** (1950), 248–255.

Gysin, W., Zur Homologietheorie der Abbildungen und Faserungen von Mannigfaltigkeiten, *Comment. Math. Helv.* **14** (1942), 61–122.

Hirsch, G., Sur les invariants attachés aux sections dans les espaces fibrés, *Colloq. Topol. Strasbourg, 1951,* No. VII, 1951.

Hirsch, G., L'anneau de cohomologie d'un espace fibré en sphères, *C. R. Acad. Sci. Paris* **241** (1955), 1021–1023.

Hopf, H., Die Coinzidenz-Cozyklen und eine Formel aus der Fasertheorie, "Algebraic Geometry and Topology," pp. 263–279. Princeton Univ. Press, Princeton, New Jersey, 1957.

Malm, D. G., The real cohomology ring of a sphere bundle over a differentiable manifold, *Trans. Amer. Math. Soc.* **102** (1962), 293–298.

Massey, W. S., On the cohomology ring of a sphere bundle, *J. Math. Mech.* **7** (1958), 265–289.

Whitney, H., On the topology of differentiable manifolds, "Lectures in Topology," pp. 101–141. Univ. of Michigan Press, Ann Arbor, 1941.

Whitney, H., On the theory of sphere bundles, *Proc. Nat. Acad. Sci. U.S.A.* **26** (1940), 148–153.

Wu, W.-T., Sur les classes caractéristiques d'un espace fibré en sphères, *C. R. Acad. Sci. Paris* **227** (1948), 582–594.

Chapter IX

Cockcroft, W. H., On the Thom isomorphism theorem, *Proc. Cambridge Philos. Soc.* **58** (1962), 206–208.

Hirsch, G., Sur un théorème de Hopf-Rueff, *Acad. Roy. Belg. Bull. Cl. Sci.* (5), **29** (1943), 516–524.

Milnor, J. W., On the immersion of n-manifolds in $(n + 1)$-space, *Comment. Math. Helv.* **30** (1956), 275–284.

Pohl, W. F., Differential geometry of higher order, *Topology* **1** (1962), 169–211.

Reinhart, B. L., Cobordism and the Euler number, *Topology* **2** (1963), 173–177.

Thom, R., Classes caractéristiques et i-carrés, *C. R. Acad. Sci. Paris* **230** (1950), 427–429.

Thom, R., Variétés plongés et i-carrés, *C. R. Acad. Sci. Paris* **230** (1950), 507–508.

Thom, R., Espaces fibrés en sphères et carrés de Steenrod, *Ann. Sci. École Norm. Sup.* (3), **69** (1952), 109–182.

Wu, G., On n-manifolds in Euclidean $2n$-space, *Sci. Rec.* (1957), Sect. 1, 35–36.

Chapter X

Hopf, H., Vektorfelder in n-dimensionalen Mannigfaltigkeiten, *Math. Ann.* **96** (1927), 225–250.

Hopf, H., Verallgemeinerung der Euler-Poincaréschen Formel, *Nachr. Akad. Wiss. Göttingen Math.-Phys. Kl.* (1928), 127–136.

Mayer, W., On products in topology, *Ann. of Math.* (2), **46** (1945), 29–57.
Mayer, W., Singular chain intersection, *Ann. of Math.* (2), **47** (1946), 767–778.
Mayer, W., Duality theorems, *Fund. Math.* **35** (1948), 188–202.
Roitberg, J., On the Lefschetz fixed point formula, *Comm. Pure Appl. Math.* **20** (1967), 139–143.
Whitney, H., On products in a complex, *Ann. of Math.* (2), **39** (1938), 397–432.

Bibliography—Books

Alexandroff, P., and Hopf, H., "Topologie." Springer-Verlag, Berlin, 1935.

Atiyah, M. F., "*K*-theory." Benjamin, New York, 1967.

Bishop, R. L., and Crittenden, R. J., "Geometry of Manifolds." Academic Press, New York, 1964.

Bredon, G. E., "Sheaf Theory." McGraw-Hill, New York, 1967.

Cairns, S., "Differential and Combinatorial Topology." Princeton Univ. Press, Princeton, New Jersey, 1965.

Cartan, E., "Leçons sur la Géométrie des Espaces de Riemann." Gauthier-Villars, Paris, 1946.

de Rham, G., "Variétés Différentiables." Hermann, Paris, 1955.

Federer, H., "Geometric Measure Theory." Springer-Verlag, Berlin and New York, 1969.

Flanders, H., "Differential Forms." Academic Press, New York, 1963.

Godement, R., "Topologie Algébrique et Théorie des Faisceaux." Hermann, Paris, 1958.

Goldberg, S., "Curvature and Homology." Academic Press, New York, 1962.

Hicks, N., "Notes on Differential Geometry" (Math. Studies). Van Nostrand-Reinhold, Princeton, New Jersey, 1965.

Hodge, W. V. D., "The Theory and Applications of Harmonic Integrals." Cambridge Univ. Press, London and New York, 1941.

Holmann, H., and Rummler, H., "Differentialformen, B. I.," Hochschultaschenbuch No. 128. To be published.

Husemoller, D., "Fibre Bundles." McGraw-Hill, New York, 1966.

Kobayashi, S., and Nomizu, K., "Foundations of Differential Geometry," Vol. I. Wiley (Interscience), New York, 1963.

Krasnosel'skiy, M. A., Perov, A. I., Povolotskiy, A. I., and Zabreiko, P. P., "Plane Vector Fields." Academic Press, New York, 1966.

Lefschetz, S., "Topology" (Colloq. Publ., Vol. XII). Amer. Math. Soc., Providence, Rhode Island, 1930.

Lefschetz, S., "Algebraic Topology" (Colloq. Publ., Vol. XXVII). Amer. Math. Soc., Providence, Rhode Island, 1942.

Lefschetz, S., "Introduction to Topology." Princeton Univ. Press, Princeton, New Jersey, 1949.

Lima, E. L., "Introdução à Topologia Differencial," (Notes de Matemática No. 23). Inst. de Mat. Pura e Aplicado de Conselho Nacional de Pesquias, Rio de Janeiro, 1961.

Milnor, J. W., "The Theory of Characteristic Classes" (Mimeographed Notes). Princeton Univ., Princeton, New Jersey, 1957.

Milnor, J. W., "Morse Theory" (Ann. of Math. Studies, No. 51). Princeton Univ. Press, Princeton, New Jersey, 1963.

Milnor, J. W., "Topology from the Differentiable Viewpoint." Univ. Press of Virginia, Charlottesville, 1965.

Morse, M., and Cairns, S., "Critical Point Theory in Global Analysis and Differential Topology." Academic Press, New York, 1969.

Munkres, J. R., "Elementary Differential Topology" (Ann. of Math. Studies, No. 54). Princeton Univ. Press, Princeton, New Jersey, 1966.

Narasimhan, R., "Analysis on Real and Complex Manifolds." North-Holland Publ., Amsterdam, 1968.

Pontrjagin, L. S., Smooth manifolds and their applications in homotopy theory, *Amer. Math. Soc. Transl.* (2), 11 (1959).

Seifert, H., and Threlfall, W., "Lehrbuch der Topologie." Teubner, Leipzig, 1934.

Spanier, E., "Algebraic Topology." McGraw-Hill, New York, 1966.

Spivak, M., "Calculus on Manifolds." Benjamin, New York, 1965.

Spivak, M., "A Comprehensive Introduction to Differential Geometry." Brandeis Univ., Waltham, Massachusetts, 1970.

Steenrod, N., "The Topology of Fibre Bundles." Princeton Univ. Press, Princeton, New Jersey, 1951.

Warner, F., "Foundations of Differentiable Manifolds and Lie Groups." Scott-Foresman, Glenview, Illinois. 1971.

Notation Index

Symbols with a fixed meaning in the text are listed below, together with the page where they first occur. Symbols defined in Chapter 0 or in the problems are *not* listed.

Index

D

Q

R

S

Pure and Applied Mathematics

A Series of Monographs and Textbooks

1: Arnold Sommerfeld. Partial Differential Equations in Physics. 1949 (Lectures on Theoretical Physics, Volume VI)
2: Reinhold Baer. Linear Algebra and Projective Geometry. 1952
3: Herbert Busemann and Paul Kelly. Projective Geometry and Projective Metrics. 1953
4: Stefan Bergman and M. Schiffer. Kernel Functions and Elliptic Differential Equations in Mathematical Physics. 1953
5: Ralph Philip Boas, Jr. Entire Functions. 1954
6: Herbert Busemann. The Geometry of Geodesics. 1955
7: Claude Chevalley. Fundamental Concepts of Algebra. 1956
8: Sze-Tsen Hu. Homotopy Theory. 1959
9: A. M. Ostrowski. Solution of Equations and Systems of Equations. Second Edition. 1966
10: J. Dieudonné. Treatise on Analysis. Volume I, Foundations of Modern Analysis, enlarged and corrected printing, 1969. Volume II, 1970.
11: S. I. Goldberg. Curvature and Homology. 1962.
12: Sigurdur Helgason. Differential Geometry and Symmetric Spaces. 1962
13: T. H. Hildebrandt. Introduction to the Theory of Integration. 1963.
14: Shreeram Abhyankar. Local Analytic Geometry. 1964
15: Richard L. Bishop and Richard J. Crittenden. Geometry of Manifolds. 1964
16: Steven A. Gaal. Point Set Topology. 1964
17: Barry Mitchell. Theory of Categories. 1965
18: Anthony P. Morse. A Theory of Sets. 1965
19: Gustave Choquet. Topology. 1966
20: Z. I. Borevich and I. R. Shafarevich. Number Theory. 1966
21: José Luis Massera and Juan Jorge Schaffer. Linear Differential Equations and Function Spaces. 1966
22: Richard D. Schafer. An Introduction to Nonassociative Alegbras. 1966
23: Martin Eichler. Introduction to the Theory of Algebraic Numbers and Functions. 1966
24: Shreeram Abhyankar. Resolution of Singularities of Embedded Algebraic Surfaces. 1966

25: François Treves. Topological Vector Spaces, Distributions, and Kernels. 1967
26: Peter D. Lax and Ralph S. Phillips. Scattering Theory. 1967.
27: Oystein Ore. The Four Color Problem. 1967
28: Maurice Heins. Complex Function Theory. 1968
29: R. M. Blumenthal and R. K. Getoor. Markov Processes and Potential Theory. 1968
30: L. J. Mordell. Diophantine Equations. 1969
31: J. Barkley Rosser. Simplified Independence Proofs: Boolean Valued Models of Set Theory. 1969
32: William F. Donoghue, Jr. Distributions and Fourier Transforms. 1969
33: Marston Morse and Stewart S. Cairns. Critical Point Theory in Global Analysis and Differential Topology. 1969
34: Edwin Weiss. Cohomology of Groups. 1969
35: Hans Freudenthal and H. De Vries. Linear Lie Groups. 1969
36: Laszlo Fuchs. Infinite Abelian Groups: Volume I. 1970
37: Keio Nagami. Dimension Theory. 1970
38: Peter L. Duren. Theory of H^p Spaces. 1970.
39: Bodo Pareigis. Categories and Functors. 1970
40: Paul L. Butzer and Rolf J. Nessel. Fourier Analysis and Approximation: Volume 1, One-Dimensional Theory. 1971
41: Eduard Prugovečki. Quantum Mechanics in Hilbert Space. 1971
42: D. V. Widder: An Introduction to Transform Theory. 1971
43: Max D. Larsen and Paul J. McCarthy. Multiplicative Theory of Ideals. 1971
44: Ernst-August Behrens. Ring Theory. 1972
45: Morris Newman. Integral Matrices. 1972
46: Glen E. Bredon. Introduction to Compact Transformation Groups. 1972
47: Werner Greub, Stephen Halperin, and Ray Vanstone. Connections, Curvature, and Cohomology: Volume I, De Rham Cohomology of Manifolds and Vector Bundles. 1972

In preparation

Ronald G. Douglas. Banach Algebra Techniques in Operator Theory

Xia Dao-xing. Measure and Integration Theory of Infinite-Dimensional Spaces: Abstract Harmonic Analysis

E. R. Kolchin. Differential Algebra and Algebraic Groups

T. Benny Rushing. Topological Embeddings

Willard Miller, Jr. Symmetry Groups and Their Applications